FIBER-OPTIC-BASED SENSING SYSTEMS

FIBER-OPTIC-BASED SENSING SYSTEMS

Lazo M. Manojlović, PhD

APPLE
ACADEMIC
PRESS

First edition published 2022

Apple Academic Press Inc.
1265 Goldenrod Circle, NE,
Palm Bay, FL 32905 USA

4164 Lakeshore Road, Burlington,
ON, L7L 1A4 Canada

CRC Press
6000 Broken Sound Parkway NW,
Suite 300, Boca Raton, FL 33487-2742 USA

2 Park Square, Milton Park,
Abingdon, Oxon, OX14 4RN UK

Library and Archives Canada Cataloguing in Publication

Title: Fiber-optic-based sensing systems / Lazo M. Manojlović, PhD.

Names: Manojlović, Lazo M., author.

Description: First edition. | Includes bibliographical references and index.

Identifiers: Canadiana (print) 20210376783 | Canadiana (ebook) 20210376813 | ISBN 9781774637241 (hardcover) | ISBN 9781774637364 (softcover) | ISBN 9781003277293 (ebook)

Subjects: LCSH: Optical fiber detectors. | LCSH: Fiber optics.

Classification: LCC TA1815 .M36 2022 | DDC 681/.25—dc23

Library of Congress Cataloging-in-Publication Data

CIP data on file with US Library of Congress

ISBN: 978-1-77463-724-1 (hbk)
ISBN: 978-1-77463-736-4 (pbk)
ISBN: 978-1-00327-729-3 (ebk)

Contents

Synopsis .... *ix*
Abbreviations .... *xi*
Preface .... *xiii*

1. The Properties and the Nature of Light 1
1.1 The Brief History of Light Phenomena Perception 1
1.2 The Wave Nature of Light 3
1.3 The Corpuscular Nature of Light 14
References 23

2. Radiometric and Photometric Measurements 25
2.1 Introduction to Radiometry and Photometry 25
2.2 Optical Radiometry 28
2.2.1 The Radiative Transfer 29
2.2.2 The Lambertian Emitters 30
2.2.3 Radiometric Measurements 31
2.2.4 Reflection, Absorption, and Transmission 34
2.2.5 Kirchhoff's Law 36
2.3 Measurement Techniques in Radiometry 38
2.3.1 Absolute Radiometer 38
2.3.2 Radiant Flux Measurement 40
2.3.3 Integrating Sphere 45
2.4 Photometry 47
2.4.1 Spectral Response of a Human Eye 47
2.4.2 Standard Photometer and the Realization of the Candela 51
References 52

3. Optical Detection 55
3.1 Photon Counting 56
3.2 Photodetection Modeling 60
3.3 Photodetectors 63
3.3.1 Photomultiplier 64
3.3.2 Photoconductors 65
3.3.3 Photodiodes 67

3.3.4 Avalanche Photodiodes 71
3.3.5 Position Sensing Photodiodes 72
3.3.6 Quadrant Photodiode....75
3.3.7 Equivalent Circuit Model of the Photodiode....92
3.3.8 Photodiode Amplifier Circuit 95
References....103

4. Coherence and Interference of Light 105

4.1 Two-Beam Interference....105
4.1.1 Wavefront Division Method....108
4.1.2 Amplitude Division Method....111
4.1.3 The Michelson Interferometer....114
4.1.4 The Mach–Zehnder Interferometer 115
4.1.5 The Sagnac Interferometer....117
4.2 COHERENCE....118
4.2.1 The Mutual Coherence Function....119
4.2.2 Spatial Coherence....122
4.2.3 Coherence Time and Coherence Length 124
4.3 White-Light Interferometry....127
4.4 Multiple-Beam Interference 135
4.5 Multilayer Thin Films 140
4.6 Interferometric Sensors 146
4.6.1 The Rayleigh Refractometer 146
4.6.2 Laser-Doppler Interferometry 147
4.6.3 Vibration Amplitudes Measurements....148
4.6.4 Michelson's Stellar Interferometer....148
References....151

5. Fiber Optics 153

5.1 Optical Fibers....155
5.1.1 Geometrical Optics and the Optical Fibers 157
5.1.2 Wave Optics and the Optical Fibers....163
5.1.3 Chromatic Dispersion....174
5.1.4 Polarization Mode Dispersion....180
5.1.5 Fiber Losses....181
5.2 Fiber-Optic Communication Systems 183
5.2.1 Point-to-Point Links 183
5.2.2 Distribution Networks 185
5.2.3 Local Area Networks....186
5.2.4 Fiber-Optic Network Design Consideration....188
5.2.5 Coherent Fiber-Optic Communication Systems....191

5.3 Basic Concepts of Fiber-Optic Sensing Systems 195
5.3.1 Fiber-Optic Sensor Basic Topologies 195
5.3.2 Basic Concepts of Interferometric Fiber-Optic Sensors 197
References 202

6. Low-Coherence Fiber-Optic Sensor Principle of Operation 205
6.1 Algorithms for Signal Processing of Low-Coherence Interferograms 209
6.1.1 Threshold Comparison Method 210
6.1.2 Envelope Coherence Function Method 212
6.1.3 Centroid Algorithm 213
6.1.4 Algorithm with Phase-Shifted Interferograms 214
6.1.5 Wavelet Transform Algorithm 215
6.2 A Modified Centroid Algorithm 217
6.2.1 Sensitivity of the Modified Centroid Algorithm with Linear Scanning 220
6.2.2 Optical Path Difference Measurement Error of Linear Scanning 228
References 242

7. Fiber-Optic Sensor Case Studies 245
7.1 Absolute Position Measurement with Low-Coherence Fiber-Optic Interferometry 245
7.2 Rough Surface Height Distribution Measurement with Low-Coherence Interferometry 258
7.3 Wide-Dynamic Range Low-Coherence Interferometry 289
7.4 Optical Coherence Tomography Technique with Enhanced Resolution 299
References 315

Author Biography* *321

Index* *323

Synopsis

Since the revival of human civilization, there was an inevitable need for some sort of simple measurements of basic physical quantities such as length, time, and mass. In order to be able to exchange goods, the first traders established some primordial measurement standards. The earliest recorded systems of weights and measures originated in the 3rd or 4th millennium BC. Therefore, one can estimate that the history of measurements and consequently the history of measurement instruments is 5 to 6 millennia old. Early measurement standard units probably have been used only in a single community or small region, where every region developed its own standards for the physical quantity of interests, such as length, area, volume, and mass.

Starting from the first industrial revolution and subsequent development of manufacturing technologies, and consequently the growing importance of trade across the globe, standardized weights and measures became critical. Hence, since the 18th century, modernized, simplified, and uniform systems of weights and measures were developed. During the last three centuries, the fundamental units were defined by ever more precise methods in the science of metrology. Today's state-of-the-art measurement systems and instruments are capable of measuring different physical quantities with the resolution and speed that was unthinkable several decades ago.

Although optical fibers are almost exclusively used in the telecommunication networks for the high capacity data transfer, there is also another specific use of the optical fibers. Namely, having in mind that optical interferometry is one of the most sensitive measuring techniques that are capable of high-precision and high-speed measurements of different physical quantities, a large number of different sensors have been developed that are based on the optical fibers.

—Dr. Lazo M. Manojlović

Abbreviations

AC	alternating current
A/D	analog/digital
APD	avalanche photodiode
BER	bit error rate
CATV	common-antenna television
CCD	charge-coupled device
CIE	Commission Internationale de l'Eclairage
CIPM	Comité International des Poids et Mesures
CW	continuous-wave
DC	direct current
DWDM	dense wavelength division multiplexing
FDDI	fiber distributed data interface
FFT	fast fourier transform
FOC	fiber-optic coupler
FWHM	full width at half maximum
HDTV	high-definition television
IMG	index matching gel
IR	infrared
LAN	local area network
LCLS	low-coherence light source
LCS	low-coherence source
LED	light emitting diode
MAN	metropolitan area networks
NA	numerical aperture
OCT	optical coherence tomography
OS	optical spectrometer
OSA	optical spectrum analyzer
PC	personnel computer
PD	photodetector
PDF	probability density function
PMD	polarization mode dispersion
PMT	photomultiplier
PSD	power spectral density

QPD	quadrant photodiode
SLD	superluminescent diode
SNR	signal-to-noise ratio
TE	transverse electric
TIA	transimpedance amplifier
TM	transverse magnetic
TV	television
UV	ultraviolet
WDM	wavelength division multiplexed
WLS	white light Source
YAG	yttrium aluminum garnet

Preface

The book *Fiber-Optic-Based Sensing Systems* deals with the applicative aspects of using optical fibers as the sensing medium as well as the medium for transmitting the corresponding optical signals to the receiving unit. Basic optical phenomena with their main emphasis on applying the optical knowledge onto solving the real engineering problems are also treated within the book. The book is aimed toward the undergraduate and graduate students who want to broaden their knowledge of fiber-optic sensing system applications to the real-life engineering problems as well as to the engineers who want to acquire the basic principles of optics, especially fiber.

The book is arranged in a way that leads the reader toward better understanding of fiber-optical phenomena and their sensing applications. Basic tools and concepts are presented in the earlier chapters, which are then developed in more detail in the later chapters. The book is organized in seven chapters covering broad range of fiber-optical sensing phenomena.

Chapter 1 gives a brief historical overview of optical phenomena perception. Although the light and light-related phenomena have always been a source of immense curiosity for ancient people, one had to wait until the late 18th century and the Maxwell's classical electromagnetic theory in order to conceive most principles of modern optics.

Chapter 2 introduces the radiometry as the substantial part in the field of optical measurement and engineering together with its measurement techniques for measuring optical power and its spectral content.

Chapter 3 shows how the light can be detected. Although in the past times, the human eye was used exclusively as an optical detector, in order to objectively measure the intensity of a light in modern optical systems, a solid-state detector has been usually used.

Chapter 4 takes into account the wave nature of light, thus introducing the corresponding interference phenomena. Being inherently short wavelength electromagnetic wave, the interference of light gives a rise of highly sensitive measurements of many physical quantities, which are also presented in this chapter.

Today's state-of-the-art communication and sensing systems are unimaginable without optical fibers. Therefore, Chapter 5 is devoted to the optical fibers and their use in communication as well as in the highly sensitive versatile measurement systems.

Due to the their possibility of absolute optical paths difference measurements, a special attention is given in Chapter 6 to the low-coherence fiber-optic sensors principle of operation, where we have treated typical algorithms for extracting the value of the optical paths difference of the sensing interferometer.

Finally, Chapter 7 presents case studies that show how the concept of low-coherence fiber-optic-based sensing systems can be used for measuring different physical parameters ranging from simple measurement of the optical paths difference to the more complicated such as surface roughness distribution estimation and high-resolution optical coherence tomography.

CHAPTER 1

The Properties and the Nature of Light

ABSTRACT

The properties and nature of light intriguing the mankind since the beginning of human civilization. Today, passing more than 25 centuries from the first known philosophical treatment of the optical phenomena, which the early Greek philosopher Pythagoras have been made, the scientist are still astonished by the dual nature of light. Maxwell's classical electromagnetic theory perfectly describes the propagation of light, whereas the quantum theory treats the interaction between light and matter or the absorption and emission of light. So, if someone ask "What is light?," there is no a simple answer to this simple questions.

1.1 THE BRIEF HISTORY OF LIGHT PHENOMENA PERCEPTION

Since the dawn of human civilization the man was astonished and puzzled by the everyday interplay between the daylight and the nightdark caused by the Sun. Being unable to understand the origins of observed optical phenomena but understanding the importance of such phenomena on everyday life, the early men ascribed to the Sun the divine imprint. For a long period of human history, the Sun, the Moon, and the stars were the only sources of light in which periodic temporal behaviors were well understood and interweaved into the human perception of time. Sometimes, this perfectly tempered clock was interrupted by the solar and Moon eclipses as well as by the lightning and the polar light that confused the early man perception of light. The first turning point in understanding and controlling the optical phenomena and being able for the first time to make a man controlled light source was the discovery of fire control. For the first time, the primordial man was able not only to heat his home but also to illuminate it.

The curious mind of the ancient Greek philosophers was the first one who tries to get a deeper insight in the nature of light. The early Greek philosopher Pythagoras (c. 582–c. 497 B.C.) believed that light, which originates from the visible objects, is carried by the tiny particles of light. Empedocles (fifth-century B.C.) believed that light came from the illuminating objects but also believed that the light rays came out from our eyes. He was the first one who postulated that light travels at a finite speed. The famous Greek mathematician Euclid (c. 325–c. 265 B.C.) thought that the eyes transmit rays of light that cause the sensation of vision. Mirrors were also studied by Euclid, where in his book, entitled *Catoptrics* and written in the third-century B.C., the law of reflection was established.

The rectilinear propagation of rays of light was the main reason for Isaac Newton in his *Treatise on Opticks* to adopt the corpuscular nature of light, more than 20 centuries after Euclid. He believed that light consists of very small bodies emanating from the shining bodies. The shadow formation behind the illuminated objects as well as the law of mirror reflections goes in the favor of corpuscular nature of light. In contrast to Newton's corpuscular theory, Christian Huygens developed a different theory where light is considered to be a wave motion that spreads from the source uniformly in all directions. The optical phenomena such as interference and diffraction, which were inexplicable in the frame of Newton's corpuscular theory, perfectly fit the wave nature of light.

In the mid–19th century, one of the milestones in understanding the nature of light was reached. Namely, due to the genius work of James Clerk Maxwell that was published in his famous paper *A Dynamical Theory of the Electromagnetic Field*, the light can be described as an electromagnetic wave. However, the beginning of 20th century brings new dilemmas. The outstanding experimental results, first announced by German physicist Philipp Lenard in 1902, showed that the energy of electrons emitted from the illuminated surface increased with the frequency of the light. The classical electromagnetic theory crashes in trying to explain this phenomenon. Therefore, a new approach was necessary. This new approach came on the wings of new quantum theory where light has been considered to consist of discrete wave packets—photons.

The presented brief history of optical phenomena perception showed that throughout the entire history of human preoccupation with the light, there was back and forth movement between the wave and corpuscular nature of light. Maxwell's classical electromagnetic theory perfectly describes the propagation of light, whereas the quantum theory treats the interaction between light and matter or the absorption and emission of light. So, if someone ask "*What is light?,*" there is no a simple answer to this simple questions.

1.2 THE WAVE NATURE OF LIGHT

Maxwell's equations together with the Lorentz force law represent the complete classical theory of the electromagnetic fields. All the phenomena where the interaction between the electromagnetic field and matter occurs can be successfully explained within the frame of classical electromagnetic theory except in those cases where the quantum effects are distinct. The set of Maxwell's equations can be written in the following forms:

$$\nabla \cdot \mathbf{E} = \frac{\rho}{\varepsilon_0}, \tag{1.1}$$

$$\nabla \cdot \mathbf{B} = 0, \tag{1.2}$$

$$\nabla \times \mathbf{E} = -\frac{\partial \mathbf{B}}{\partial t}, \tag{1.3}$$

$$\nabla \times \mathbf{B} = \mu_0 \left(\mathbf{J} + \varepsilon_0 \frac{\partial \mathbf{E}}{\partial t} \right) \tag{1.4}$$

where **E** is the electric field vector (or electric field strength), **B** is the magnetic induction, ρ is the electric charge density, **J** is the electric current density, $\varepsilon_0 = 8.85 \times 10^{-12}$ F/m is the vacuum permittivity, and $\mu_0 = 4\pi \times 10^{-7}$ H/m is the vacuum permeability. The force **f** per unit volume acting on the electric charge density ρ and the electric current density **J** is given by the Lorentz force law:

$$\mathbf{f} = \rho\mathbf{E} + \mathbf{J} \times \mathbf{B}. \tag{1.5}$$

If the electromagnetic field is observed in the regions where there are no electric charges ($\rho = 0$) and no electric currents ($\mathbf{J} = 0$), such as in a vacuum, Maxwell's equations reduce to:

$$\nabla \cdot \mathbf{E} = 0, \tag{1.6}$$

$$\nabla \cdot \mathbf{B} = 0, \tag{1.7}$$

$$\nabla \times \mathbf{E} = -\frac{\partial \mathbf{B}}{\partial t}, \tag{1.8}$$

$$\nabla \times \mathbf{B} = \varepsilon_0 \mu_0 \frac{\partial \mathbf{E}}{\partial t}. \tag{1.9}$$

By taking curl ($\nabla\times$) of the curl equations and by using the identity ($\nabla\times (\nabla \times \mathbf{X}) = \nabla(\nabla \cdot \mathbf{X}) - \nabla^2\mathbf{X}$, one can obtain the following wave equations:

$$\nabla^2 \mathbf{E} = \frac{1}{c^2}\frac{\partial^2 \mathbf{E}}{\partial t^2}. \tag{1.10}$$

$$\nabla^2 \mathbf{B} = \frac{1}{c^2}\frac{\partial^2 \mathbf{B}}{\partial t^2}, \tag{1.11}$$

where c is the speed of light in vacuum identified as follows:

$$c = \frac{1}{\sqrt{\varepsilon_0 \mu_0}} = 299{,}792{,}458 \mathrm{m/s}. \tag{1.12}$$

Although the speed of light in vacuum has the exact value, the speed of light v in matter depends on the medium characteristics such as relative permittivity ε_r and relative permeability μ_r in the following way:

$$v = \frac{1}{\sqrt{\varepsilon_r \varepsilon_0 \mu_r \mu_0}} = \frac{c}{n}, \tag{1.13}$$

where $n = \sqrt{\varepsilon_r \mu_r}$, is the material index of refraction. The electric field **E** and magnetic induction **B** of an electromagnetic wave are mutually perpendicular to each other and to the direction of wave propagation and are in phase with each other, that is, the wave propagates in the direction of $\mathbf{E} \times \mathbf{B}$ as it is presented in Figure 1.1.

If we take into consideration the Cartesian coordinate system and resolve the vector wave eqs 1.10 and 1.11 into components, one can observe that both components **E** and **B** satisfy the general scalar wave equation:

$$\frac{\partial^2 U}{\partial x^2} + \frac{\partial^2 U}{\partial y^2} + \frac{\partial^2 U}{\partial z^2} = \frac{1}{v^2}\frac{\partial^2 U}{\partial t^2}, \tag{1.14}$$

where U represents any of the electric field components E_x, E_y, and E_z, as well as any of the magnetic induction components E_x, E_y, and E_z. Let further consider the simplest case, where the spatial variation of U occurs only along some particular direction, for example, along the z-axis, that is, $U = U(z, t)$. In this particular case, eq 1.14 simplifies to the one-dimensional wave equation:

$$\frac{\partial^2 U}{\partial z^2} = \frac{1}{v^2}\frac{\partial^2 U}{\partial t^2} \tag{1.15}$$

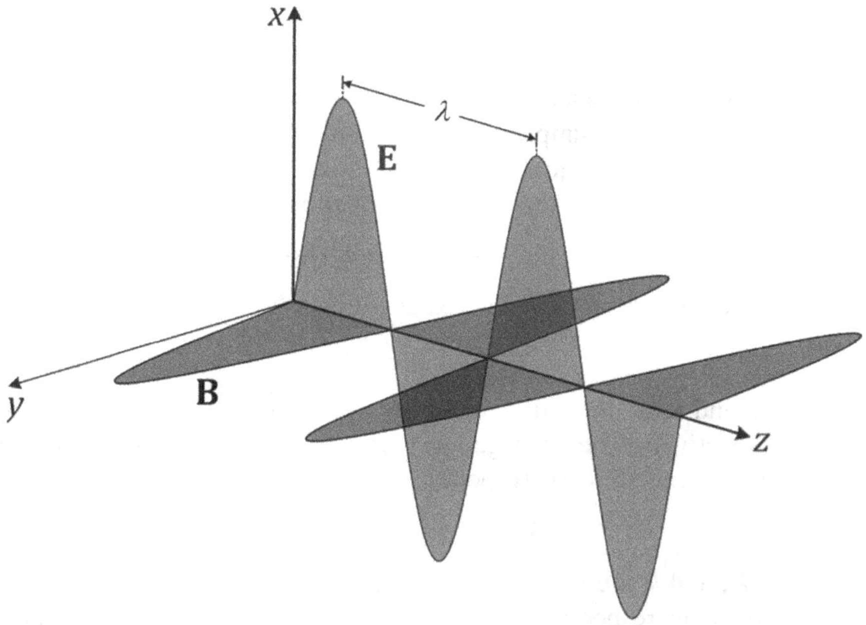

FIGURE 1.1 The electric field **E** and the magnetic induction **B** are mutually perpendicular to each other and to the direction of wave propagation and are in phase with each other.

A simple harmonic solution of this one-dimensional wave equation can be obtained by the variable separation in the following form:

$$U(z, t) = U_0\cos(\omega t - kz), \tag{1.16}$$

where U_0 is the corresponding electromagnetic wave amplitude and where wave number k and angular frequency ω are related as:

$$v = \frac{\omega}{k}, \tag{1.17}$$

where the ratio of the angular frequency ω and the wave number k represents the phase velocity v. The presented particular solution, given by eq 1.16, is fundamental to the study of optics and represents the plane harmonic wave. The wavelength λ is defined as the distance, measured along the direction of the wave propagation, where the wave function completes one whole cycle, that is, the distance between two maximums (or minimums) of the wave function. The wavelength is determined by the wave number as:

$$\lambda = \frac{2\pi}{k}. \tag{1.18}$$

If we make a step back to the three-dimensional wave eq 1.14, it is easy to show that a similar simple harmonic solution is one of the particular solutions of this wave equation. The three-dimensional plane harmonic wave function is given by the following equation:

$$U(x, y, z, t) = U_0 \cos(\omega t - \mathbf{k} \cdot \mathbf{r}), \tag{1.19}$$

where the position vector r is defined as follows:

$$\mathbf{r} = \hat{\mathbf{i}}x + \hat{\mathbf{j}}y + \hat{\mathbf{k}}z, \tag{1.20}$$

where $\hat{\mathbf{i}}$, $\hat{\mathbf{j}}$, and $\hat{\mathbf{k}}$ are the unit vectors ($|\hat{\mathbf{i}}| = |\hat{\mathbf{j}}| = |\hat{\mathbf{k}}| = 1$) along x-, y-, and z-axis, respectively. The propagation vector or the wave vector $\mathbf{k}$ can be also given in terms of its corresponding components as follows:

$$\mathbf{k} = \hat{\mathbf{i}}k_x + \hat{\mathbf{j}}k_y + \hat{\mathbf{k}}k_z, \tag{1.21}$$

where k_x, k_y, and k_z are the corresponding wave vector components along x-, y-, and z-axis, respectively. The magnitude of the wave vector is equal to the wave number, which is given in terms of the corresponding wave vector components as follows:

$$|\mathbf{k}| = k = \sqrt{k_x^2 + k_y^2 + k_z^2}. \tag{1.22}$$

If we further inspect eq 1.19 and its cosine function argument $\omega t - \mathbf{k} \cdot \mathbf{r}$, one can observe that the constant values of this argument define a set of planes in space named surfaces of constant phase (equiphase surfaces) given by the following equation:

$$\omega t - \mathbf{k} \cdot \mathbf{r} = \omega t - (k_x x + k_y y + k_z z) = \text{const.} \tag{1.23}$$

By interpreting eq 1.23, it is obvious that wave vector $\mathbf{k}$ is normal to the surfaces of constant phase as shown in Figure 1.2. Moreover, the constant-phase surfaces are moving in the direction of the wave vector with the velocity equal to the phase velocity:

$$v = \frac{\omega}{k} = \frac{\omega}{\sqrt{k_x^2 + k_y^2 + k_z^2}}. \tag{1.23}$$

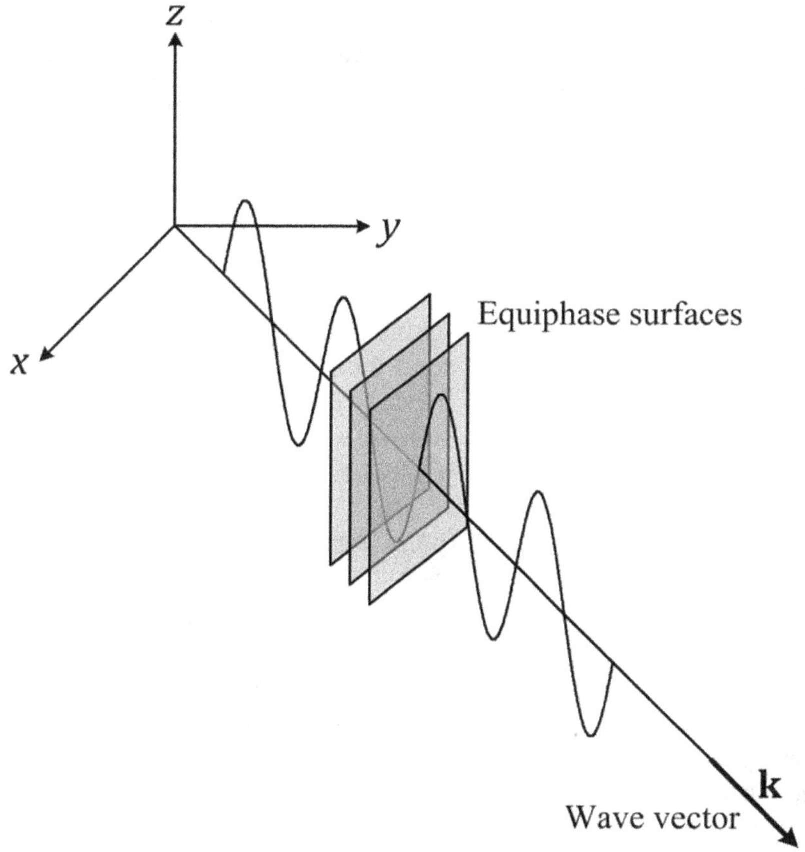

FIGURE 1.2 Equiphase surfaces of a plane wave.

Electromagnetic waves are produced by oscillating electric charged particles, so the frequency of oscillation determines the type of emitted radiation. Such produced electromagnetic disturbance propagates through space as a monochromatic wave, which is characterized by a single frequency, or polychromatic wave, which consists of many frequencies, either discrete or in a continuum. The energetic content of an electromagnetic wave, which is distributed among the various constituent waves, is called the spectrum of the radiation. Typically, to the different regions of the electromagnetic spectrum, specific names are given, such as radio waves, microwaves, optical waves (infrared, visible, and ultraviolet light), X-rays, gamma rays, and cosmic rays due to the different way of its production or detection. The common names, descriptions as well as typical frequency ranges are presented in Figure 1.3 where the electromagnetic spectrum is given in both terms frequency v and wavelength λ. One can recall that these two quantities, frequency and wavelength, are linked as follows:

$$v\lambda = c. \tag{1.24}$$

In the case when all the charges oscillate unison, for the emitted electromagnetic radiation, it is said to be coherent. In the case when all the charges oscillate independently and/or randomly, for the emitted electromagnetic radiation, it is said to be incoherent. Typical radiation sources in the optical domain such as flames, incandescent light bulbs, fluorescent lamps, and light-emitting diodes (LED) are incoherent light sources. In contrast to these incoherent sources of radiation, typical man-made radiation sources in the low-frequency range such as radio waves and microwaves are coherent. The only light source in optical domain that emits coherent radiation is laser.

Sometimes, it is convenient to present the wave function in the complex domain by using Euler representation:

$$U(x,y,z,t) = U_0 \exp\left[j(\omega t - \mathbf{k}\cdot\mathbf{r})\right]. \tag{1.25}$$

The real part of the complex wave function represents the actual physical quantity, where the real part is equal to the actual wave function given by eq 1.19. The reason why the complex representation of the wave function is often used is the simplicity of the algebra with the complex representation than with the trigonometric representation of the wave function. Besides simple harmonic solution of the plain waves that satisfies the wave equation in Cartesian coordinate system, given by eq 1.14, there is also a particular solution in the form of the spherical waves, which can be easily obtained if we rearrange the wave equation given by eqs 1.10 and 1.11 in the polar coordinate system. The wave function of the spherical waves is given by the following equation:

$$U(r,t) = \frac{U_0}{r}\cos(\omega t - kr) \text{ or } U(r,t) = \frac{U_0}{r}\exp\left[j(\omega t - kr)\right]. \tag{1.26}$$

where r is the radial distance from the radiation source to any particular point in the space.

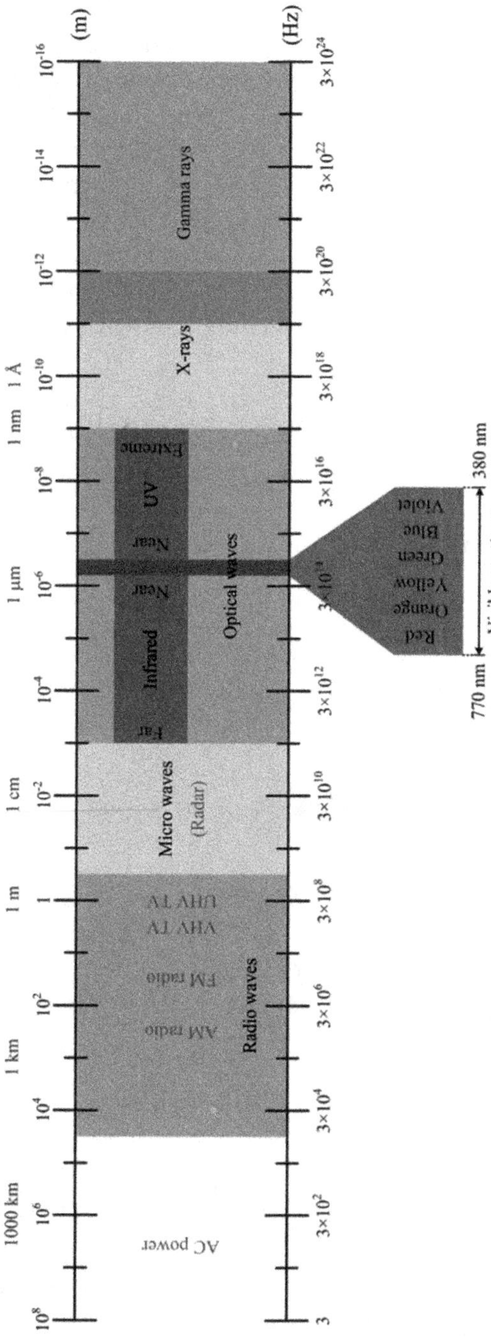

FIGURE 1.3 Electromagnetic spectrum arranged by wavelength in meters (m) and frequency in hertz (Hz).

As it was mentioned earlier, the phase velocity is the rate at which the wave phase propagates in space and it is defined as $v_p = \omega/k$. On the other side, the group velocity is defined as the rate at which the wave envelope propagates through space. To find the relation for the group velocity, we will observe two waves with the angular frequencies $\omega + \mathrm{d}\omega$ and $\omega - \mathrm{d}\omega$ and wave numbers $k + \mathrm{d}k$ and $k - \mathrm{d}k$. In a simple case, which will be considered here, we will assume without losing the generality of the analysis that both waves have the same amplitude U_0 and propagate in the same direction, let say along the z-axis. The superposition of both waves will result in a wave in which complex wave function is given by the following equation:

$$U = U_0 \exp\left\{j\left[(\omega + \mathrm{d}\omega)t - (k + \mathrm{d}k)r\right]\right\} \\ + U_0 \exp\left\{j\left[(\omega - \mathrm{d}\omega)t - (k - \mathrm{d}k)r\right]\right\}. \tag{1.27}$$

After the rearrangement of eq 1.27, we have:

$$U = 2U_0 \exp\left[j(\omega t - kr)\right]\cos(t\mathrm{d}\omega - r\mathrm{d}k). \tag{1.28}$$

The last equation can be interpreted as a single wave with the wave function $2U_0 \exp\,[j(\omega t - kr)]$ that is modulated by the envelope function $\cos(t\mathrm{d}\omega - r\mathrm{d}k)$. The envelope travels not with the phase velocity but with the group velocity equal to:

$$v_{\mathrm{g}} = \frac{\mathrm{d}\omega}{\mathrm{d}k}. \tag{1.29}$$

Electromagnetic wave when propagates throughout the space transfers a certain amount of energy and momentum. Therefore, there is a certain electromagnetic energy that is stored in the space through which wave travels. The energy density, stored into the free space, is given by the following equation:

$$w_E = \frac{\varepsilon_0}{2} E^2, \tag{1.30}$$

$$w_B = \frac{1}{2\mu_0} B^2, \tag{1.31}$$

where w_E and w_B are the energy densities of electric and magnetic fields, respectively. As $E = cB$ is fulfilled, we have $w_E = w$. So, the total energy density is given by the following equation:

$$w = w_E + w_B = \varepsilon_0 E^2 = \frac{1}{\mu_0} B^2 \tag{1.32}$$

When propagates, electromagnetic wave fills the space with the energetic content. If we observe an electromagnetic wave for a very short period of time Δt, the electromagnetic energy that is brought into the observed elementary volume is equal to:

$$W = cAw\Delta t, \tag{1.33}$$

where A is the cross section of the elementary volume. By inspecting eq 1.33, a particular physical quantity can be defined that represents the energy transfer through the unit area and unit time as:

$$S = \frac{W}{A\Delta t} = cw. \tag{1.34}$$

This physical quantity represents the magnitude of the Poynting vector, which is defined as follows:

$$S = \frac{1}{\mu_0} \mathbf{E} \times \mathbf{B}. \tag{1.35}$$

Besides the energy transfer, electromagnetic radiation carries also the momentum, so when electromagnetic wave impinges the surface of some object, it exerts force. As Maxwell showed, the radiation pressure is equal to the energy density of the electromagnetic wave. The radiation pressure exerted on the object that perfectly absorbs the radiation can be found by using virtual work principle. Therefore, we will observe the electromagnetic radiation that impinges the perfectly absorbing body, as it is presented in Figure 1.4. If we allow force, which is generated by the radiation, to virtually move the object for an infinitesimally short distance δz, the work done by this force will, according to the first principle, be equal to the change in stored electromagnetic energy. Thus, the energy conservation law gives:

$$\mathrm{PA}\, \delta z = wA\, \delta z. \tag{1.36}$$

where P is the radiation pressure and A is the cross section of the observed elementary volume. From eq 1.36, it is simply obtained $P = w$.

Referring further to Figure 1.5, where it is presented the radiation that impinges the perfectly absorbing object, the total momentum of the electromagnetic radiation that is stored in the elementary volume with the cross section A and thickness Δz will be transferred to the object in period of time equal to $\Delta t = \Delta z/c$. The overall momentum M that will be transferred to the object is equal to:

$$M = p'A\Delta z, \tag{1.37}$$

where p' is the volume density of electromagnetic momentum. According to the third Newton's law of motion, we have:

$$PA = M\Delta t. \tag{1.38}$$

that is, the force exerted on the object is caused by the overall momentum transfer during the period of time Δt. Therefore, the electromagnetic momentum volume density is given by the following equation:

$$p' = \frac{w}{c} = \frac{S}{c^2}. \tag{1.39}$$

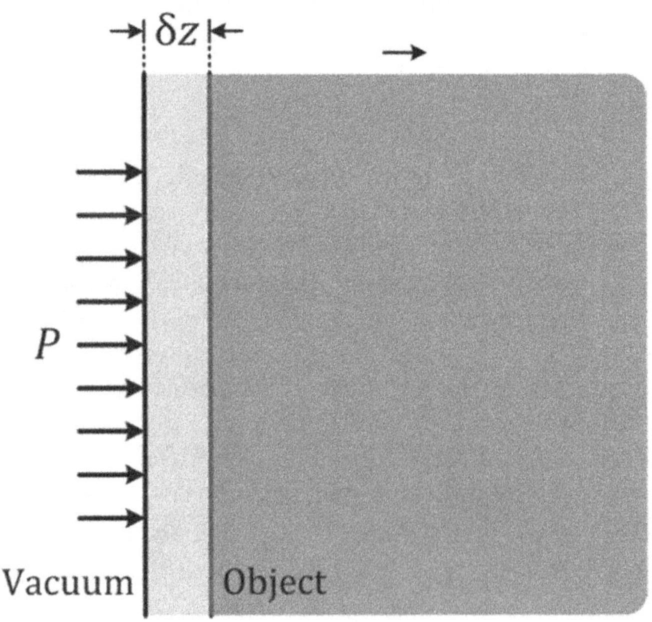

FIGURE 1.4 Electromagnetic radiation impinges the perfectly absorbing object.

One of the important physical quantities, especially in the radiometry, is the time average of the energy that is received by the unit area and unit time and is named irradiance. By definition, we can write:

$$I = \langle S \rangle, \tag{1.40}$$

where $\langle \bullet \rangle$ denotes the time average operator. So, the irradiance can be interpreted as the time-averaged Poynting vector.

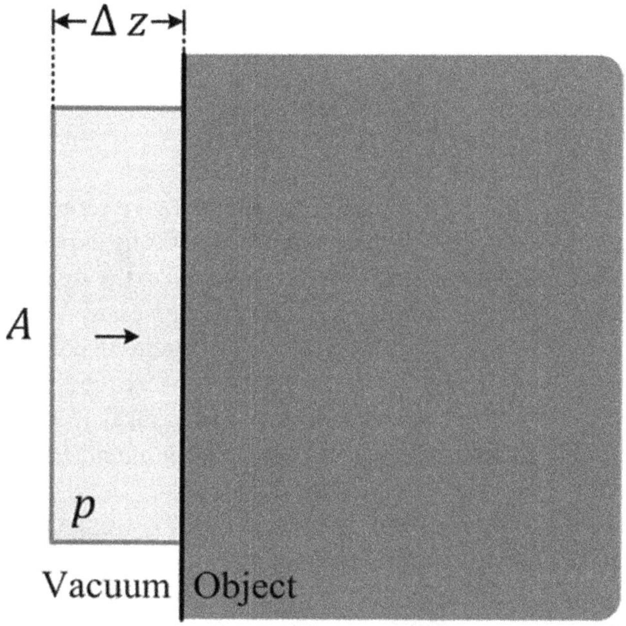

FIGURE 1.5 Electromagnetic radiation impinges the perfectly absorbing object and transfers the momentum.

If a monochromatic source of electromagnetic radiation is moving uniformly with a constant velocity toward the motionless observer and simultaneously transmits the electromagnetic waves with the particular frequency ν, then, in the observer's frame of reference, according to the relativistic Doppler effect, there will be a change in the received radiation frequency in the following way:

$$\nu' = \nu\sqrt{\frac{c \mp u}{c \pm u}}, \tag{1.41}$$

where ν' is the received frequency, u is the source velocity relative to the observer, and where upper signs stand for the case when the source is moving away from the observer and lower signs when the source is approaching the observer. Equation 1.41 is obtained by taking into consideration the Lorentz transformation, that is, eq 1.41 is valid for any velocity. Typically, the source (or the observer) velocity is much smaller than speed of light ($u \ll c$), so eq 1.41 simplifies to:

$$\nu' \approx \nu\left(1 \mp \frac{u}{c}\right) \tag{1.42}$$

1.3 THE CORPUSCULAR NATURE OF LIGHT

As mentioned earlier, Lenard's experiment had the far-reaching consequences on the deeper understanding of the nature of light. The experiment showed that the energy of emitted electrons is proportional to the frequency of light. The classical electromagnetic theory failed in this case. According to the classical theory, the energy of an emitted electron must be proportional to the intensity of the light but not to its frequency. Moreover, a very dim light would be sufficient to trigger the electron emission that was in opposite to the observed frequency threshold, that is, the minimal frequency where photoelectric effects start. Below the threshold, electrons aren't emitted regardless the light intensity or the exposure time. To bridge this gap, Albert Einstein suggested that a light beam is not a wave that propagates through space, but rather a set of discrete wave packets (photons), each with energy $\varepsilon = h\nu$, with $h = 6.626 \times 10^{-34}$Js being the Planck constant. There has been passed almost two decades before Einstein's light quanta hypothesis had been accepted. The period of transition between the hypothesis rejection and acceptance would undoubtedly have been longer if Compton's experiment had been less striking. Based on the thermodynamic approach to the black-body radiation, Einstein showed that monochromatic black-body radiation in Wien's law spectral region behaves with respect to thermal phenomena as if it consists of independently moving particles or quanta of radiant energy each with the energy proportional to its frequency.

There is also an alternative way to find the relation between the photon energy and the relevant physical quantities of the macroscopic electromagnetic wave. In the analysis shown next, the only assumption that will be taken into account is that electromagnetic radiation is quantized and the photon is the quantum of such an electromagnetic radiation. Further, it will be taken into consideration that an arbitrary monochromatic wave motion $u(\mathbf{r},t)$ at a particular point in space, defined by its position vector $\mathbf{r}$, is fully determined by its amplitude $U(\mathbf{r})$, wave number $\mathbf{k}$, and angular frequency ω as:

$$u(\mathbf{r},t) = U(\mathbf{r})\cos(\omega t - \mathbf{k}\cdot\mathbf{r} + \varphi), \tag{1.43}$$

where, without losing the generality of the analysis, the influence of the initial wave phase φ will be neglected, we will assume that the photon, as an integral part of the electromagnetic wave (radiation), has an average energy ε that is dependent on these three abovementioned parameters, as well as potentially dependent on some other physical parameters $\mathbf{x} = [x_1\ x_2\ \ldots\ x_n]$, that is, $\varepsilon = \varepsilon(U, \mathbf{k}, \omega, \mathbf{x})$, where $x_i \neq x_i(U, \mathbf{k}, \omega)$ $(i = 1,2, \ldots, n)$ must be satisfied. In the opposite case, where we have $x_i = x_i(U, \mathbf{k}, \omega)$ fulfilled, the average photon energy will be dependent only on the electromagnetic wave parameters and not on this particular physical quantity. Also, if the initial phase affects the average photon energy, the average photon energy will be dependent on the choice of the initial time instant, which is in the collision with the energy conservation law and thus absurd. For the wave number, we have $|\mathbf{k}| = 2\pi/\lambda$, where λ is the wavelength of the monochromatic electromagnetic wave in vacuum and $\lambda = 2\pi c/\omega$, is valid, where c is the speed of light in vacuum, so the wave number is only dependent on the frequency $\mathbf{k} = \mathbf{k}(\omega)$. Therefore, for the photon energy, we have $\varepsilon = \varepsilon(U, \omega, \mathbf{x})$.

To find out what is the dependence of the photon average energy on the wave amplitude, we will observe the unvarying, static, and monochromatic sources S of the electromagnetic radiation, as it is depicted in Figure 1.6. Looking at the distance z_1 from the source within a small solid angle $\Delta\Omega$ ($\Delta\Omega \ll 1$), the amplitude of the radiation is equal to U_1 whereas at the distance z_2 from the source, the amplitude is equal to U_2. Since $z_1 \neq z_2$, we have fulfilled $U_1 \neq U_2$. Taking into consideration that the source is unvarying, it emits a constant radiant flux $\Delta\Phi$ of the electromagnetic radiation within the observed solid angle $\Delta\Omega$, or equivalently, it emits a constant number of photons per unit time q within the corresponding solid angle having the average photon energy $\varepsilon = \varepsilon(U, \mathbf{k}, \omega, \mathbf{x})$. The photon velocity has been considered to be exactly the same as the speed of light in vacuum. If we consider the case where the photon velocity is smaller than the speed of light in vacuum, some parts of the electromagnetic wave that was first broadcasted by the transmitter will be not quantized at large distances from the transmitter and the others will be, which is naturally absurd. In the opposite case, where we consider the photon velocity to be greater than the speed of light in vacuum, we have also absurd situation since it is in the collision with the basic principle of the special relativity. Therefore, after the time interval $t_1 = z_1/c$ from the moment of their emission, photons reach the distance z_1 from the source S. The total number of photons, contained in a thin spherical shell, that is, located within the slab $A_1B_1C_1D_1$, having the solid angle $\Delta\Omega$, cross-section area ΔS_1 and with the thickness Δz_1 at the distance z_1, whereby $\Delta z_1 \ll \Delta z_1$ is satisfied, is $N_1 = q\,\Delta z_1$, where $\Delta t_1 = \Delta z_1/c$, with the assumption that there are no photons absorbed, scattered, and/or generated on their way since they are propagating in vacuum. Similarly, after time interval $t_2 = z_2/c$ from the moment of their emission, photons reach the distance z_2 from the source. Also, the total number of photons, contained in a thin spherical shell, that is, located within the slab $A_2B_2C_2D_2$, having the solid angle $\Delta\Omega$, cross-section area ΔS_2 and with the thickness Δz_2 at the distance z_2, whereby $\Delta z_2 \ll \Delta z_2$ is satisfied, is $N_2 = q\,\Delta t_2$ where $\Delta t_2 = \Delta z_2/c$, with the same assumption that there are no photons absorption, scattering, and/or generation on their way.

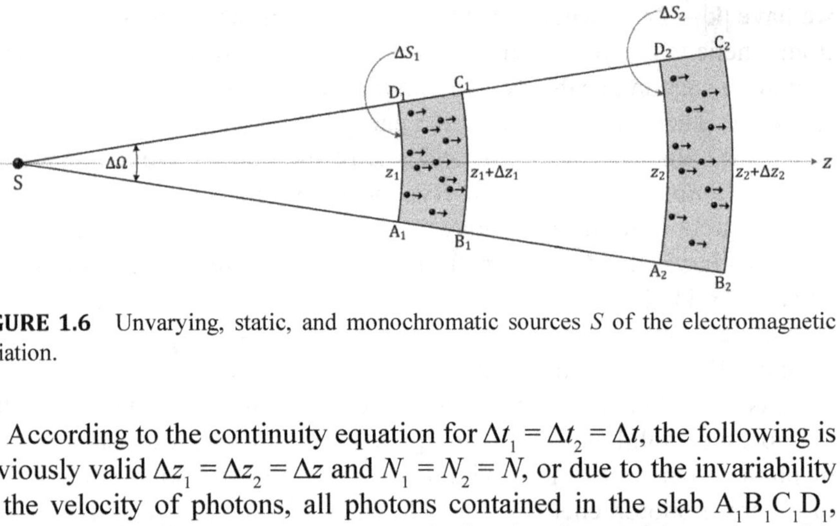

FIGURE 1.6 Unvarying, static, and monochromatic sources S of the electromagnetic radiation.

According to the continuity equation for $\Delta t_1 = \Delta t_2 = \Delta t$, the following is obviously valid $\Delta z_1 = \Delta z_2 = \Delta z$ and $N_1 = N_2 = N$, or due to the invariability of the velocity of photons, all photons contained in the slab $A_1B_1C_1D_1$, specified by the distances $[z_1 = \Delta z_1 = \Delta z_1]$ after time interval $\Delta t_{12} = (z_2 - z_1)/c$ are contained in the slab $A_2B_2C_2D_2$, specified by the distances $[z_2, z_2 + \Delta z_2]$, where $\Delta t = \Delta z/c$ is valid. The energy stored in the slab $A_1B_1C_1D_1$ or the energy of all photons that are located in the slab $A_1B_1C_1D_1$ at the distance z_1 is $W_1 = N\varepsilon(U_1, \omega, \mathbf{x}_1)$, while the energy stored in the slab $A_2B_2C_2D_2$ or the energy of all photons that are located in the slab $A_2B_2C_2D_2$ at a distance z_2 is $W_2 = N\varepsilon(U_2, \omega, \mathbf{x}_2)$, where $\mathbf{x}_1$ and $\mathbf{x}_2$ are the corresponding other physical parameters' values that may influence the photon energy at the position of $A_1B_1C_1D_1$ and $A_2B_2C_2D_2$ slabs, respectively. Further, if we simply apply the energy conservation law to the photons contained within the cone

$A_1A_2D_1D_2$ during their movement and having in mind that there are no external forces exerted on the slab surfaces A_1D_1 (B_1C_1) and A_2D_2 (B_2C_2) and thus there is no mechanical work done over the distances Δz_1 and Δz_2, the energy stored in the cone $A_1A_2D_1D_2$ must be the same as the energy stored in the cone $B_1B_2C_1C_2$. Furthermore, this leads to the conclusion that the energy content of the slab $A_1B_1C_1D_1$ is the same as the energy content of the slab $A_2B_2C_2D_2$, so we have $W_1 = W_2$ fulfilled, or equivalently:

$$\varepsilon(U_1, \omega, \mathbf{x}_1) = \varepsilon(U_2, \omega, \mathbf{x}_2). \tag{1.44}$$

Now, if we, without losing the generality of the analysis, assume that these two slabs are close enough, that is, infinitesimally close where $z_1 = z$ and $z_2 = z + \mathrm{d}z$ are fulfilled, the corresponding amplitudes of the electromagnetic wave have the following values $U_1 = U$ and $U_2 = U + \mathrm{d}U$, within the slabs $A_1B_1C_1D_1$ and $A_2B_2C_2D_2$, respectively, with $\mathrm{d}U$ being the infinitesimal change of the electromagnetic wave amplitude between these two slabs, and where the other influencing physical parameters have the following values $\mathbf{x}_1 = \mathbf{x}$ and $\mathbf{x}_2 = \mathbf{x} + \mathbf{dx}$, with $\mathbf{dx} = [\mathrm{d}x_1\ \mathrm{d}x_2\ \ldots\ \mathrm{d}x_n]$ being the infinitesimal change of the physical parameters' values between these two slabs. Having this in mind, eq 1.44 becomes:

$$\varepsilon(U_2,\omega,\mathbf{x}_2) - \varepsilon(U_1,\omega,\mathbf{x}_1) = \mathrm{d}\varepsilon(U,\omega,\mathbf{x}) = \frac{\partial\varepsilon}{\partial U}\mathrm{d}U + \sum_{i=1}^{n}\frac{\partial\varepsilon}{\partial x_i}\mathrm{d}x_i = 0. \tag{1.45}$$

with $\mathrm{d}\varepsilon(U, \omega, \mathbf{x})$ being the total derivative of the photon energy. After rearranging eq 1.45, one obtains:

$$\frac{\partial\varepsilon}{\partial U} = -\sum_{i=1}^{n}\frac{\partial\varepsilon}{\partial x_i}\frac{\mathrm{d}x_i}{\mathrm{d}U}. \tag{1.46}$$

Since $x_i \neq x_i(U, \mathbf{k}, \omega)$, where $i = 1,2, \ldots, n$ is valid, one obtains $\mathrm{d}x_i / \mathrm{d}U \equiv 0$. Therefore, from eq 1.46, we have:

$$\frac{\partial\varepsilon}{\partial U} \equiv 0 \rightarrow \varepsilon \neq \varepsilon(U), \tag{1.47}$$

or the photon energy is not dependent on the electromagnetic wave amplitude, so one has $\varepsilon = \varepsilon(\omega, \mathbf{x})$, that is, the photon energy potentially depends on the angular frequency of the electromagnetic radiation and potentially on various other physical parameters.

To obtain the mathematical relation between the photon energy and the angular frequency of the electromagnetic wave, as well as between the photon energy and other physical parameters that may influence photon energy, one will observe the moving emitter that emits plane, unvarying, and monochromatic electromagnetic waves in the direction of the motionless absorber, as shown in Figure 1.7. The moving emitter in its inertial frame of reference emits the total radiant flux of Φ. Since the emitted flux is constant, the photon rate is also constant and equals q in the moving emitter frame of reference. The measured flux and the photon rate are related by the following relation:

$$\Phi = q\varepsilon(\omega, \mathbf{x}). \tag{1.48}$$

Now, we will consider the thought experiment where the perfect, plane, motionless absorber, in which surface is parallel to the wavefront, absorbs each photon that is emitted by the moving emitter, as it is presented in Figure 1.7. Due to the Doppler effect, in the case of the motionless absorber, the absorbed radiant flux and angular frequency will differ from case of moving emitter emitted radiant flux and angular frequency. In the motionless absorber inertial frame of reference, the absorbed radiant flux and the angular frequency are $\Phi'$and ω', respectively.

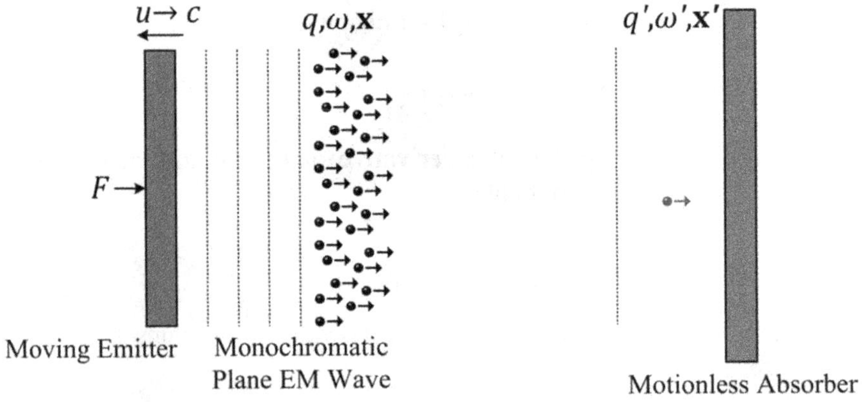

FIGURE 1.7 Moving emitter emits planar, unvarying, and monochromatic electromagnetic waves in the direction of the motionless absorber.

Due to the time dilation, caused by the motion of the moving emitter in the motionless absorber frame of reference, the number of photons per unit time that reaches the motionless absorber is q', where $q' \neq q$. In this case, the average time between observing two photons is $\tau = 1/q$, while the average time between absorbing two photons is $\tau' = 1/q'$, where $\tau' \neq \tau$

is also valid. The abovementioned three parameters Φ', q', and ω' that correspond to the motionless absorber are related by the following relation Φ', $= q'\varepsilon(\omega',\mathbf{x}')$. Based on the Lorentz transformations, the following is valid for the angular frequency ω' of the electromagnetic wave and the number of photons per unit time q' that have been absorbed:

$$\omega' = \omega\sqrt{\frac{c-u}{c+u}} \tag{1.49}$$

$$\tau' = \frac{\tau}{\sqrt{1-\left(u^2/c^2\right)}} \Rightarrow q' = q\sqrt{1-\frac{u^2}{c^2}} \tag{1.50}$$

According to the energy conservation law in the motionless absorber frame of reference, we have:

$$\Phi' = \Phi - Fu, \tag{1.51}$$

where F is the braking force that must be applied to the moving emitter in the opposite direction of its motion to provide the uniform motion of the emitter with constant velocity, or the emitter must not accelerate due to the repulsive force of the electromagnetic radiation that it captures. Further, eq 1.51 becomes:

$$q'\varepsilon(\omega',\mathbf{x}') = q\varepsilon(\omega,\mathbf{x}) - Fu. \tag{1.52}$$

By combining eqs 1.49, 1.50, and 1.52, we have:

$$q\sqrt{1-\frac{u^2}{c^2}}\,\varepsilon\left(\omega\sqrt{\frac{c-u}{c+u}},\mathbf{x}'\right) = q\varepsilon\left(\omega,\mathbf{x}\right) - Fu. \tag{1.53}$$

If the moving emitter velocity is very close to the speed of light in vacuum, that is, $u \rightarrow c$, according to eq 1.50, it is valid $q' \rightarrow 0$, that is, the average time between absorbing two consecutive photons tends to infinity $\tau' \rightarrow +\infty$ and $\varepsilon(\omega', x') \rightarrow \varepsilon(0, x')$, so regardless the value of $\varepsilon(0, x')$, eq 1.53 becomes:

$$F = q\frac{\varepsilon\left(\omega,\mathbf{x}\right)}{c}, \tag{1.54}$$

where it is assumed $\varepsilon(0, x') < +\infty$, while otherwise an indefinitely large energy would be needed to establish a time invariant electromagnetic field even in a limited volume. Based on the third Newton's law, the force F is also the repulsive force of the electromagnetic radiation that acts on the emitter. According to eq 1.54, the photon momentum $p(\omega, x)$ is given by

$p(\omega, x) = \varepsilon(\omega, x)/c$, which is the well-known relationship for the photon momentum.

The goal of the above-presented analyses was to find the photon momentum as a function of its energy. In the next step of the analysis, we will consider the case when the emitter is moving steadily toward the motionless absorber with constant velocity u, as shown in Figure 1.8. To overcome the repulsive force of the emitted electromagnetic radiation and to ensure uniform motion of the emitter, the force F must be exerted on the moving emitter in the direction of its motion.

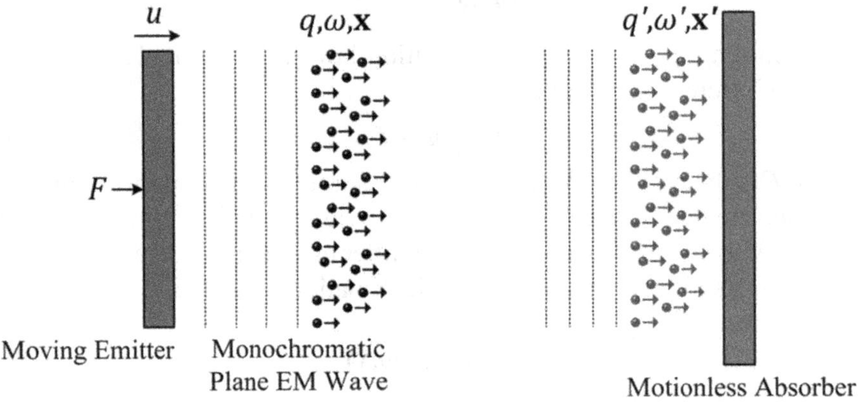

FIGURE 1.8 The moving emitter is moving uniformly toward the motionless absorber with a constant velocity.

According to the energy conservation law in the motionless absorber frame of reference, the following is valid:

$$\Phi' = \Phi - Fu. \tag{1.55}$$

Further, taking into consideration eq 1.54, the following is valid for eq 1.55:

$$q'\varepsilon(\omega', \mathbf{x}') = q\varepsilon(\omega, \mathbf{x}) + q\varepsilon(\omega, \mathbf{x})\frac{u}{c}. \tag{1.56}$$

By combining eqs 1.49, 1.50, and 1.56, we obtain the following functional equation:

$$\varepsilon\left(\omega\sqrt{\frac{c+u}{c-u}}, \mathbf{x}'\right)\sqrt{1-\frac{u^2}{c^2}} = \varepsilon(\omega, \mathbf{x}) + \varepsilon(\omega, \mathbf{x})\frac{u}{c}. \tag{1.57}$$

Equation 1.57 can be rearranged in the following way:

$$\varepsilon\left(\omega\sqrt{\frac{c+u}{c-u}},\mathbf{x}'\right)=\sqrt{\frac{c+u}{c-u}}\varepsilon(\omega,\mathbf{x}). \tag{1.58}$$

If we take $a=\sqrt{(c+u)/(c-u)}$, eq 1.58 becomes:

$$\varepsilon(a\omega,\mathbf{x}')=a\varepsilon(\omega,\mathbf{x}), \tag{1.59}$$

where obviously $a \geq 0$ is satisfied. By taking partial derivative of both sides of eq 1.59 with respect to the variable ω (angular frequency), one obtains:

$$\frac{\partial}{\partial\omega}\varepsilon(a\omega,\mathbf{x}')=a\frac{\partial}{\partial\omega}\varepsilon(\omega,\mathbf{x}). \tag{1.60}$$

After rearranging eq 1.60, we have:

$$\frac{\partial}{\partial(a\omega)}\varepsilon(a\omega,\mathbf{x}')=\frac{\partial}{\partial\omega}\varepsilon(\omega,\mathbf{x}), \tag{1.61}$$

which can be further written in the following form:

$$\varepsilon'_{\omega}(\omega',\mathbf{x}')=\varepsilon'_{\omega}(\omega,\mathbf{x}), \tag{1.62}$$

with ε'_{ω} representing the partial derivative of photon energy ε with respect to the angular frequency and $\omega' = a\omega$. Now, if we, without losing the generality of the analysis, assume that the angular frequencies and the values of other influencing physical parameters differ for very small, infinitesimal values in two inertial frames of reference (inertial frames of reference of the moving emitter and of the motionless absorber), thus having the following values $\omega' = \omega + \mathrm{d}\omega$, with $\mathrm{d}\omega$ being the infinitesimal change of the angular frequency in the abovementioned inertial frames of reference and $\mathbf{x}' = \mathbf{x} + \mathbf{dx}$, with $\mathbf{dx} = [\mathrm{d}x_1\ \mathrm{d}x_2 \ldots \mathrm{d}x_n]$ being the infinitesimal change of the physical parameters' values also in the abovementioned inertial frames of reference. Having this in mind, eq 1.62 becomes:

$$\varepsilon'_{\omega}(\omega',\mathbf{x}')-\varepsilon'_{\omega}(\omega,\mathbf{x})=\mathrm{d}\varepsilon'_{\omega}(\omega,\mathbf{x})=\frac{\partial\varepsilon'_{\omega}}{\partial\omega}\mathrm{d}\omega+\sum_{i=1}^{n}\frac{\partial\varepsilon'_{\omega}}{\partial x_i}\mathrm{d}x_i=0, \tag{1.63}$$

with $\mathrm{d}\varepsilon'_{\omega}(\omega, \mathbf{x})$ being the total derivative of the function $\varepsilon'_{\omega}(\omega,\mathbf{x})$. After rearranging eq 1.63, one obtains:

$$\frac{\partial\varepsilon'_{\omega}}{\partial\omega}=-\sum_{i=1}^{n}\frac{\partial\varepsilon'_{\omega}}{\partial x_i}\frac{\mathrm{d}x_i}{\mathrm{d}\omega}. \tag{1.64}$$

Since $x_i \neq x_i(U, \mathbf{k}, \omega)$, where $i = 1,2, \ldots, n$ is valid, one obtains $\mathrm{d}x_i / \mathrm{d}\omega \equiv 0$ and eq 1.64 becomes:

$$\frac{\partial \varepsilon'_\omega}{\partial \omega} = \frac{\partial^2}{\partial \omega^2} \varepsilon(\omega, \mathbf{x}) \equiv 0. \tag{1.65}$$

The solution of the second order partial differential equation, shown in eq 1.65, is given by the following equation:

$$\varepsilon(\omega, \mathbf{x}) = \omega f(\mathbf{x}) + g(\mathbf{x}), \tag{1.66}$$

where $f(\mathbf{x})$ and $g(\mathbf{x})$ are arbitrary functions of $\mathbf{x}$. After substituting eq 1.66 into eq 1.59, one has:

$$a\omega f(\mathbf{x}') + g(\mathbf{x}') = a[\omega f(\mathbf{x}) + g(\mathbf{x})]. \tag{1.67}$$

Equation 1.67 can be rearranged in the following form:

$$a\omega[f(\mathbf{x}') - f(\mathbf{x})] + g(\mathbf{x}') - ag(\mathbf{x}) = 0. \tag{1.68}$$

Since eq 1.68 must be satisfied for any value of the angular frequency, it is obvious that $f(\mathbf{x}') - f(\mathbf{x})$ (where $\mathbf{x}' \neq \mathbf{x}$) must be fulfilled, which further leads to the conclusion that the value of $f(\mathbf{x})$ must be constant as it takes the same value for an arbitrary value of $\mathbf{x}$, that is, $f(\mathbf{x}) = \text{const.} = \hbar$. Consequently, we have $g(\mathbf{x}') = ag(\mathbf{x})$ also fulfilled. The photon energy is then given by $\varepsilon(\omega, \mathbf{x}) = \hbar\omega + g(\mathbf{x})$. For $\omega = 0$, we have $\varepsilon(0, \mathbf{x}) = g(\mathbf{x})$, or the photon will have the energy even though there are no electromagnetic radiation and thus no photons. This is naturally absurd, so we have $g(\mathbf{x}) \equiv 0$, which finally leads us to the conclusion that the photon energy is only dependent on the angular frequency of the radiation and that there are no other physical quantity that may have an effect on the photon energy. Finally, the photon energy is given by the following equation:

$$\varepsilon(\omega) = \hbar\omega, \tag{1.69}$$

or equivalently the photon energy–frequency relationship is given by the following equation:

$$\varepsilon = h\nu, \tag{1.70}$$

where $\nu = \omega/(2\pi)$ is the frequency of the electromagnetic radiation and where the constant h must be experimentally determined. The relationship given in eq 1.70 represents the well-known relationship between the energy and the frequency of the photon, where h is also the well-known Planck's constant $h = 6.626 \times 10^{-34}$ Js that was first measured by Millikan a century ago. Equation 1.70 confirms our earlier assumption that $\varepsilon(\omega = 0, \mathbf{x}) < +\infty$

since according to eq 1.70, we have $\varepsilon(\omega = 0, \mathbf{x}) = 0$. To show the uniqueness of $\varepsilon(\omega) = \hbar\omega$ as a solution to eq 1.59, we will assume that there is another solution $\varepsilon'(\omega)$, which, in general case, can be represented as $\varepsilon'(\omega) = l(\omega) \cdot \varepsilon(\omega)$ where $l(\omega)$ is an arbitrary function of ω. This solution must also satisfy eq 1.59, so we have $\varepsilon'(a\omega) = a\varepsilon'(\omega)$ and consequently $l(a\omega) \cdot \varepsilon(a\omega) = al(\omega) \cdot \varepsilon(\omega)$. As $\varepsilon(\omega)$ is the solution to eq 1.59, we have $\varepsilon(a\omega) = a\varepsilon(\omega)$, which further leads to $l(a\omega) = l(\omega)$. Therefore, function $l(\omega)$ takes the same value for any value of ω, so it has a constant value $l(\omega) = b$ or consequently $\varepsilon'(\omega) = b\varepsilon(\omega)$ is satisfied, where b is an arbitrary but positive constant. In conclusion, both solutions $\varepsilon'(\omega)$ and $\varepsilon(\omega)$ are of the same form, given by eq 1.59, that is, they depend linearly on the frequency. Hence, the solution given by eq 1.69 is the unique solution to the functional equation given by eq 1.69.

If we go backward to the equation $p(\omega) = \varepsilon(\omega)/c$, one can see that the photon momentum in vacuum is given by $p = h\nu/c$, or in the vector notation as:

$$\mathbf{p} = \hbar\mathbf{k}, \tag{1.71}$$

where $\hbar = h/2\pi$ is the reduced Planck constant.

KEYWORDS

- **dual nature of light**
- **corpuscular nature of light**
- **optical phenomena**
- **nature of light**
- **properties of light**
- **wave nature of light**

REFERENCES

Beeson, S.; Mayer, J. W. *Patterns of Light*; Steven Beeson and James W. Mayer: New York, NY, 2008.

Greene, D. *Light and Dark*; Institute of Physics Publishing Ltd: Bristol and Philadelphia, 2003.

Hakfoort, C. *Optics in the Age of Euler*; Cambridge University Press: Cambridge, 1995.

Hecht, E. *Optics*, 5th ed. Pearson Education Limited: Harlow, 2017.

Kenyon, I. R. *The Light Fantastic: A Modern Introduction to Classical and Quantum Optics*; Oxford University Press: Oxford, 2008.

Manojlović, L. M. *Is It Possible That Photon Energy Depends on Any Other Physical Parameter Except Its Frequency?* Optik, accepted for publication.

Montwill, A.; Breslin, A. *Let There Be Light*; Imperial College Press: London, 2008.

Overduin, J. M.; Wesson, P. S. *The Light/Dark Universe: Galactic Light, Dark Matter and Dark Energy*; World Scientific Publishing Co. Pte. Ltd.: Danvers, 2008.

Pedrotti, F. L.; Pedrotti, L. M.; Pedrotti, L. S. *Introduction to Optics*, 3rd ed; Pearson Education Limited: Harlow, 2014.

Römer, H. *Theoretical Optics: An Introduction*; WILEY-VCH Verlag GmbH & Co. KGaA: Weinheim, 2005.

Roychoudhuri, C.; Kracklauer, A. F.; Creath, K., Eds. *The Nature of Light: What Is a Photon?* Taylor & Francis Group, LLC: Boca Raton, FL, 2008.

Smith, F. G.; King, T. A.; Wilkins, D. *Optics and Photonics: An Introduction*, 2nd ed; John Wiley & Sons Ltd: Chichester, 2007.

Weiss, R. J. *A Brief History of Light and Those That Lit the Way*; World Scientific: Singapore, 1995.

CHAPTER 2

Radiometric and Photometric Measurements

ABSTRACT

The word "radiometry" means the radiation measurement. The term radiometer is usually exclusively applied to the devices that measure radiance. However, this term is used in a more general sense for a device that measures one of several power-dependent optical quantities. Radiance, defined as the optical power emitted from an area within particular solid angle, is one of many optical terms that will be defined and treated in this chapter. The corresponding mathematical techniques for calculating the radiative transfer as well as the measurement techniques for characterizing of fluxes and radiometric properties of different kinds will be included in this chapter.

2.1 INTRODUCTION TO RADIOMETRY AND PHOTOMETRY

Radiometry is the substantial part in the field of optical measurement and engineering, where the measurement techniques for power, polarization, spectral content, and other relevant parameters for an electromagnetic radiation interrogation, which are important in characterization of optical sources and detectors, have been studied. An instrument that measures optical radiation is named a radiometer. Radiometer is typically used to detect the radiation and in many applications, where it is essential, to measure the amount of the captured radiation. As a matter of fact, the word "radiometry" itself means the radiation measurement. The term radiometer is usually exclusively applied to the devices that measure radiance. However, this term will be used in a more general sense for a device that measures one of several power-dependent optical quantities.

For example, radiance, defined as the optical power emitted from an area within particular solid angle, is one of many optical terms that will be defined and treated in this chapter. Moreover, this chapter will include both the corresponding mathematical techniques for calculating the radiative transfer as well as the measurement techniques for characterizing of fluxes and radiometric properties of different kinds.

Before starting introducing radiometric quantities, there are some important geometrical parameters that must be first introduced. One of these parameters is surely the solid angle. A solid angle can be understood as the two dimensional equivalent of a linear angle or as the two-dimensional angle in three-dimensional space. This angle is the measure of the size of the object that appears to an observer looking at it from a certain point. The solid angle can be equivalently defined as the projection of an area onto a sphere surface divided by the square of the radius of the sphere, as is presented in Figure 2.1:

$$\Omega = \frac{S}{r^2}, \tag{2.1}$$

where S is the projected area onto the sphere surface and r is the radius of the sphere. The maximal value of the solid angle is equal to 4π sr (steradians).

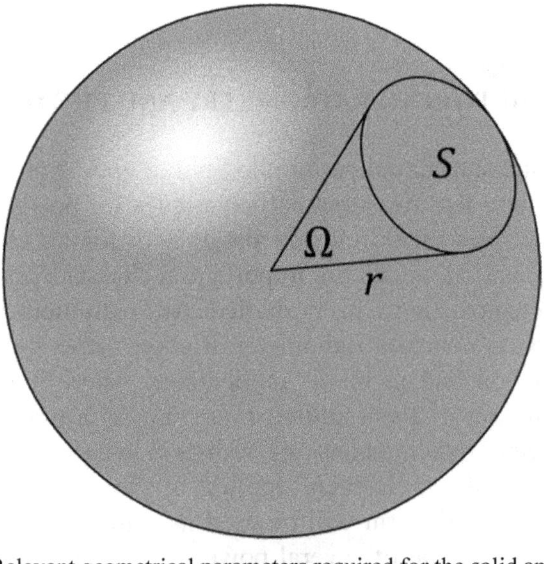

FIGURE 2.1 Relevant geometrical parameters required for the solid angle definition.

The way that two bodies exchange the radiated powers or energies to and from each other is described by their mutual radiative transfer. There are several physical quantities that can give more insight into this process, such as the energy, the power, and certain geometric characteristics of the power, like its density, that is, the power emitted by the unit area.

To understand these transfer processes next, it will be given some important definitions. The radiometric and photometric quantities, which are listed in Table 2.1 together with their typical symbols and units, can be introduced as presented in Figure 2.2.

TABLE 2.1 Radiometric and Photometric Quantities and Their Symbols and Units.

Radiometric quantity	Symbol	Units	Photometric quantity	Symbol	Units
Radiant energy	Q	J	Luminous energy	Q_V	lm s
Radiant flux (power)	Φ, P	W	Luminous flux	Φ_V	lm
Irradiance	E	W/m^2	Illuminance	E_V	(lm/m^2) = lx
Radiance	L	W/(m^2 sr)	Luminance	L_V	lm/(m^2 sr)
Radiant intensity	I	W/sr	Luminous intensity	I_V	(lm/sr) = cd
Radiant exitance	M	W/m^2	Luminous exitance	M_V	lm/m^2
Radiant exposure	H	W s/m^2	Luminous exposure	H_V	Lx s
Radiance temperature	T	K	Color temperature	T_C	K

Optical power or radiant flux Φ, emitted by the light source or detected by the detector, is the primary radiometric quantity measured by radiometers. In some cases, flux is the radiometric parameter to be measured, but in many other radiometric measurements, radiance of the source and irradiance incident upon a surface is the quantities of interest. According to Figure 2.2, an elementary area dA_S acts as a source thus emitting the radiation toward the elementary area dA_S acting as the receiver. To simplify the description, it will be assumed that the area dA_S emits uniformly in all directions and that both surfaces are mutually parallel and centered on and perpendicular to the centerline. For a source of radiation that emits uniformly in all directions, it is said that it acts as a Lambertian source. Following this approach, the radiant intensity can be presented as the radiant flux from a center point of the elementary area dA_S into the cone of light with the base located onto the elementary area dA_S. Therefore, the radiant intensity can be defined as the elementary radiant flux emitted into the elementary solid angle. This radiometric quantity is typically used with the point sources, but in general, it designates the radiant flux per unit solid angle from an entire area:

$$I = \frac{\mathrm{d}\Phi}{\mathrm{d}\Omega}, \tag{2.2}$$

where I represents the radiant intensity and dΦ represents the elementary radiant flux emitted in the elementary solid angle dΩ. The radiance of the source is defined as the radiant flux from the elementary area $\mathrm{d}A_S$ of the source within the space defined by the truncated cone of radiation emitted from the entire elementary area $\mathrm{d}A_S$ that impinge the receiver elementary area $\mathrm{d}A_R$. Therefore, the radiance L can be defined as the elementary radiant flux emitted into the elementary solid angle from the elementary area as:

$$L = \frac{\mathrm{d}\Phi}{\mathrm{d}A\mathrm{d}\Omega}, \tag{2.3}$$

where $\mathrm{d}A$ is the elementary area. If it is assumed that the radiant flux, which impinges the receiver elementary surface $\mathrm{d}A_R$, is uniformly distributed over the entire area, the irradiance E is defined as the elementary radiant flux per elementary area, that is, as:

$$E = \frac{\mathrm{d}\Phi}{\mathrm{d}A}. \tag{2.4}$$

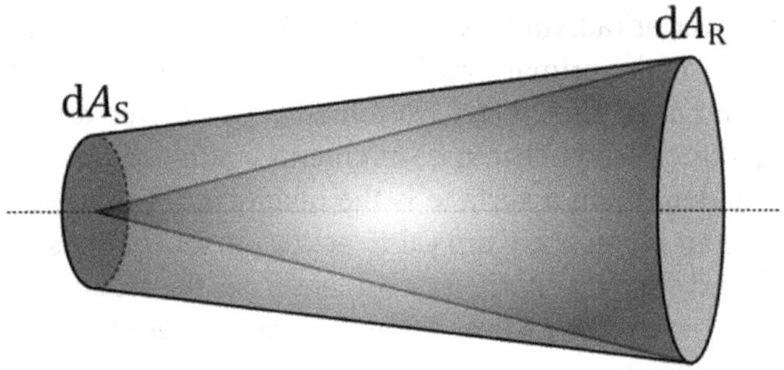

FIGURE 2.2 Geometric representation of basic radiometric quantities.

2.2 OPTICAL RADIOMETRY

Radiative transfer between the bodies is of the paramount importance in the optical radiometry as it explains the concepts of energy and power transfer between the surfaces as well as in the optical systems. To determine the transfer of radiation between the arbitrary surfaces, the radiometric quantities such as radiant flux, radiant intensity, radiance, and irradiance will be taken into account.

2.2.1 THE RADIATIVE TRANSFER

To determine the amount of the radiant flux that originates from the elementary area on one body that reaches the elementary area of the other body, we will observe the position of two corresponding radiating objects in the space with all relevant geometrical parameters for the analysis that are presented in Figure 2.3. Radiance is the typical parameter that describes the radiating characteristics of a surface. Therefore, it will be assumed that the radiance of the first object at the place of the elementary area $\mathrm{d}A_1$ is equal to L_1. Consequently, the radiance of the second object at the place of the elementary area $\mathrm{d}A_2$ is equal to L_2. The distance between the elementary surfaces is R. The angles between the unit vectors $\mathbf{n}_1$ and $\mathbf{n}_2$, normal to the elementary areas $\mathrm{d}A_1$ and $\mathrm{d}A_2$, respectively, and the line that connects elementary areas centers are θ_1 and θ_2, respectively.

The elementary fluxes $\mathrm{d}\Phi_1$ and $\mathrm{d}\Phi_2$, radiated by the elementary areas $\mathrm{d}A_1$ and $\mathrm{d}A_2$, respectively, in the direction of the elementary areas $\mathrm{d}A_2$ and $\mathrm{d}A_1$, respectively, are given by:

$$\mathrm{d}\Phi_1 = L_1 \cos\theta_1 \,\mathrm{d}A_1 \,\mathrm{d}\Omega_2, \tag{2.5}$$

$$\mathrm{d}\Phi_2 = L_2 \cos\theta_2 \,\mathrm{d}A_2 \,\mathrm{d}\Omega_1. \tag{2.6}$$

Based on the geometry presented in Figure 2.3, for the solid angles $\mathrm{d}\Omega_1$ and $\mathrm{d}\Omega_2$, that is, the solid angles under which one elementary has been seen from another, the following is valid:

$$\mathrm{d}\Omega_1 = \frac{\cos\theta_1}{R^2}\mathrm{d}A_1, \tag{2.7}$$

$$\mathrm{d}\Omega_2 = \frac{\cos\theta_2}{R^2}\mathrm{d}A_2. \tag{2.8}$$

By combining eqs 2.5, 2.6, 2.7, and 2.8, we have:

$$\mathrm{d}\Phi_1 = \frac{L_1\cos\theta_1\cos\theta_2}{R^2}\mathrm{d}A_1\mathrm{d}A_2, \tag{2.9}$$

$$d\Phi_2 = \frac{L_2 \cos\theta_1 \cos\theta_2}{R^2} dA_1 dA_2. \quad (2.10)$$

The radiant fluxes $d\Phi_1$ and $d\Phi_2$, given by eqs 2.9 and 2.10, respectively, represent also the radiant fluxes originating from the elementary areas dA_1 and dA_2, respectively, that is captured by the elementary areas dA_2 and dA_1, respectively.

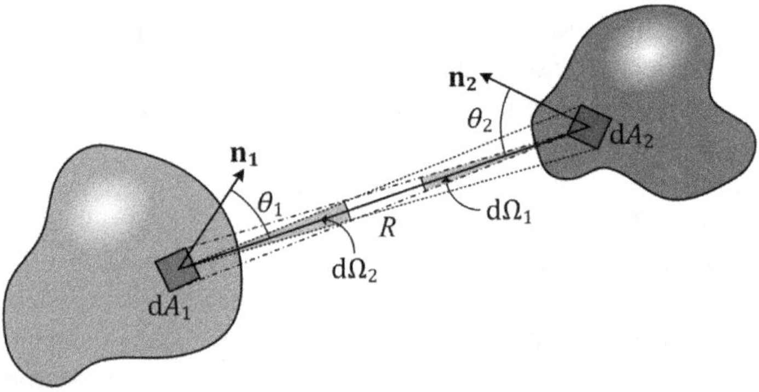

FIGURE 2.3 Position of two radiating objects in the space with all relevant geometrical parameters for the analysis.

2.2.2 THE LAMBERTIAN EMITTERS

The Lambertian radiator has the angle independent radiance, that is, its radiance does not depend on the angle into which the radiation is directed, so $L \neq L(\theta)$ is valid. As the Lambert emitter is the isotropic radiator, the determination of the radiant exitance will be presented on the base of the geometrical parameters given in Figure 2.4. An elementary area dS, positioned onto the radiating body that obeys Lambert law, radiates with the constant radiance L into the overlying hemisphere with the radius R. According to eqs 2.9 and 2.10, the elementary radiant flux $d\Phi$, emitted into the cone limited by the azimuth angles $[\varphi, \varphi + d\varphi]$ and by the elevation angles $[\theta, \theta + d\theta]$, is given by:

$$d\Phi = \frac{L\cos\varphi\cos\theta}{R^2} dSdA \quad (2.11)$$

where $\varphi = 0$ since the normal to the elementary area dA, positioned onto the hemisphere surface, is directed toward its center, that is, toward the elementary area d*S*. Based on the geometrical parameters presented in Figure 2.4, the elementary area d*A* is equal to $\mathrm{d}A = \mathrm{R}^2 \sin\theta \mathrm{d}\theta \mathrm{d}\varphi$. Therefore, for the elementary radiant flux dΦ, the following is valid:

$$\mathrm{d}\Phi = L\sin\theta\cos\theta\mathrm{d}S\mathrm{d}\theta d\varphi. \tag{2.12}$$

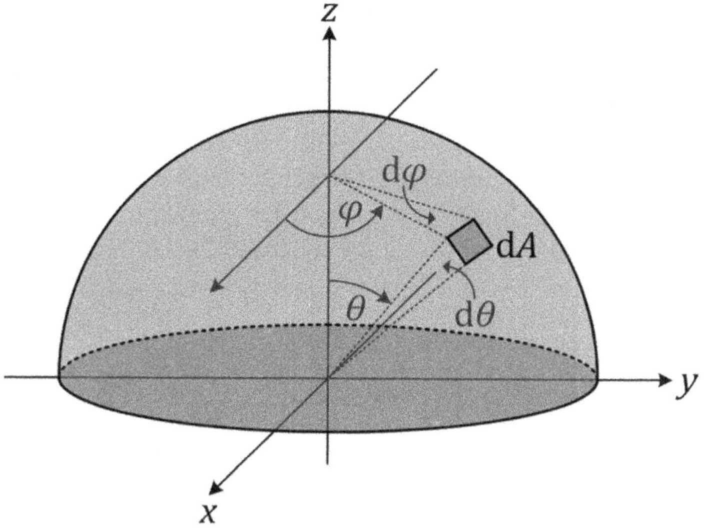

FIGURE 2.4 Lambertian emitter relevant geometry.

The radiant exitance *M*, or equivalently the radiant flux per unit area, is thus given as:

$$M = \frac{\mathrm{d}\Phi}{\mathrm{d}S} = L\int_{\grave{e}=0}^{\grave{e}=\pi/2}\int_{\varphi=0}^{\varphi=2\pi} \sin\theta\cos\theta\mathrm{d}\theta\mathrm{d}\varphi \tag{2.13}$$

Integration of eq 2.13 gives:

$$M = \pi L. \tag{2.14}$$

The radiant exitance is π times the radiance, or as usually said, the radiance is the radiant exitance divided by π.

2.2.3 RADIOMETRIC MEASUREMENTS

All radiometric quantities that have been introduced, such as energy, radiant flux, and radiance, are integral radiometric quantities. Measurement of such energetic quantities can be performed by using some kind of thermal detectors that converts the captured energy into the heat and subsequently into the temperature increase of the detector, such as thermocouples and bolometers. However, the sensitivity of such detectors is rather low so the measurements of such parameters are usually performed by the optical detectors that have different sensitivities for different wavelengths such as solid-state photodetectors (photodiodes, avalanche photodiodes, photoconductors) and photomultipliers. Typically, silicon photodiode are sensitive in the range from 200 to 1100 m with the peak sensitivity in the near-infrared range (800 – 900 nm). As the radiometric measurement can be spectrally sensitive, one can define the corresponding spectral radiometric quantities. The spectral radiant flux Φ_λ or Φ_v is the radiant flux emitted by the source in the wavelength range $[\lambda,\lambda + d\lambda]$ or in the frequency range $[v,v + dv]$ given by:

$$\Phi_\lambda = \frac{d\Phi}{d\lambda}, \tag{2.15}$$

$$\Phi_v = \frac{d\Phi}{dv}. \tag{2.16}$$

Bearing in mind that $\lambda v = c$, we have:

$$\Phi_\lambda = \frac{d\Phi}{d\lambda} = \frac{d\Phi}{d(c/v)} = -\frac{v^2}{c}\frac{d\Phi}{dv} = -\frac{v^2}{c}\Phi_v. \tag{2.17}$$

Typical configuration of the radiometric measurement is given in Figure 2.5, where the photodetector with spectrally dependent sensitivity is used to characterize the source of radiation. A source of optical radiation emits the radiation through the source aperture of area A_S. The detector aperture of the area A_D is positioned at the distance r from the source aperture. The area of the detector aperture limits the flux that after passing through the filter F impinge the detector D. In the case of a solid-state detector, the output signal is an electrical current i_D that is proportional to the detector incident flux Φ_D with the conversion factor $\Re$, that is, $i_D = \Re\ \Phi_D$. To convert the current signal into the voltage signal a transimpedance amplifier A is used. The relation that describes the overall signal response of the measurement chain is given by:

$$V(\lambda) = R(\lambda)\ \Phi(\lambda). \tag{2.18}$$

where $V(\lambda)$ is the output voltage signal for a given wavelength λ and $R(\lambda)$ is the overall responsivity of the measurement system.

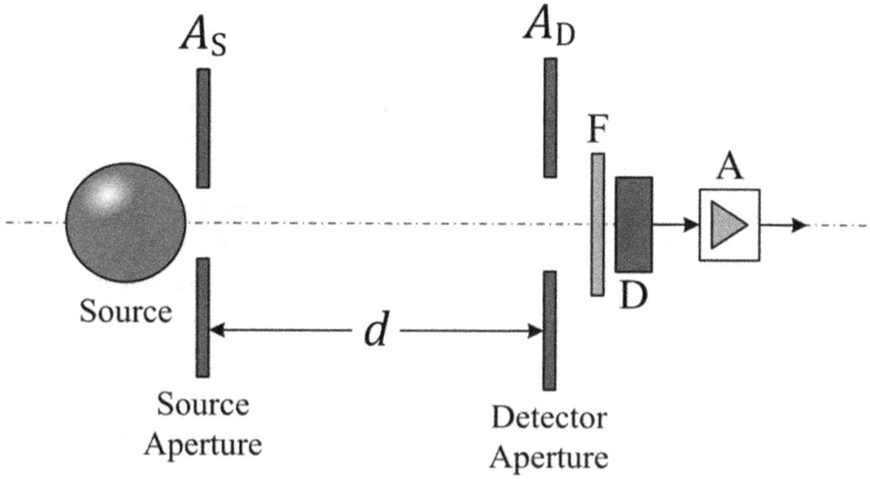

FIGURE 2.5 Typical configuration of the radiometric measurement system.

The relation between the output signal of the entire measurement system, given in Figure 2.5 and the relevant parameters of the same will be derived in the case of the radiation source that follows the Lambertian law with the spectral radiance of $L_\lambda(\lambda)$. Typically, the apertures are of the circular shape, normal to the line of sight, with the radiuses of r_S and r_D for source and detector apertures, respectively. The distance between the apertures is d. In the case of uniform spatial distribution of the detector sensitivity, the output voltage signal V is given by:

$$V = \frac{A_S A_D (1+c)}{r_S^2 + r_D^2 + d^2} \int_\lambda L_\lambda(\lambda)\tau(\lambda)R(\lambda)\mathrm{d}\lambda, \tag{2.19}$$

where c is the correction factor that depends on the system geometry and $\tau(\lambda)$ is the filter transmittance. Typical employed radiant sources have the spectrum located at the wavelength λ_0 with the spectral width $\Delta\lambda$ that is much smaller than the central frequency $\Delta\lambda \gg \lambda_0$, or the bandpass of the filter is narrow compared to the central frequency. In this case, eq 2.19 can be simplified into:

$$V = \frac{A_S A_D (1+c)}{r_S^2 + r_D^2 + d^2} L_\lambda(\lambda_0)\tau(\lambda_0)R(\lambda_0)\Delta\lambda. \tag{2.20}$$

The last equation is often used in radiometry but it is important to be aware of the assumptions that have been made to derive it.

Wavelength selective measurement of the optical radiation is performed by a spectral radiometer. The measurement system, presented in Figure 2.5, although aimed for radiometric measurement at the particular wavelength, defined by the optical filter, cannot be referred as the spectral radiometer as the measurement is performed at the fixed wavelength. The measurement principle of the spectral radiometer is given in Figure 2.6 aimed for spectral radiance measurement. The measurement system is presented in such a way that with minimal changes in the design it can be adapted to measure spectral irradiance or spectral flux. Similarly to the measurement system in Figure 2.5, the wavelength selective measurement systems measure the radiation from the source that passes through the source aperture of area A_S, detector aperture of area A_D. After passing the lens L, the optical radiation is focused on the monochromator M that at its entrance has a slit S_1 and an exit slit S_2. A monochromator is an optical device that can transmit a selectable narrow band of wavelengths of optical radiation from a wider range of wavelengths that illuminates its entrance. Due to the dispersive element of the monochromator, the optical radiation with the selected wavelength impinges the detector D that is further connected to the signal amplifier A. The overall spectral responsivity $R_\lambda(\lambda_0)$ of the imaging system, the monochromator, and the detector is a function of wavelength where the monochromator is adjusted to the wavelength λ_0. The dependence of the transmittance, the reflectance of the used optical components can also influence the overall spectral responsivity.

Internationally recognized designations for the various wavelength regions and its wavelength intervals that are commonly used in radiometric measurements are presented in Table 2.2.

2.2.4 REFLECTION, ABSORPTION, AND TRANSMISSION

As mentioned in the previous chapter, transmission, reflection, and absorption, which can be caused by various factors, can significantly influence the overall responsivity and spectral responsivity of the radiometric measurement system. Generally, a material may transmit, reflect, emit, or absorb radiation and typically does more than one of these at a time. Spectral and geometric parameters of an optical component influence all of these properties. The corresponding processes are graphically presented in Figure 2.7 for a light incident on a plane parallel plate where it is shown that the incident light can be reflected, absorbed, or transmitted through the material.

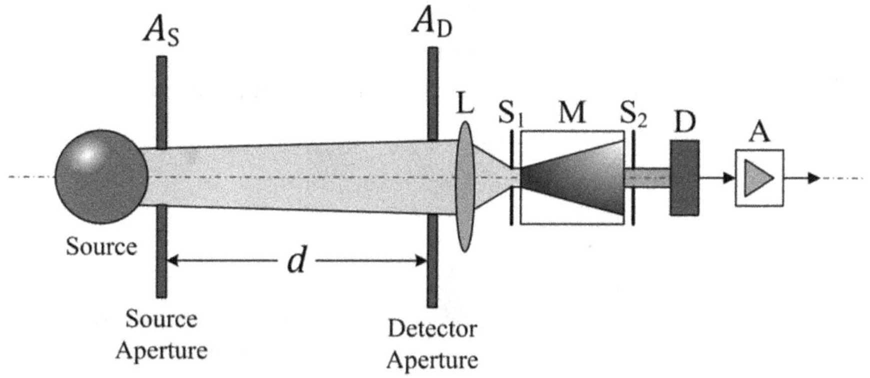

FIGURE 2.6 Measurement principle of the spectral radiometer.

TABLE 2.2 Wavelength Regions and Wavelength Intervals.

Wavelength regions	Wavelength intervals
UV-C	100–280 nm
UV-B	280–315 nm
UV-A	315–400 nm
Visible	380–780 nm
IR-A	780–1400 nm
IR-B	1.4–3 μm
IR-C	– 1000 μm

Based on the energy conservation law, the sum of reflected Φ_R, absorbed Φ_A, and transmitted Φ_T flux must be equal to the incident flux Φ_I or:

$$\Phi_R + \Phi_A + \Phi_T = \Phi_I, \tag{2.21}$$

$$\frac{\Phi_R}{\Phi_I} + \frac{\Phi_A}{\Phi_I} + \frac{\Phi_T}{\Phi_I} = \rho + \alpha + \tau = 1, \tag{2.22}$$

where the reflectivity ρ, absorptivity α, and transmissivity τ are defined by this equation. If there are no nonlinear processes within the material and thus no change in the radiation frequency (there is no frequency doubling, no fluorescence, etc.), the energy conservation law can be also applied to any particular wavelength, so we have:

$$\frac{\Phi_R(\lambda)}{\Phi_I(\lambda)}+\frac{\Phi_A(\lambda)}{\Phi_I(\lambda)}+\frac{\Phi_T(\lambda)}{\Phi_I(\lambda)}=\rho(\lambda)+\alpha(\lambda)+\tau(\lambda)=1 \tag{2.23}$$

A similar argument can be applied to each of the linear or circular polarization components, as long as the polarization is conserved.

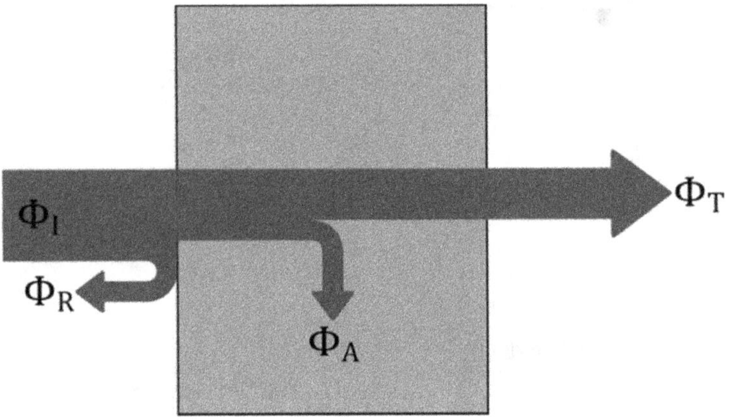

FIGURE 2.7 Transmittance, absorptance, and reflectance.

2.2.5 KIRCHHOFF'S LAW

Kirchhoff's law of heat radiation states that under certain conditions, the emissivity is equal to the absorptivity of the isothermal bodies. If we consider an isothermal cavity that is also in the thermal equilibrium, as shown in Figure 2.8, which every part is at the same constant temperature *T* and the walls are made of the same material, meaning that optical properties of the inner cavity walls are uniform, then according to the energy conservation law, the cavity walls radiant flux of the emitted radiation must be equal to the flux that is captured by the cavity walls. The equilibrium is maintained only if the radiation is perfectly reflected back. Moreover, the photon that is contained within the cavity walls must be only emitted or reflected from the cavity walls. Therefore, one can say that the probability that the photon, within the cavity, is emitted by the cavity walls or it is reflected from the cavity wall is equal to unity, thus giving $\varepsilon + \rho = 1$, where emissivity ε

represents at the same time the probability of the photon emission from the cavity walls surface and reflectivity ρ represents at the same time the probability of the photon emission from the cavity walls surface. If the radiation doesn't perfectly reflect, then any deficiency in the reflectivity must be compensated exactly by emissivity, or:

$$\varepsilon = 1 - \rho. \tag{2.24}$$

Then, by combining eqs 2.22 and 2.24 and taking into considerations that the cavity walls are opaque ($\tau = 0$), we have:

$$\varepsilon = \alpha, \tag{2.25}$$

or the emissivity must be equal to absorptivity. Similarly, it can be stated that for each wavelength the emissivity is equal to absorptivity:

$$\varepsilon(\lambda_0) = \alpha(\lambda), \tag{2.26}$$

while in the opposite case $\varepsilon(\lambda_0) \neq \alpha(\lambda)$, the energy conservation law would be broken.

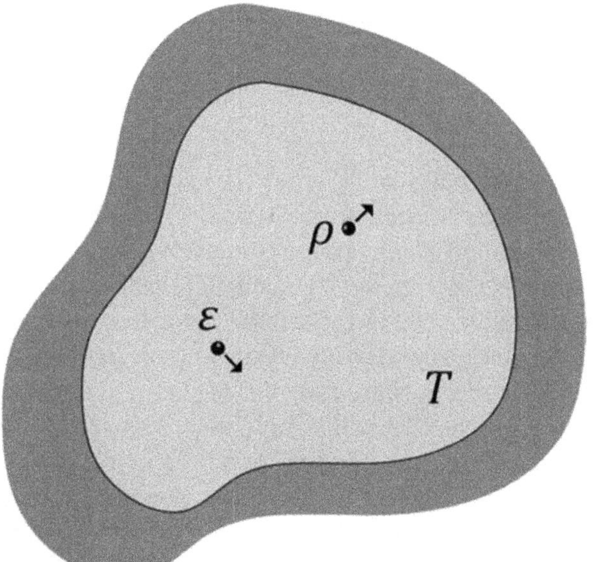

FIGURE 2.8 Isothermal cavity.

However, there are some cases where the application of Kirchhoff's law is not straightforward, for example, when bodies are covered with the paints with metallic particles, layered optical materials, or semi-infinite bodies that have a large thermal gradient at the surface.

2.3 MEASUREMENT TECHNIQUES IN RADIOMETRY

2.3.1 ABSOLUTE RADIOMETER

Absolute radiometer is an instrument that can not only measure the level of captured optical radiation but also cannot be simply calibrated by comparing it with another instrument or reference source, that is, it must be a self-calibrated device. The absolute radiometer is typically based on one of two basic physical principles: electrical substitution (thermal effects occur) and quantum physics (photon detection). To make a solid-state detector an absolute radiometer, one can perform several measurements to find the quantum efficiency of the solid-state detector (photodiode), that is, the number of charge carriers that arrive at the inputs of the measuring electrical circuitry per incident photon, and thus to estimate its spectral response. The photodiode responsivity $\Re_\lambda$ at a particular wavelength λ is given as:

$$\Re_\lambda = \frac{e}{hc}\eta\lambda, \tag{2.27}$$

where e is the electron charge and η is the photodiode quantum efficiency. The basic principle that lies behind the photodiode as the absolute radiometer is the fact that quantum efficiency of the photodiode can be determined from a number of simple experiments. To accurately determine the photodiode efficiency, one must measure its reflectance ρ as it is satisfied $\eta = 1 - \rho$. The photodiode reflectance can be easily measured very accurately.

Beside the radiometers based on the quantum effects, as mentioned previously, there are also absolute radiometers based on the electrical substitution principles. The operating principle of electrical substitution radiometers is to compare the heating effect of optical radiation with the heating effect of the corresponding electrical heater. Figure 2.9 presents the operating principle of the electrical substitution radiometer. A typical electrical substitution radiometer consists of an optical absorbing element, in which the simplest construction can be in the shape of a metal plate, connected by the poorly conducting heat link, to a reference heat sink, which is kept at a constant temperature. The maintenance of the constant temperature of the heat sink can be accomplished actively by a temperature control loop, or passively through its thermal mass.

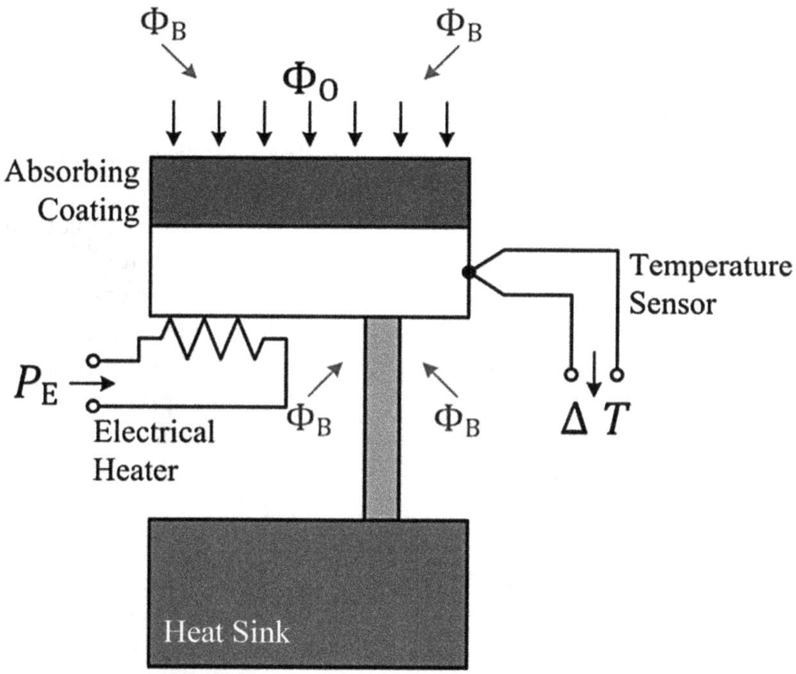

FIGURE 2.9 Block schematic representation of the absolute radiometer based on the electrical substitution principle.

The main physical principle is to keep the same temperature at the reference heat sink whether it is exposed to the optical radiation with the radiant flux Φ_O or heated by means of an electrical heater that delivers power P_E to the reference heat sink. The temperature difference ΔT between the ambient and the reference heat sink is given by:

$$\Delta T_O = K(\Phi_O + \Phi_B), \tag{2.28}$$

$$\Delta T_E = K(P_E + \Phi_B), \tag{2.29}$$

where ΔT_O is the temperature difference caused by the optical radiation, ΔT_E is the temperature difference caused by the electrical heater, K is the constant that depends on the radiometer construction, and Φ_B is the radiant flux of the background radiation. By keeping $\Delta T_E = \Delta T_O$, one can simply measure the radiant flux by measuring the delivered electrical power as $\Phi_O = P_E$.

2.3.2 RADIANT FLUX MEASUREMENT

The radiometric quantities measurements, such as radiant flux, irradiance, radiance, and intensity, are made with the help of a radiometer, that is, the instrument that measures the optical flux captured by its surface. Typically, the flux that impinges the detector surface generates an electrical signal, which is proportional to the captured flux. By calibration, it is possible to make a relation between the measured radiant flux and the corresponding electrical signal. It is important to notice that due to the relation between the radiant flux and certain geometrical parameters, it is possible to measure other radiometric quantities by involving appropriate adjustments of the measurement setup. For example, if the detector measures a certain radiant flux that is captured by the detector active surface, the irradiance or the flux density can be measured by dividing the measured flux by the active area of the detector surface. Of course, the measured irradiance is averaged over the area of the entrance aperture.

Exitance is similar to the irradiance but it is the property of the radiation source. However, it can be also determined by measuring the radiant flux that is captured by the detector. If the radiant flux is known, then, based on the geometry of the measurement setup, the exitance of the source can be determined if the distance from the source to the detector and the size of the source are known.

To measure the radiant intensity of a source, the radiant flux angular distribution must be determined. Typically, in the case of a point source, radiant intensity can be measured by measuring the optical flux that impinges on the detector and certain geometrical parameters of the measurement setup. First of all, the distance between the source and the detector must be much larger than the source dimensions thus "forcing" the source to be a point-like source. If the distance is large enough, as it is presented in Figure 2.10, the radiation emitted in the solid angle Ω defined by the detector aperture area A and the distance d as $\Omega = A/d^2$ will be captured by the detector D and amplified by the signal amplifier A, so the measured flux will be Φ. According to the presented geometry of the measurement setup, the measured radiant intensity will be:

$$I = \frac{\Phi}{\Omega} = \frac{d^2}{A}\Phi \tag{2.30}$$

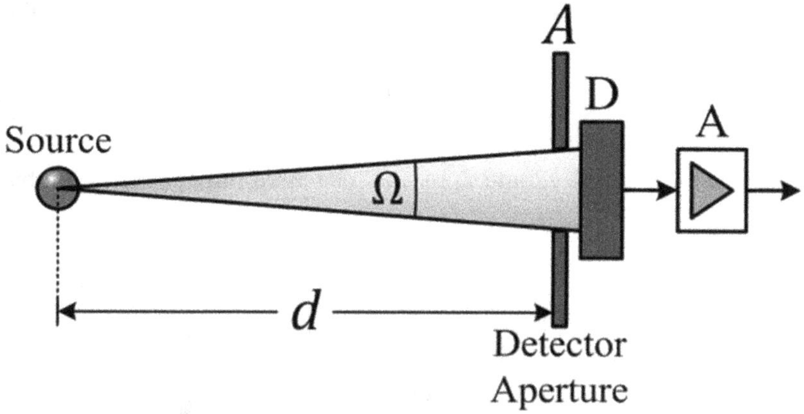

FIGURE 2.10 Measurement setup for measuring the radiant intensity.

The radiance measurement can be performed according to the measurement setup presented in Figure 2.11. The radiation source and the detector D are positioned in such a way the between them, in the simplest measurement arrangement, is an aperture with the entrance opening area A_1. This aperture serves to limit the solid angle to a certain value to enable that all of the radiations that pass through the aperture impinges the detector. This condition is fulfilled if it is satisfied:

$$\frac{A_1}{d_1^2} < \frac{A_2}{(d_1 + d_2)^2}, \tag{2.31}$$

where d_1 is the distance between the aperture and the source, d_2 is the distance between the aperture and the detector, and A_2 is the entrance opening area of the detector aperture.

To avoid the averaging processes $d_1^2 \gg A_1$ and $(d_1 + d_2)2 \gg A_2$ must be fulfilled, that is, all solid angles must be much smaller than unity. If the detector D and signal amplifier A measure the radiant flux of Φ, then the corresponding radiant intensity of the source at a given surface point is given by:

$$I = \frac{\Phi d_1^2}{A_1}. \tag{2.32}$$

Due to very small opening of the aperture between the source and the detector and the corresponding geometrical parameters of the measurement setup, the detector receives the radiation only from a small portion of the source surface. The area of the source that emits the light toward the detector is given by:

$$A_S = \frac{d_1}{d_2} A_2. \tag{2.33}$$

According to eqs 2.32 and 2.33, the measured radiance is given by:

$$L = \frac{I}{A_S} = \frac{d_1 d_2}{A_1 A_2} \Phi. \tag{2.34}$$

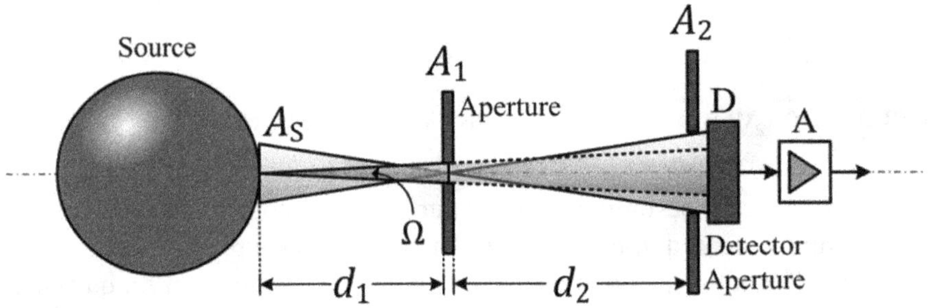

FIGURE 2.11 Measurement setup for measuring the radiance.

The presented measurement schematic, aimed for the radiance measurements, is given in its simplest form, that is, there are no needs for additional optical elements. For example, if instead of a simple aperture between the source and the detector a lens with much higher aperture area is placed, then a much higher level of captured optical power can lead to the higher precision measurement. One must pay attention that all solid angles still must be kept well below unity to avoid averaging. Typical measurement setup with the lens in the middle is given in Figure 2.12.

The measurement of low radiant fluxes emanating from faint radiation sources can be a problem if the measurement has been performed with the sensing setups shown earlier. Due to the very low light levels, the influences of other parasitic effects can interfere and thus hide the useful signal. For example, relative high levels of background light, voltage offset of the signal amplifier, and its thermal and time drifts as well as high levels of flicker noise of the amplifier can significantly deteriorate the measurement accuracy and precision. However, there is a solution in moving the useful signal toward the higher frequencies and perform the phase-sensitive detection with the help of lock-in amplifier. The typical measurement setup is presented in Figure 2.13. Between the source and the main detector D_1, an optical chopper has been installed. Optical chopper is a device that periodically interrupts a light beam. There are several different types of optical choppers. One of them is presented in Figure 2.13 and consists of an opaque rotating wheel with symmetrically positioned transparent parts (holes) that is driven by a motor *M*. The signal from the detector is amplified by the signal amplifier A_1 in which output has been brought to the input port of the lock-in amplifier LA. To perform a phase-sensitive detection, there is a need for the corresponding reference signal that must be brought to the reference input of the lock-in amplifier. The reference signal is obtained with the help of auxiliary light source LS and detector D_2 that are separated by the same optical chopper wheel. The reference signal is obtained at the output of the second signal amplifier A_2. The lock-in amplifier output signal will be proportional to the detector captured radiant flux but most of the disruptive influences will be significantly suppressed.

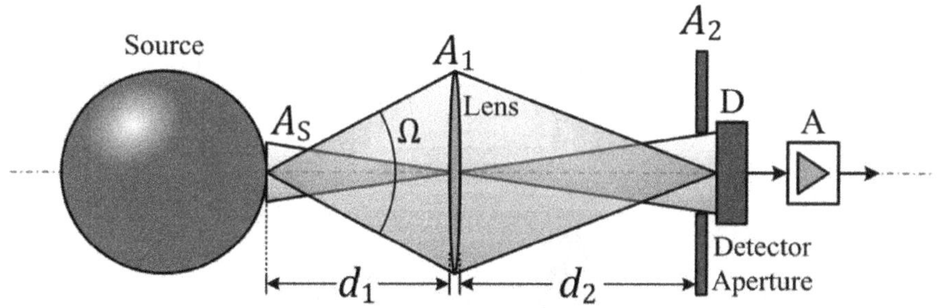

FIGURE 2.12 Measurement setup for measuring the radiance with incorporated simple lens-based optical system.

Accurate and high precision measurement of very low light levels or radiant fluxes draws a lot of attention in the last two decades as one of the most powerful measurement techniques to detect and find the extrasolar planets. The main principle of an application of the above-presented measurement technique is given in Figure 2.14. When the planet transits the star disk in the direction A–B–C–D–E, there is a small drop in the measured radiant flux emanating from the star, as presented on a diagram given in the lower part of Figure 2.14. The reason for this phenomenon is the shadow that the planet makes. The measured flux when the planet is far away from the star Φ_0 is just slightly higher than in the case when the planet transits the star disk $\Phi_0 - \Delta\Phi$, where $\Delta\Phi$ is the flux drop that is typically much smaller than the star flux $\Delta\Phi \gg \Delta\Phi_0$. As the transition lasts for a very long period of time, the measurement technique that is capable of detecting extrasolar planets must be very precise to be able to measure very small flux changes and very stable (low output signal drift) for a very long period of time (suppressed flicker noise).

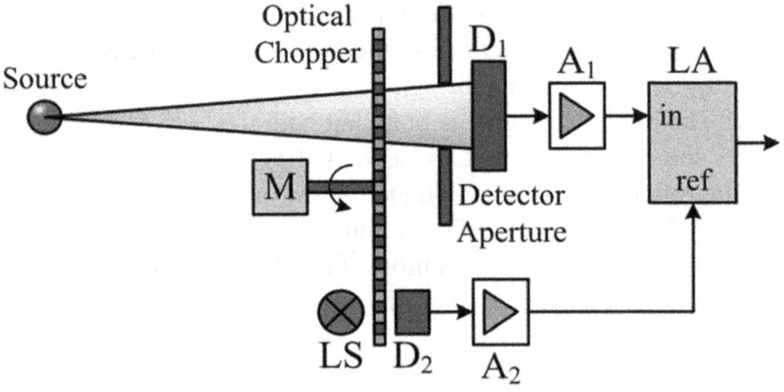

FIGURE 2.13 Measurement setup aimed for measuring low radiant fluxes.

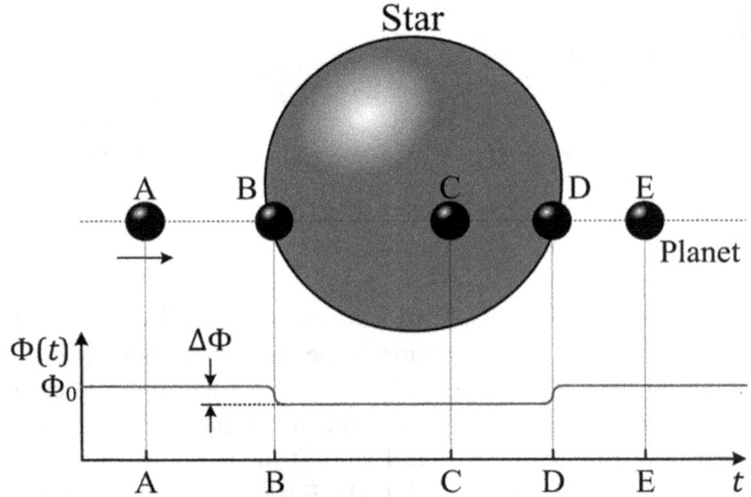

FIGURE 2.14 Method for detecting the extrasolar planet.

2.3.3 INTEGRATING SPHERE

An integrating sphere is an optical device that consists of a spherical shell in which inner walls are covered with a diffuse, white, and highly reflective coating. To be able to perform the measurements, there are small holes at the sphere for entrance and exit ports. The main function of an integrating sphere is to spatially integrate the radiant flux by a uniform scattering or diffusing effect that occurs at its inner walls. Light rays, which are incident at any point on the integrating sphere inner walls, are due to the multiple scatterings, distributed evenly to the all other points at the inner sphere surface. An integrating sphere is typically used with some light source and a detector for radiant flux measurement and it has become a standard instrument in optical radiometry and photometry. It has a unique function of the total flux measurement of any light source in a single measurement. Typical integrating sphere arrangement is presented in Figure 2.15. The light, which radiant flux Φ_I is to be measured, enters the integrating sphere through the input hole with the area of entrance opening of A_I. Due to the multiple scatterings at the inner highly reflective and diffusive walls, the light emanating from the source illuminates the inner walls with uniform irradiance E.

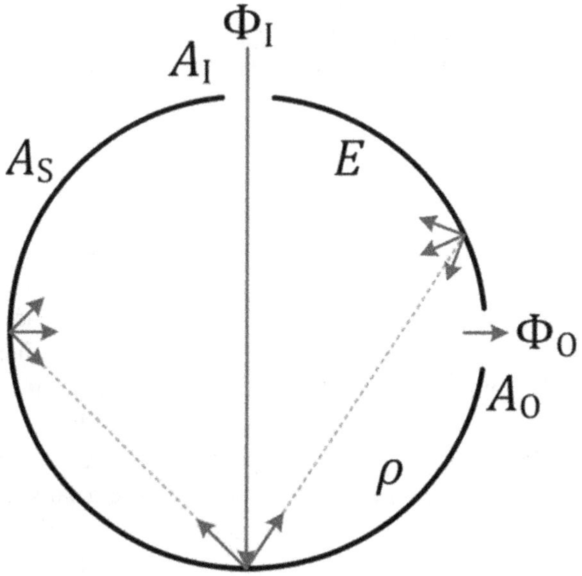

FIGURE 2.15 Integrating cavity with relevant parameters.

Based on the energy conservation law, we have:

$$\Phi_I = E(A_I + A_O) + E(1 - \rho)(A_S - A_I - A_O), \tag{2.35}$$

where A_O represents the area of the exit opening, ρ is the reflectance of the inner walls, and A_S is the overall area of the inner walls. One can see that the incoming flux is divided into two parts. The first one, which is represented by the first argument at the right side of eq 2.35, is the flux loss due to the entrance and exit holes where the light can leave the integrating sphere. The second one, which is represented by the second argument at the right side of eq 2.35, is the flux loss due to the wall absorption. If the detector is positioned at the exit hole and captures all the flux leaving the sphere at that point, the detector measures the flux Φ_O that is given by:

$$\Phi_O = EA_O. \tag{2.36}$$

By combining the last two equations, we have:

$$\Phi_O = \frac{\Phi_I}{(1+(A_I/A_O))+(1-\rho)\left[(A_S/A_O)-(1+(A_I/A_O))\right]} \tag{2.37}$$

that is, the measured flux is proportional to the incoming flux (light source flux). Typically, the integrating sphere is calibrated by the known radiant flux Φ_{IC}, where the measured flux is Φ_{OC}, so, based on the measured flux Φ_O, one can determine the light source flux as:

$$\Phi_I = \frac{\Phi_{IC}}{\Phi_{OC}}\Phi_O = K_C\Phi_O, \tag{2.38}$$

where $K_C = \Phi_{IC}/\Phi_{OC}$ is the calibration constant. One can notice that integrating sphere parameters such the areas of holes, the area of integrating sphere inner walls, and the reflectance of the walls, although figure in the measurement, are not necessary to know as the calibration process eliminates them from the equation. As mentioned previously, an integrating sphere is capable of measuring the total radiant flux of an arbitrary light source. Simply, by placing the source inside the integrating sphere and hiding it behind the baffles that are covered with the same material as the inner sphere walls, the measured output flux will be proportional to the total source flux. The baffles serve to stop the source radiation to left the integrating sphere before its reflection from the walls.

2.4 PHOTOMETRY

In photometry, the light is measured in such a way that the measurement result is in accordance with the human vision. A large number of devices are aimed for human use such as computer displays, TVs, and traffic signal lights, therefore they must be tested and evaluated based on the spectral responsivity of the average human eye. Radiometry covers all optical spectral regions, ranging from ultraviolet to infrared, while the photometry only the range of light that is visible for average human eye, that is, in the spectral range from 380 to 770 nm. In the last times, human eye was used exclusively as an optical detector. To measure the intensity of a test light, the test light source was moved and placed at the varied distances from an eye or a screen and compared with the intensity of a standard light source positioned at a fixed position by a visual observation. The distance of the test light source was set so that both light sources, test and standard, appear equally bright. The estimated intensity of the test light source is calculated as the ratio of the intensity of the standard source and the ratio of the distances squared. Such a measured light intensity was named candle power, or in present terminology luminous intensity. This photometric quantity was the first defined photometric quantity. Since the 1940s the human eye, which was predominantly used as an optical detector, has been replaced by a light-sensitive detector (solid-state detector, photomultiplier tube).

Modern photometry measurements have been performed with the help of photodetectors, thus referring to the physical photometry. Physical photometry characterizes the optical radiation with an optical detector that mimics the spectral response of the human eye, that is, it has incorporated weighted spectral response same as the spectral response of a human eye. According to Table 2.1, the photometric units include the lumen or the luminous flux, the candela or the luminous intensity, the lux or the illuminance, and the candela per square meter or the luminance. Each photometric quantity is the spectrally integrated corresponding radiometric quantities weighted by the human eye spectral response.

2.4.1 SPECTRAL RESPONSE OF A HUMAN EYE

In 1924, the relative spectral response of the human eye was specified by the Commission Internationale de l'Eclairage (CIE), (the International

Commission on Illumination). Such a defined relative spectral response of the human eye was called the spectral luminous efficiency for photopic vision with a joined symbol $V(\lambda)$. The relative spectral response is defined in the spectral domain ranging from 360 to 830 nm and it is normalized to its peak value at 555 nm. The values were republished by CIE in 1983 and adopted by Comité International des Poids et Mesures (CIPM) (the International Committee on Weights and Measures) in the same year. Tabular presentation of the spectral luminous efficiency for photopic vision values at 5 nm increments is presented in Table 2.3. One can notice from the tabular values that outside the spectral range ranging from 380 to 770 nm, the values of the spectral luminous efficiency for photopic vision are smaller than 10^{-4} thus negligible for the calculations.

A photometric quantity X_V can be related to the corresponding radiometric spectral quantity X_λ in the following way:

$$X_V = K_m \int_{\lambda=360\ \text{nm}}^{\lambda=830\ \text{nm}} X_\lambda V(\lambda)\,d\lambda, \tag{2.39}$$

where K_m is the constant that relates photometric and radiometric quantities and it is named the maximum spectral luminous efficiency of radiation for photopic vision. The value of this constant is defined as the spectral luminous efficiency of radiation at the frequency of 540×10^{-12}Hz (or at the wavelength of 555.016 nm in the standard air) to be $K_m = 683$ lm/W. This value isn't defined strictly for the wavelength of 555 nm, where the peak value should appear, but for the wavelength that is slightly shifted from the maximum value. The exact maximum value of the constant K_m can be accurately calculated as 683 lm/W × $V(\lambda = 555\ \text{nm})/V(\lambda = 555.016\ \text{nm})$ = 683.002 lm/W, which can be simply rounded to 683 lm/W, with the negligible error.

The function $V(\lambda)$, the values of which are given in Table 2.3 and which is also graphically presented in Figure 2.16, is defined for the standard photometric observer for photopic vision at relatively high luminance levels (higher than 1 cd/m^2). The spectral responsivity of an average human eye significantly differs at very low luminance levels (at luminance levels lower than 10^{-3} cd/m^2) when the eye rods are the predominant receptors. This type of vision is named scotopic vision and its spectral luminous efficiency values $V'(\lambda)$ are presented in Table 2.4 and graphically in Figure 2.16. The scotopic vision spectral luminous efficiency has the maximum value at 507 nm.

TABLE 2.3 Definitive Values of the Spectral Luminous Efficiency Function for Photopic Vision $V(\lambda)$.

λ[nm]	V(λ) [a. u.]	λ [nm]	V(λ) [a. u.]	λ [nm]	V(λ) [a. u.]	λ [nm]	V(λ) [a. u.]
360	3.9170×10^{-6}	480	0.13902	600	0.63100	720	1.0470×10^{-3}
365	6.9650×10^{-6}	485	0.16930	605	0.56680	725	7.4000×10^{-4}
370	1.2390×10^{-5}	490	0.20802	610	0.50300	730	5.2000×10^{-4}
375	2.2020×10^{-5}	495	0.25860	615	0.44120	735	3.6110×10^{-4}
380	3.9000×10^{-5}	500	0.32300	620	0.38100	740	2.4920×10^{-4}
385	6.4000×10^{-5}	505	0.40730	625	0.32100	745	1.7190×10^{-4}
390	1.2000×10^{-4}	510	0.50300	630	0.26500	750	1.2000×10^{-4}
395	2.1700×10^{-4}	515	0.60820	635	0.21700	755	8.4800×10^{-5}
400	3.9600×10^{-4}	520	0.71000	640	0.17500	760	6.0000×10^{-5}
405	6.4000×10^{-4}	525	0.79320	645	0.13820	765	4.2400×10^{-5}
410	1.2100×10^{-3}	530	0.86200	650	0.10700	770	3.0000×10^{-5}
415	2.1800×10^{-3}	535	0.91485	655	8.1600×10^{-2}	775	2.1200×10^{-5}
420	4.0000×10^{-3}	540	0.95400	660	6.1000×10^{-2}	780	1.4990×10^{-5}
425	7.3000×10^{-3}	545	0.98030	665	4.4580×10^{-2}	785	1.0600×10^{-5}
430	1.1600×10^{-2}	550	0.99495	670	3.2000×10^{-2}	790	7.4657×10^{-6}
435	1.6840×10^{-2}	555	1.00000	675	2.3200×10^{-2}	795	5.2578×10^{-6}
440	2.3000×10^{-2}	560	0.99500	680	1.7000×10^{-2}	800	3.7029×10^{-6}
445	2.9800×10^{-2}	565	0.97860	685	1.1920×10^{-2}	805	2.6078×10^{-6}
450	3.8000×10^{-2}	570	0.95200	690	8.2100×10^{-3}	810	1.8366×10^{-6}
455	4.8000×10^{-2}	575	0.91540	695	5.7230×10^{-3}	815	1.2934×10^{-6}
460	6.0000×10^{-2}	580	0.87000	700	4.1020×10^{-3}	820	9.1093×10^{-7}
465	7.3900×10^{-2}	585	0.81630	705	2.9290×10^{-3}	825	6.4153×10^{-7}
470	9.0980×10^{-2}	590	0.75700	710	2.0910×10^{-3}	830	4.5181×10^{-7}
475	0.11260	595	0.69490	715	1.4840×10^{-3}		

TABLE 2.4 Definitive Values of the Spectral Luminous Efficiency Function for Scotopic Vision $V'(\lambda)$.

λ [nm]	V′ (λ) [a.u.]	λ [nm]	V′ (λ) [a.u.]	λ [nm]	V′ (λ) [a.u.]	λ [nm]	V′ (λ) [a.u.]
380	5.890×10^{-4}	485	8.510×10^{-1}	590	6.550×10^{-2}	695	2.501×10^{-5}
385	1.108×10^{-3}	490	9.040×10^{-1}	595	4.690×10^{-2}	700	1.780×10^{-5}
390	2.209×10^{-3}	495	9.490×10^{-1}	600	3.315×10^{-2}	705	1.273×10^{-5}
395	4.530×10^{-3}	500	9.820×10^{-1}	605	2.312×10^{-2}	710	9.140×10^{-6}
400	9.290×10^{-3}	505	9.980×10^{-1}	610	1.593×10^{-2}	715	6.600×10^{-6}
405	1.852×10^{-2}	510	9.970×10^{-1}	615	1.088×10^{-2}	720	4.780×10^{-6}
410	3.484×10^{-2}	515	9.750×10^{-1}	620	7.370×10^{-3}	725	3.482×10^{-6}
415	6.040×10^{-2}	520	9.350×10^{-1}	625	4.970×10^{-3}	730	2.546×10^{-6}
420	9.660×10^{-2}	525	8.800×10^{-1}	630	3.335×10^{-3}	735	1.870×10^{-6}
425	1.436×10^{-1}	530	8.110×10^{-1}	635	2.235×10^{-3}	740	1.379×10^{-6}
430	1.998×10^{-1}	535	7.330×10^{-1}	640	1.497×10^{-3}	745	1.022×10^{-6}
435	2.625×10^{-1}	540	6.500×10^{-1}	645	1.005×10^{-3}	750	7.600×10^{-7}
440	3.281×10^{-1}	545	5.640×10^{-1}	650	6.770×10^{-4}	755	5.670×10^{-7}
445	3.931×10^{-1}	550	4.810×10^{-1}	655	4.590×10^{-4}	760	4.250×10^{-7}
450	4.550×10^{-1}	555	4.020×10^{-1}	660	3.129×10^{-4}	765	3.196×10^{-7}
455	$5.130e \times 10^{-1}$	560	3.288×10^{-1}	665	2.146×10^{-4}	770	2.413×10^{-7}
460	5.670×10^{-1}	565	2.639×10^{-1}	670	1.480×10^{-4}	775	1.829×10^{-7}
465	6.200×10^{-1}	570	2.076×10^{-1}	675	1.026×10^{-4}	780	1.390×10^{-7}
470	6.760×10^{-1}	575	1.602×10^{-1}	680	7.150×10^{-5}		
475	7.340×10^{-1}	580	1.212×10^{-1}	685	5.010×10^{-5}		
480	7.930×10^{-1}	585	8.990×10^{-2}	690	3.533×10^{-5}		

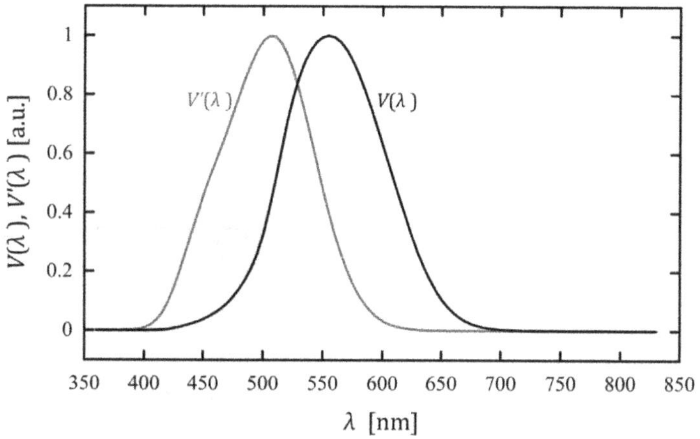

FIGURE 2.16 Spectral luminous efficiency function for photopic vision $V(\lambda)$ and for scotopic vision $V'(\lambda)$.

2.4.2 STANDARD PHOTOMETER AND THE REALIZATION OF THE CANDELA

Typically, a standard photometer consists of a silicon photodiode, a $V(\lambda)$ correction filter, and a photometer head with the precisely defined aperture opening, as presented in Figure 2.17. If $s(\lambda)$ is the absolute spectral power responsivity of the whole photometer, then the illuminance responsivity s_V of the photometer is given by:

$$s_V = \frac{A\int_\lambda S(\lambda)s(\lambda)\mathrm{d}\lambda}{K_m\int_\lambda S(\lambda)V(\lambda)\mathrm{d}\lambda}, \tag{2.40}$$

where A is the photometer head area of aperture opening and $S(\lambda)$ is the relative power spectral distribution of the light source to be measured. If the light source with the power spectral distribution of $S(\lambda)$ is measured with the photometer at a distance d, the illuminance at the photometer aperture is given by:

$$E_V = \frac{i_P}{s_V}, \tag{2.41}$$

where i_P is the of the photometer output current. If the distance d, measured from the light source to the photometer, is accurately known, then the luminous intensity of the source is given by:

$$I_V = \frac{E_V d^2}{\Omega_0} = \frac{i_P d^2}{s_V \Omega_0}, \tag{2.42}$$

where $\Omega_0 = 1$ sr is the unit solid angle.

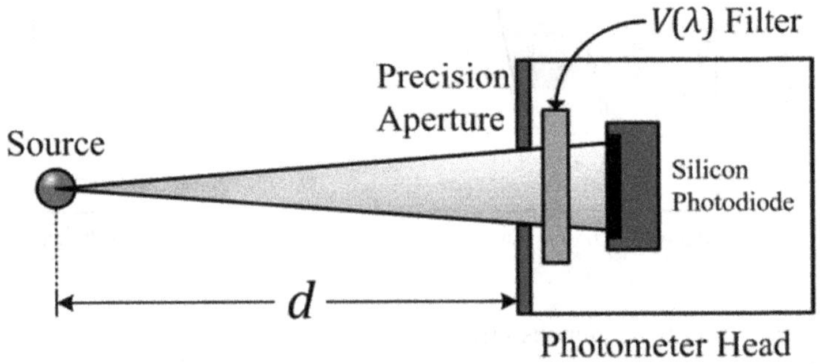

FIGURE 2.17 Detector-based candela realization.

KEYWORDS

- **absorption**
- **Lambertian emitter**
- **photometry**
- **radiometry**
- **reflection**
- **transmission**

REFERENCES

Budding, E.; Demircan, O. *Introduction to Astronomical Photometry*, 2nd ed.; Cambridge University Press: Cambridge, 2007.

Bukshtab, M. *Applied Photometry, Radiometry, and Measurements of Optical Losses*; Springer Science + Business Media B.V.: Dordrecht, 2012.

Celotta, R.; Lucatorto, T., Eds. *Experimental Methods in the Physical Sciences: Volume 41 Optical Radiometry*; Elsevier Inc.: Amsterdam, 2005.

Grant, B. G. *Field Guide to Radiometry*; Society of Photo-Optical Instrumentation Engineers (SPIE): Bellingham, 2011.

International Commission on Illumination. *The Basis of Physical Photometry*. Publication CIE No 18.2 (TC-I .2) 1983 Commission Internationale De L'éclairage 52, Boulevard Malesherbes–75008 Paris, France.

Lucatorto, T.; Parr, A. C., Eds. *Experimental Methods in the Physical Sciences: Volume 43 Radiometric Temperature Measurements*; Elsevier Inc.: Amsterdam, 2010.

Mahajan, V. N. *Fundamentals of Geometrical Optics*; Society of Photo-Optical Instrumentation Engineers (SPIE): Bellingham, 2014.

Milone, E. F.; Sterken, C., Eds. *Astronomical Photometry: Past, Present, and Future*; Springer Science+Business Media, LLC: New York, NY, 2011.

Warner, B. D. *A Practical Guide to Lightcurve Photometry and Analysis*; Springer Science+Business Media, Inc.: New York, NY, 2006.

Willers, C. J. *Electro-Optical System Analysis and Design: A Radiometry Perspective*; SPIE – The International Society for Optical Engineering: Bellingham, 2013.

Wolfe, W. L. *Introduction to Radiometry*; The Society of Photo-Optical Instrumentation Engineers: Bellingham, 1998.

CHAPTER 3

Optical Detection

ABSTRACT

Optical detector converts an optical signal into more convenient form for signal extraction and further processing. As an example of an imaging optical system can serve the camera where at the end one has a two-dimensional array of optical detector that converts the image into electrical signal that is more convenient for further processing. In the case of a nonimaging optics, one can refer to the collectors of the solar radiation where at the end there is a photovoltaic solar cell that converts the captured optical energy into electrical energy that is, once again, more convenient for further distribution. Fiber-optic communication and optical sensing systems always at their end have an optical detector where the optical data stream is converted into an electrical data stream or an electrical signal proportional to the measured physical quantity, respectively.

Roughly speaking, we can place an arbitrary optical system into several categories such as imaging optics, nonimiaging optics, and fiber-optics. Regardless of the purpose of the particular optical system, at the end, there is an optical detector. Typically, optical detector converts an optical signal into more convenient form for signal extraction and further processing. As an example of an imaging optical system can serve the camera where at the end one has a two-dimensional array of optical detector that converts the image into electrical signal that is more convenient for further processing. In the case of a nonimaging optics, we can refer to the collectors of the solar radiation where at the end there is a photovoltaic solar cell that converts the captured optical energy into electrical energy that is, once again, more convenient for further distribution. Fiber-optic communication and optical sensing systems always at their end have an optical detector where the optical data stream is converted into an electrical data stream or an electrical signal proportional to the measured physical quantity, respectively. However, the captured light isn't always transformed in its electrical counterpart; yet again, most detectors employ this type of optical detection. Table 3.1 lists the most common detection techniques that are associated with the detection of optical signals.

TABLE 3.1 Photodetection Techniques.

Thermal effects	Wave interaction effects	Photon (quantum) effects
Thermoelectric effect	Parametric downconversion	Photoconductors
Pyromagnetic effect	Parametric upconversion	Photoemissives
Pyroelectric effect	Parametric amplifiers	Photovoltaics
Liquid crystals		
Bolometers		

Optical detectors based on the thermal effect employ the detector temperature change when absorbing the optical radiation. With the detector temperature change alters some other device parameter, usually its resistance that can be externally sensed. Typical example of the thermoelectric detector is the bolometer where the bolometer resistance changes with the temperature change. Also, we have the pyroelectric detectors where the detector capacitive charge alters with the temperature changes.

Optical detectors that are based on the wave interaction effect employ the interaction of the lightwave and a nonlinear material to form the sum or difference frequency of the lightwaves. The wave interaction is the basis for the construction of optical parametric amplifiers and frequency doubles. However, efficient optical parametric amplifiers are relatively complex to implement. Therefore, they haven't been widely used in the optical communication systems.

The third group of optical detectors that are based on the photon (quantum) effects is the most popular one. These types of detectors employ the absorption of the photons to convert them into the corresponding photocarrier stream. The generation of a photocarrier corresponds to the electron–hole pair formation within the photodetector material. The photocarriers transport the charge thus forming an electrical photocurrent that can be further processed by using conventional electronics.

3.1 PHOTON COUNTING

Optical signal that impinges an optical detector can be modeled as a stream of incident photons. In the case of a monochromatic lightwave, each photon carries the same amount of energy, which is independent of the incident beam. However, the photon rate varies with the power of the optical signal. High energetic content of an optical signal corresponds to the high photon arrival rate. Therefore, from the receiver point of view, the most important parameter is the number of the received photons during particular time interval. Essentially, the detection process is an events counting process known as the photon counting process as illustrated in Figure 3.1.

A constant power, monochromatic beam, which consists of a photon stream with photon each carrying energy $E = hv$, illuminates an ideal photon to photon carrier converter, as is presented in Figure 3.1(a). The photons arrive at discrete points in time as presented in Figure 3.1(b). Such formed point process is random.

An ideal photon to photocarrier converter is a device with infinite bandwidth that converts every incident photon without an exception into a photoelectron. Further, an ideal electric counter counts the generated photocarriers. When the counter reset occurs after specified time interval, the number of photons has been counted by the detector within each time interval, as illustrated in Figure 3.1(c). If the light source power is modulated, the number of counted received photons within the defined time interval also varies in the same manner.

It is well known that the photon generation of an optical source (transmitter) is a random process. Therefore, the point process at the receiver site is as well random process. The number of photocarriers generated within an ideal photodetector illuminated by a constant power monochromatic light is described by the Poisson distribution as:

$$P(n|t) = \frac{(rt)^n \exp(-rt)}{n!}. \tag{3.1}$$

where $P(n|t)$ is the probability of detecting n photons within the time interval t and r is the mean photon arrival rate.

For the fixed observation time t and the mean photon arrival rate r, we have:

$$P(n) = \frac{s^n \exp(-s)}{n!} \tag{3.2}$$

where $s = rt$ is valid. The examples of the Poisson distributions are presented in Figure 3.2 for the mean arrival rates of $s = 5$, $s = 20$, and $s = 80$ photons per observation time interval.

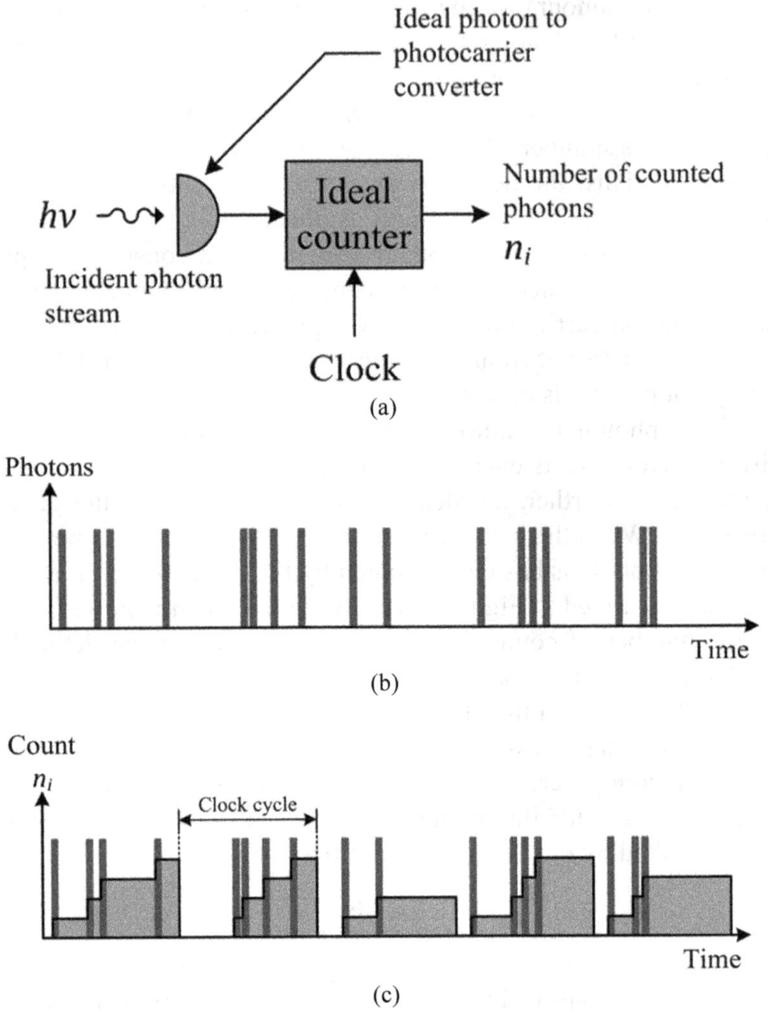

FIGURE 3.1 Photon counting: (a) ideal photon detector and counter, (b) photon arrivals, and (c) photon count.

Poisson distribution represents an accurate description of many processes with the common features that they appear as a series of random independent events and where the probability that the event will occur during certain time interval is proportional to the time interval length. This distribution as well describes the process of the photocarrier generation within the photodetector when illuminated with coherent, monochromatic light such as is generated by laser. However, the photons of the light, generated by the incoherent sources such as incandescent and fluorescent light bulbs, the Sun, and the amplified spontaneous emission from the optical amplifiers don't obey the Poisson distribution. The distribution that describes such random process, where the photodetector is illuminated by the incoherent narrowband light, is known as the Bose–Einstein distribution which is given by:

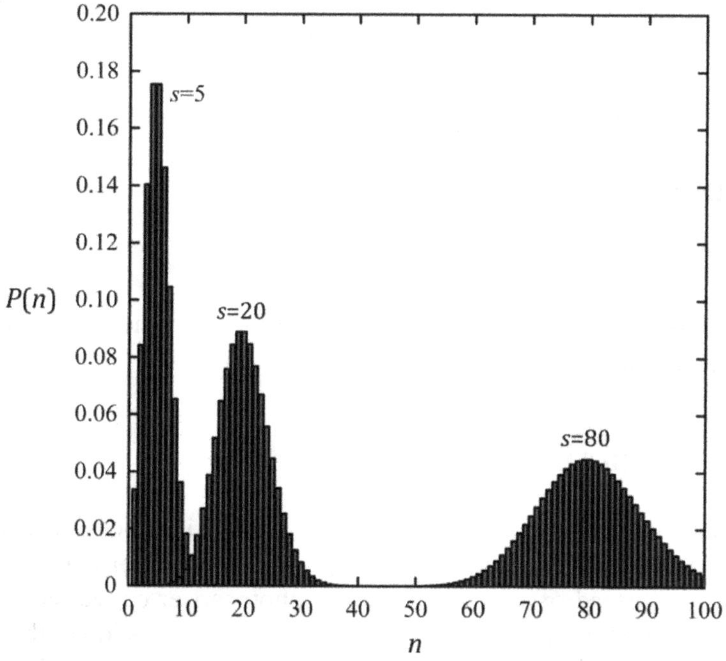

FIGURE 3.2 Poisson distribution examples.

$$P(n) = \left(\frac{1}{1+b}\right)\left(\frac{b}{1+b}\right)^n, \tag{3.3}$$

where b is again the mean number of incoherent background photons which is defined as:

$$b = \frac{N}{h\nu}, \tag{3.4}$$

where N is the spectral density of incoherent light specified as:

$$N = \frac{\Phi}{B}, \tag{3.5}$$

where Φ is the radiant flux of the incoherent source and B is the optical bandwidth of the incoherent source. If the t is the photon counting time interval of the incoherent source, the narrowband condition that must be met is $t \leq 1/B$.

In the case when the photodetector is illuminated by both sources a coherent signal and a narrowband background component, the photon counting statistics obeys the Laguerre distribution given by:

$$P(n) = \left(\frac{1}{1+b}\right)\left(\frac{b}{1+b}\right)^n \exp\left(\frac{s}{1+b}\right) L_n\left[-\frac{s}{b(1+b)}\right], \tag{3.6}$$

where the mean photon rate of the coherent signal s can be presented as:

$$s = \frac{\Phi_s}{h\nu_s}, \tag{3.7}$$

where Φ_s is the flux of the coherent signal of the frequency ν_s, and $L_n(x)$ is the Laguerre polynomial.

3.2 PHOTODETECTION MODELING

The ideal photon counter, as presented in Figure 3.1 is impossible to realize in practice. A more realistic model of a real photodetector is presented in Figure 3.3. Instead of an ideal photon to photocarrier converter, here we have the converter with conversion efficiency η, which is lower than unity and an electrical low-pass filter to include into account the finite time response of a real photodetector.

Consequently, Figure 3.4 represents the corresponding current outputs from the real photon to photocarrier converter. The generated photocurrent is a series of current pulses each carrying the charge of a single electron.

How the inherent low-pass filter of the photodetector acts on the current pulses is presented in Figure 3.4(b). The obtained current pulses have a finite time duration that depends on the filter bandwidth. Each pulse still carries a charge equal to the charge of a single electron. When the number of photons per observation time interval increases, a continuous photocurrent is obtained as presented in Figure 3.4(c). Bearing in mind that the photodetector represents a linear system, in the general case, we can model the photodetector current as a sum of individual photodetector impulse responses as:

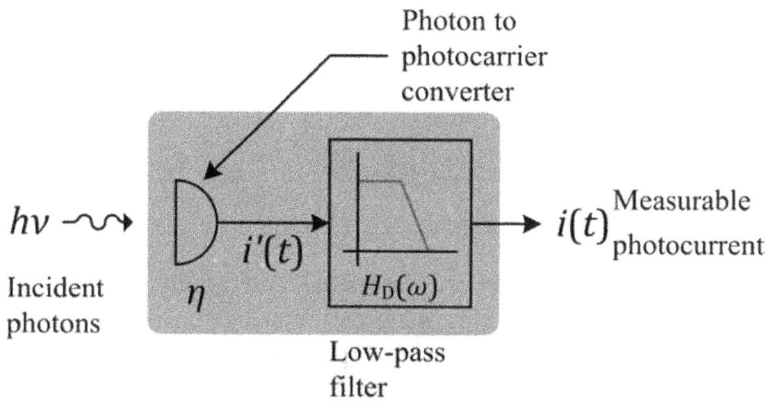

FIGURE 3.3 Photodetector model.

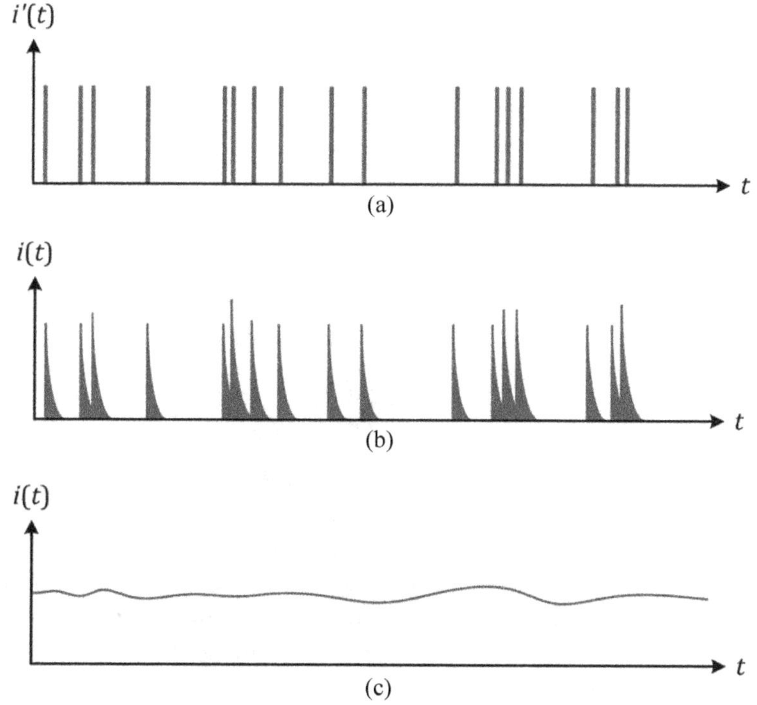

FIGURE 3.4 Photodetector current characteristics.

$$i(t) = \sum_{k=1}^{n} h_D(t - \tau_k), \tag{3.8}$$

where $h_D(t)$ represents the photodetector impulse response, n is the total number of generated photocarriers, and τ_k is the random time at which the k th photocarrier has been generated. Each current pulse carries a charge of a single electron thus we have:

$$\int h_D(t)\,dt = q, \tag{3.9}$$

where q is the electron charge.

The number of photocarriers generated during a particular time interval depends on the optical power (flux) captured by the photodetector. By including the variation of the captured optical power with time, the photon rate also changes with the time as:

$$r(t) = \frac{\eta}{h\nu} p(t), \tag{3.10}$$

where $p(t)$ is the captured optical power (flux) and η is the so-called quantum efficiency defined as the ration of the produced photocarriers number and the number of incident photons for which is valid $0 \leq \eta \leq 1$. Bearing in mind that the photocurrent is proportional to the photon rate as $i(t) = qr(t)$ and taking into consideration eq 3.10, one obtains:

$$r(t) = \frac{\eta q}{h\nu} p(t) = \Re p(t). \tag{3.11}$$

The term is given by:

$$\Re = \frac{\eta q}{h\nu}, \tag{3.12}$$

is known as the photodetector responsivity. As presented in Figure 3.5, the responsivity varies with the wavelength. Responsivity increases with wavelength because there are more photons per unit power at long wavelengths than there are at short wavelengths. Since the amount of photocurrents depends on the photon number not the photon energy, longer wavelengths generate more photocurrent per unit power than do the short wavelengths.

3.3 PHOTODETECTORS

The photon effect–based photodetector directly generates the photocurrent from the interactions between the photons and the atoms within the detector material. After the penetration into a photodetector, the probability that the particular atom will absorb a photon and thus generate free carrier is rather small. However, the number of atoms within the detector is huge so the probability that the photon will be absorbed by some atom is quite high. Photon-effect photodetectors can be designed to fulfill two major requirements such as high sensitivity and wide bandwidth (fast response time). Such photodetectors are typically grouped into four major categories:

- Photomultipliers (PMTs),
- Photoconductors,
- Photodiodes, and
- Avalanche photodiodes (APDs).

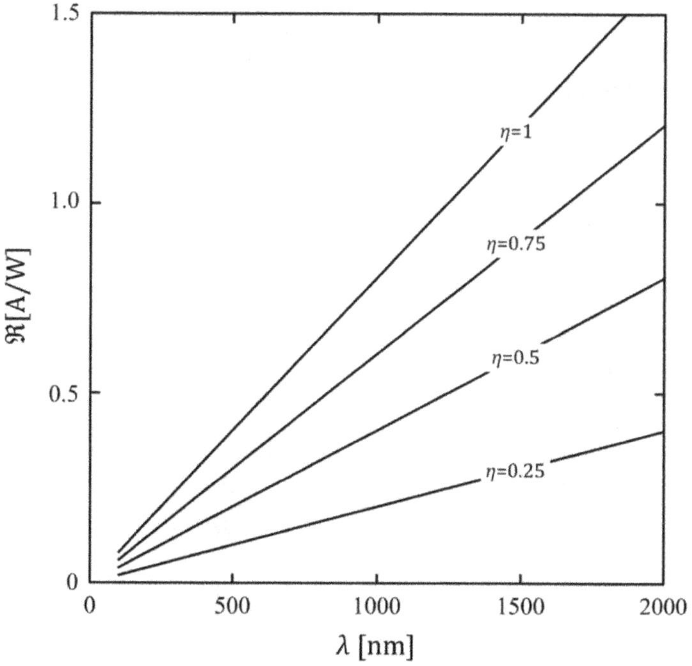

FIGURE 3.5 Wavelength dependence of the responsivity.

Each of these four photon effect–based photodetectors will be described in the following sections.

3.3.1 PHOTOMULTIPLIER

A PMT is a vacuum tube that employs the photoelectric effect as well as the secondary emission of electrons to provide high current gain. A schematic representation of the PMT tube is given in Figure 3.6. When photon reaches the photocathode, which is made of a low work-function metal or semiconductor, then if the photon energy exceeds the work function, the electron will be ejected with a certain probability. After the ejection, electron is accelerated by the electric field formed between the series of electrodes called dynodes. The dynodes are coated with a material that is prone to secondary emission of electrons and are connected in series between the resistors. The resistors R that are connected in series between the dynodes form the gradient of voltage from the high voltage source $+V_{HV}$ to the ground. The dynodes generate additional electrons by the secondary emission. Finally, all the electrons that are generated by the dynodes are captured by the anode. The current multiplication process continues for each dynode in the PMT tube, thus providing the current gain that can exceed 10^6.

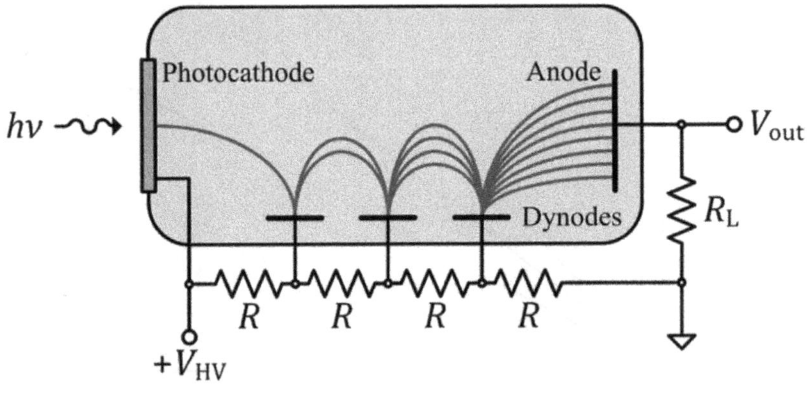

FIGURE 3.6 Photomultiplier tube.

The probability of detecting incident photon is equal to the quantum efficiency of the PMT tube. Typical photocathode can reach a quantum efficiency of 30% meaning that in average approximately every third photon is detected.

PMTs are usually used for the laboratory instrumentation. Due to their very high current gain, the PMT tubes are used to detect extremely low optical signals. PMTs are physically large, require high voltage power supplies, and are mechanically fragile.

3.3.2 PHOTOCONDUCTORS

Photoconductors employ the photon absorption in semiconductor material. There are three principal absorption processes that can be found in semiconductors, as presented in Figure 3.7. The intrinsic band-to-band absorption appears in the cases when the bandgap energy E_G of the material is lower than the photon energy. Therefore, the photon is able to excite the electron from the valence to the conduction band. Such obtained hole and electron constitute a charge carrier. If an external electric field is applied, the hole and electron propagate throughout the semiconductor material to the external circuit thus causing photocurrent flow.

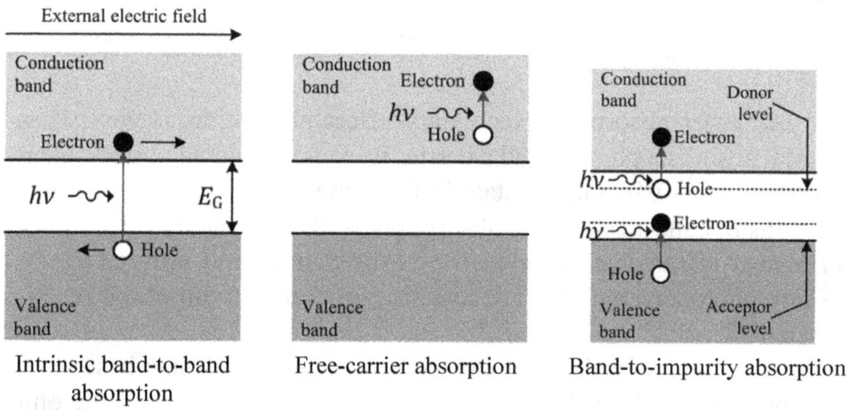

FIGURE 3.7 Photon absorption mechanisms in semiconductors.

Intrinsic band-to-band absorption represents a dominant absorption mechanism in most semiconductors which are used for the photodetection purposes. To trigger the intrinsic band-to-band absorption, the following condition must be fulfilled:

$$hv > E_G, \qquad (3.13)$$

$$\frac{hc}{\lambda} > E_G, \qquad (3.14)$$

where hv is the photon energy. According to eq 3.14, there is a maximum wavelength where the intrinsic band-to-band absorption still can occur. This maximum wavelength is given by:

$$\lambda_{max} = \frac{hc}{E_G}. \qquad (3.15)$$

The bandgap energies and the corresponding maximum wavelengths for typical semiconductor materials used in the photodetection are presented in Table 3.2.

TABLE 3.2 Typical Semiconductor Materials Characteristics.

Material	Bandgap energy (eV)	Maximum wavelength (nm)	Operating spectral range (nm)
Si	1.12	1110	500–900
Ge	0.67	1850	900–1300
GaAs	1.43	870	750–850
$In_xGa_{1-x}As_yP_{1-y}$	0.38–2.25	550–3260	1000–1600

Free-carrier absorption, secondary effect absorption, occurs when a material absorbs a photon, and a carrier is excited from an already-excited state to another, unoccupied state in the same band. The third absorption mechanism, band-to-impurity absorption, is also secondary effect absorption at near-infrared band. Typically, it is used in the mid-infrared region to develop photodetectors that are sensitive up to the 30 µm of the radiation wavelength.

A photoconductor detector can have a very simple design based on the semiconductor slab of the bulk semiconducting material with the ohmic (metal) contacts at the ends, as presented in Figure 3.8(a) or it may have an interdigitated structure of the ohmic contacts on the top as presented in Figure 3.8(b). To bias the photoconductor, a simple DC circuit can be implemented, which generates the corresponding photocurrent i_p proportional to the photoconductor conductance and thus the captured optical power.

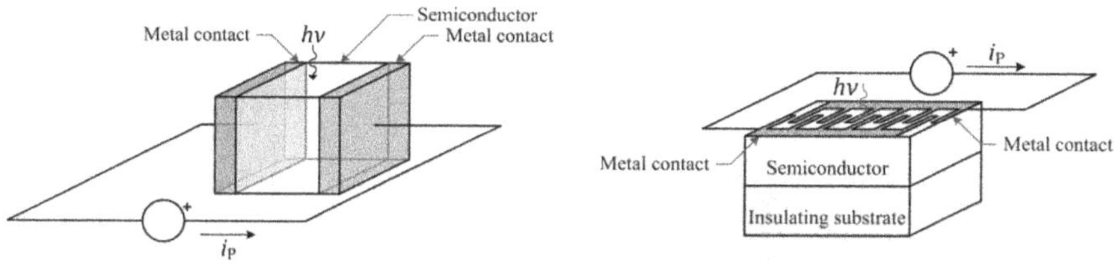

FIGURE 3.8 Photoconductor design: (a) bulk semiconductor and (b) semiconductor film with an interdigitated electrode structure.

The photocurrent of an illuminated photoconductor is influenced by the carrier lifetime and the time required to make their way through the device:

$$i_P = \frac{\mu q}{h\nu} \frac{\tau_C}{\tau_T} p + i_D, \tag{3.16}$$

where τ_C is the mean carrier lifetime, τ_T transit time between the electrical contacts, p is the captured optical power, and i_D is the dark current.

3.3.3 PHOTODIODES

A semiconductor p-n junction (diode) can be used as a photodetector which is frequently called a photodiode. Similarly like in the case of the photoconductors, photodiode relies on the photon absorption in a semiconductor material. Such generated photocarriers are separated by an electric field thus resulting in a photocurrent proportional to the captured optical power. The p-n junction is illustrated in Figure 3.9(a) together with a biasing circuitry. The junction doping profiles are shown in Figure 3.9(b) where the p-type side of the photodiode contains a surplus of holes and which has been doped with the acceptor atoms of a concentration of N_A. The other side of a photodiode is the n-type semiconductor containing a surplus of electrons. The concentration of the donor atoms at the n-side is N_D.

When the junction has been formed, the electrons from the n-type material and holes from the p-type material diffuse across the junction to equalize the free-carrier density throughout the semiconductor material. Electrons and holes continue to flow until the charge buildup, as shown in Figure 3.9(c), is sufficient to prevent any additional carriers from crossing the junction.

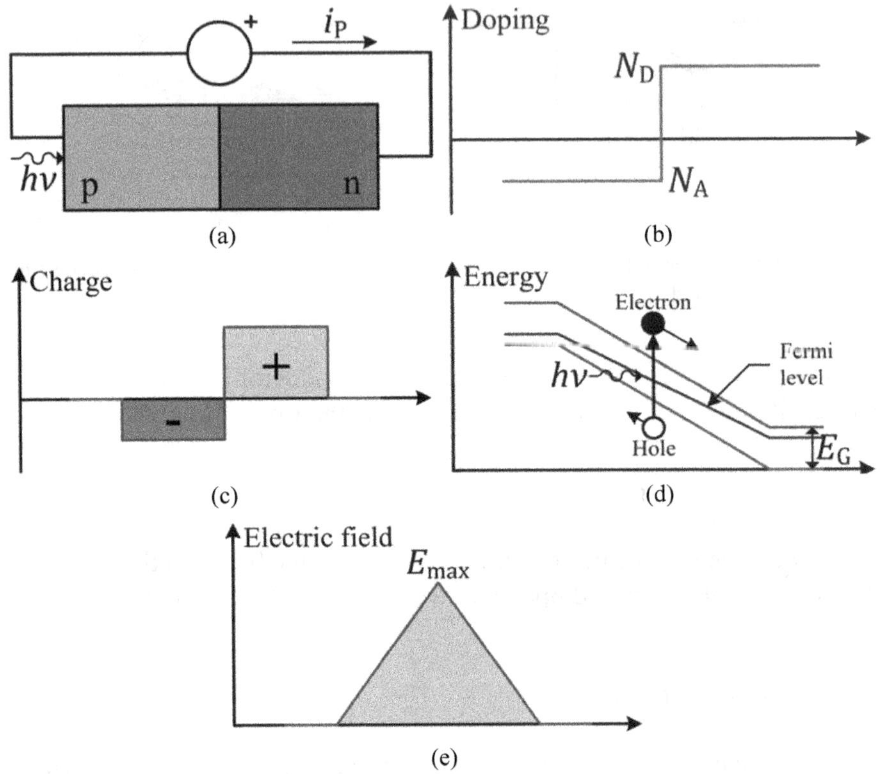

FIGURE 3.9 Reverse-biased p-n diode as a photodetector.

The region in the junction vicinity where the charge buildup occurs is known as the space-charge region or the depletion region. The corresponding energy diagram is shown in Figure 3.9(d) where the energy band diagram illustrates the relationship between the top of the valence band and the bottom of the conductance band across the photodiode. The Fermi level, which defines the energy level with a 50% probability of occupancy, is also shown. Finally, in Figure 3.9(e) is shown the electric field profile in the depletion region.

When a photon is absorbed within the depletion region an electron–hole pair will be formed. The electron and hole will be transported throughout the depletion region by the inherent electric field to the edges of the depletion region. After leaving the depletion region, the carriers travel to the terminals of the photodiode, thus forming a photocurrent that further flows into the external circuitry.

The built-in electric filed is usually supplemented by an external field that biases the photodiode reversely. The stronger the external electric field the wider the depletion region and the more rapidly the free carriers will move toward the region edges. The average time required for a photocarrier to travel across the depletion region is called the transit time and is given by:

$$\tau_T = \frac{l_D}{v_S}, \tag{3.17}$$

where l_D is the depletion region length and v_S is the average carrier saturation velocity. The shorter the transit time of the photodiode, the faster is its response to the abrupt illumination.

As it is well known, the photodiode is still a diode. The diodes can operate in three modes, unbiased, forward biased, and reverse biased. Figure 3.10 illustrates these three operating modes with three corresponding bias schematics in which the photodiode can operate when connected to the external circuit. These three modes are usually referred to as open circuit or the photovoltaic mode, short circuit or photoconductive mode, and the reverse-biased photoconductive mode.

Regardless of the operation mode the photodiode current vs. voltage characteristics given by:

$$i_{\text{PD}} = I_S\left[\exp\left(\frac{V_{\text{PD}}}{V_T}\right) - 1\right] + i_L - i_P, \tag{3.18}$$

where I_S is the reverse saturation current, i_L is the photodiode leakage current, V_{PD} is the voltage applied across the photodiode, V_T is the thermal voltage given by:

$$V_T = \frac{k_B T}{q}, \tag{3.19}$$

where k_B is the Boltzmann constant, T is the photodiode absolute temperature, and q is elementary positive charge. The photocurrent i_p from eq 3.18 is given as:

$$i_P = \frac{\eta q}{h\nu} p, \tag{3.20}$$

with p being the photodiode captured optical power. The photodiode dark current, which flows through the photodiode when it isn't illuminated, is given by:

$$i_D = I_S\left[\exp\left(\frac{V_{\text{PD}}}{V_T}\right) - 1\right] + i_L. \tag{3.21}$$

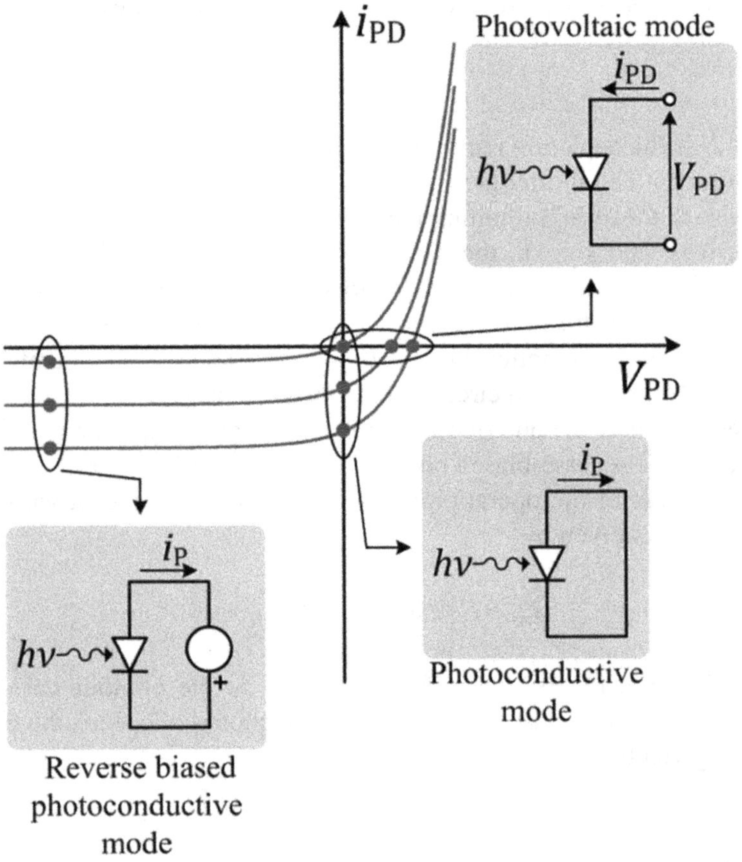

FIGURE 3.10 Photodiode operating regions.

When the photodiode is in the photovoltaic mode, that is, when its terminals are open, the photodiode is then typically used as a solar cell. The corresponding voltage at the diode open ends can be obtained from eq 3.18 for $i_{\text{PD}} = 0$, that is, generated voltage is:

$$V_{\text{PD}} = V_T \ln\left(1 + \frac{i_P}{I_S}\right). \tag{3.22}$$

3.3.4 AVALANCHE PHOTODIODES

Unlike the conventional photodiode, which can generate single electron–hole pair from a single absorbed photon, an APD can produce many electron–hole pairs from a single absorbed photon. The internal gain of the APD enables its use for constructing the highly sensitive optical receivers. The typical structure of an APD is presented in Figure 3.11(a).

As presented in Figure 3.11(a), the APD consists of four sections, which doping profile is presented in Figure 3.11(b). The leftmost section is a p-type semiconductor that is doped with the acceptor atoms with the concentration N_A and the rightmost section is a n-type semiconductor that is doped with the donor atoms with the concentration N_D. In between, there is a middle section that is lightly doped semiconductor with a second p-type section. The corresponding charge density is presented in Figure 3.11(c), the energy diagram in Figure 3.11(d), and the electric field profile along the APD in Figure 3.11(e).

In the case of high reverse-biased p-n junction, there are two breakdown mechanisms that can occur. The first mechanism implies that the atoms are directly ionized by the applied electric field. This is well-known Zener breakdown that is commonly used for making the voltage-regulating diodes. The second mechanism is caused by the high carrier velocity that causes the impact ionization within the semiconductor. This mechanism is called avalanche breakdown. The photons are absorbed within the lightly doped semiconductor material thus generating the electron–hole pairs. The generated carriers are quickly transported by the applied electric filed to the absorption region edges. After reaching the gain region, the carriers are accelerated by the very high electric field thus achieving sufficient velocity (kinetic energy) to trigger the generation of additional electron–hole pairs. This process is called the impact ionization. These additional carriers can then undergo other impact ionizations thus causing an avalanche effect. Therefore, an APD is similar to the PMT where a single absorbed photon can generated a shower of the carriers. The corresponding current gain or multiplication factor of an APD is given by:

$$M = \frac{i_{\text{APD}}}{i_p}, \tag{3.23}$$

where i_{APD} is the APD photocurrent and i_p is the internal photocurrent before the multiplication that is given by eq 3.20. The typical values of the

APD gain range somewhere between few tens and few hundreds, although the gains as high as 10^3 or 10^4 can be reached.

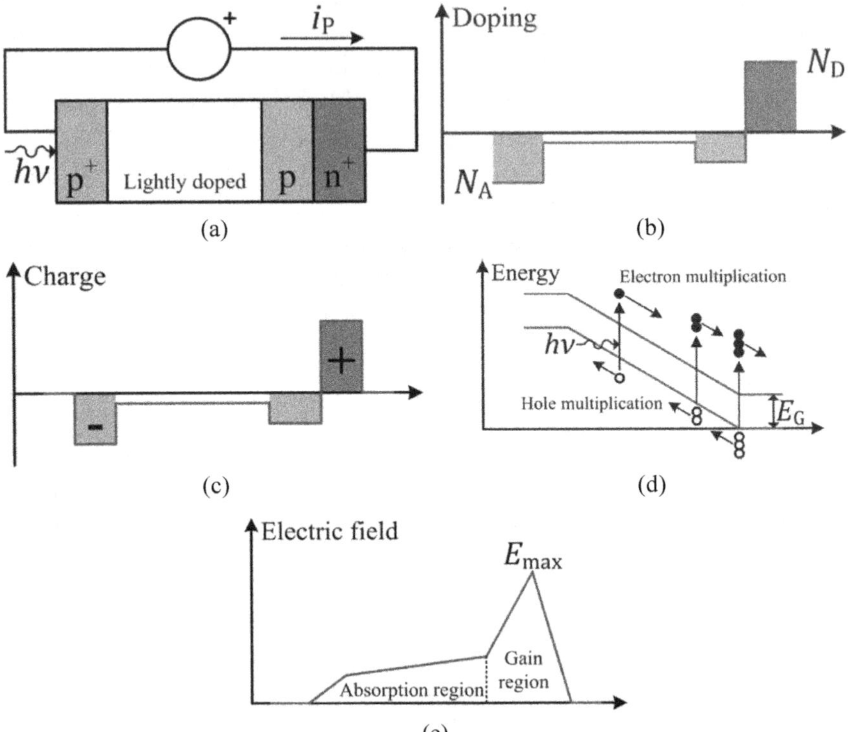

FIGURE 3.11 Avalanche photodiode as a photodetector.

3.3.5 POSITION SENSING PHOTODIODES

To measure the position of the light spot in two dimensions, one can employ the multiple elements and lateral effect photodiodes. Such photodiodes are capable of generating signals that indicate the light spot position that impinges the diode's surface in two dimensions. Figure 3.12 illustrates the duo-lateral structure of the lateral effect photodiode with two pairs of electrodes that are in contact with diode's p and n layers. The electrical contact stripes, that is, the electrodes, marked with X_1 and X_2 and positioned at the opposite side of the chip, collect the photocurrent supplied to the p layer. The bottom electrodes, also in the form of the

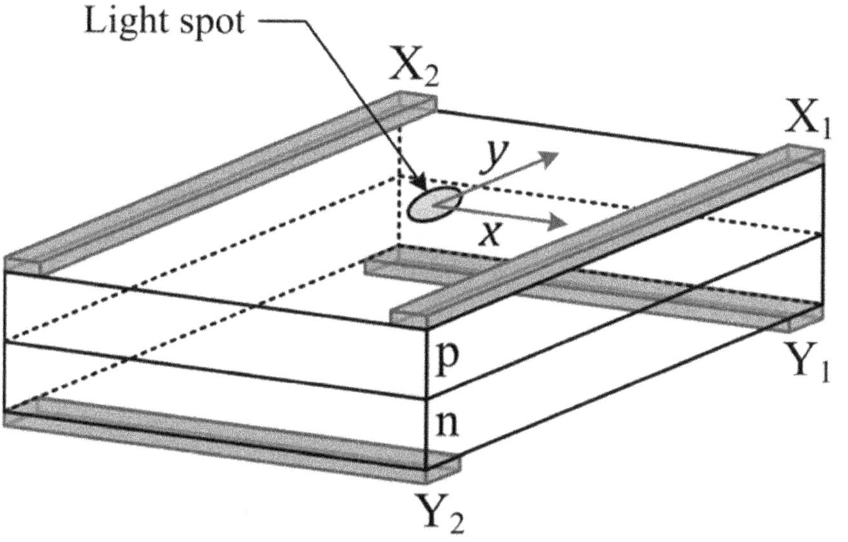

FIGURE 3.12 Structure of a duo-lateral photodiode.

contact strips, Y_1 and Y_2 divide the photocurrent through the same lateral effect as for the top electrodes. The X electrodes receive the portion of the photocurrent as determined by the resistive divider presented by the p region and similarly, the Y electrodes receive currents appointed by the resistivity of the n region. Therefore, the corresponding X and Y current components can define the light spot position in the following way:

$$x = \frac{i_{X_1} - i_{X_2}}{i_{X_1} + i_{X_2}} L_X, \tag{3.24}$$

$$y = \frac{i_{Y_1} - i_{Y_2}}{i_{Y_1} + i_{Y_2}} L_Y, \tag{3.25}$$

where x and y represent the spot positions on the photodiode surface, L_X and L_Y represent the interelectrode spacing for the top and bottom electrode pairs, respectively, and $i_{X_{1,2}}$ and $i_{Y_{1,2}}$ the corresponding electrode currents. Figure 3.13 represents the equivalent circuit of a duo-lateral photodiode where the potentiometer wiper arms represent the actual position of the light spot along X and Y axes. One must take into consideration that the wiper arms positions are independent due to the orthogonal orientation of the electrode pairs. The relations between the partial potentiometer resistances and the actual position of the light spot are given in the followings:

$$R_{1X} = \frac{L_X - x}{2L_X} R_{PX}, \tag{3.26}$$

$$R_{2X} = \frac{L_X + x}{2L_X} R_{PX}, \tag{3.27}$$

and

$$R_{1Y} = \frac{L_Y - y}{2L_Y} R_{PY}, \tag{3.28}$$

$$R_{2Y} = \frac{L_Y + y}{2L_Y} R_{PY}, \tag{3.29}$$

where R_{PX} and R_{PY} represent the interelectrode resistances.

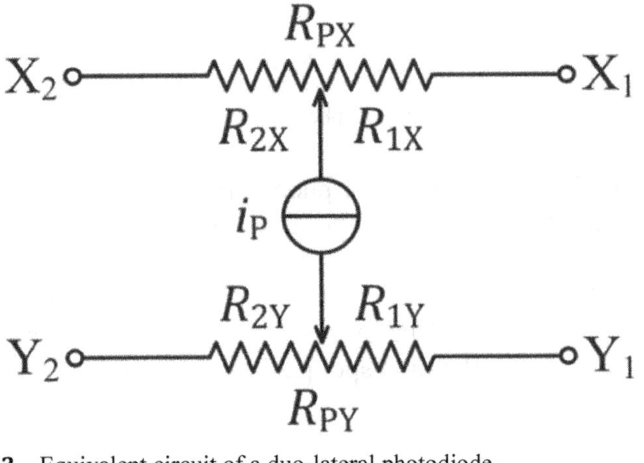

FIGURE 3.13 Equivalent circuit of a duo-lateral photodiode.

The tetra-lateral photodiode has all four electrodes placed at the illuminating surface of the photodiode, as is depicted in Figure 3.14. The electrodes are positioned at the ends of the top surface whereas the continuous electrode is located at the bottom side of the photodiode. At the upper diode surface, the corresponding resistances divide the photocurrent into the component that flow toward various output electrodes. However, the two-dimensional current flow complicates the current division. To linearize the tetra-lateral photodiode transfer function, one must pay a special attention into the electrode and photodiode geometry. Therefore, typically a pincushion-shaped diode area with point-like diode contacts has been used to compensate the distortion.

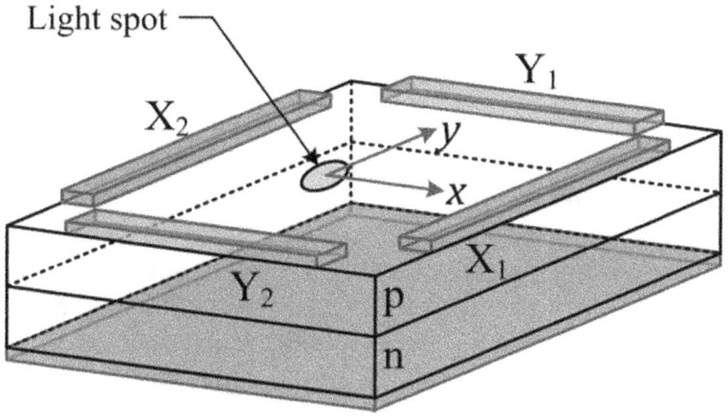

FIGURE 3.14 Structure of a tetra-lateral photodiode.

Figure 3.15 represents the equivalent circuit of a tetra-lateral photodiode with a four-way output current divider. Due to its geometry, the tetra-lateral photodiode may use essentially the same resistance equations as there were given in the case of a duo-lateral photodiode in eqs 3.26–3.29. By holding, in some way, all four electrodes at the same potential one can take into consideration that all resistors are connected in parallel and the photocurrent is effectively divided into two currents that correspond to two orthogonal axes, into i_X and i_Y where is fulfilled $i_p = i_X + i_Y$. These two currents, i_X and i_Y, correspond to the resistors $R_X = R_{1X} \| R_{2X}$ and $R_Y = R_{1Y} \| R_{2Y}$, respectively, that is, the following is valid:

$$i_X = \frac{R_Y}{R_X + R_Y} i_P, \tag{3.30}$$

$$i_Y = \frac{R_X}{R_X + R_Y} i_P. \tag{3.31}$$

Further, the currents i_X and i_Y have been divided into their respective subcomponents in the same way as it has been described in the case of duo-lateral photodiode.

3.3.6 QUADRANT PHOTODIODE

Quadrant photodiode (QPD) is probably one of the most commonly used sensors aimed for very high precision measurement of the lateral position and the angle of the incoming light beam. The possibility of reaching subnanometer resolution for a very broad frequency range, very simple analog signal processing circuitry that follows up every QPD, as well as wide operating temperature range make them an almost optimal solution for the position sensing in the nanometer range. Therefore, QPDs become an integral part of many atomic force microscopes where they are used to sense the microcantilever position. Nevertheless, QPDs are not exclusively aimed for the position measurements in the micro and nano world. Since being low-noise, high sensitive and very robust, QPDs have found, also many other applications in the macro world. Lidars, laser-guided weapons, free-space optic communications, satellite communications, civil and maritime engineering, mining industry, etc., are just a few possible applications among unlimited number of them.

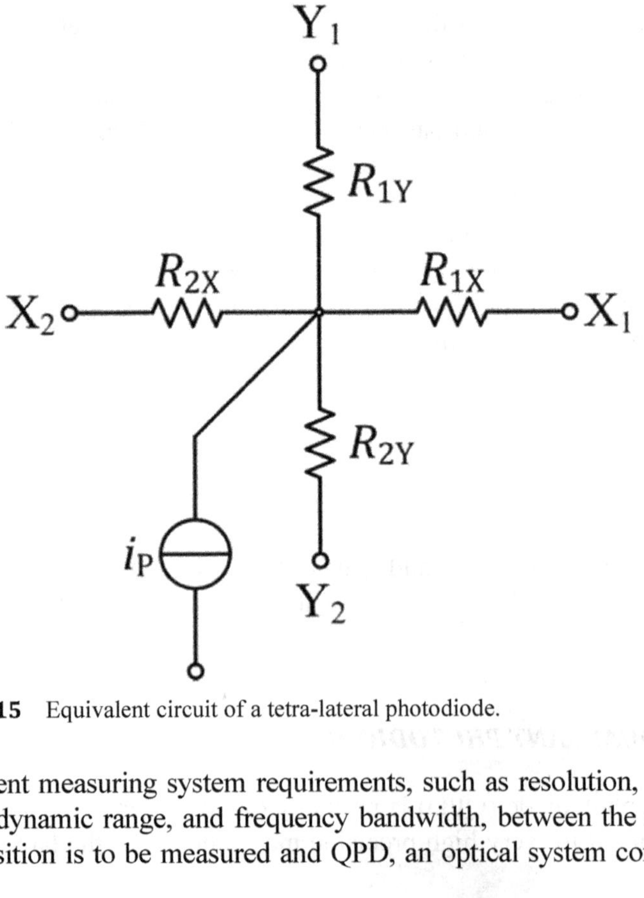

FIGURE 3.15 Equivalent circuit of a tetra-lateral photodiode.

Different measuring system requirements, such as resolution, accuracy, linearity, dynamic range, and frequency bandwidth, between the object, in which position is to be measured and QPD, an optical system consisted of light source (usually a laser or the light coming out of the single-mode optical fiber is used) and a few passive optical components are used to fulfill the requirements. The role of this optical system is to make a relation between the object position and the position of the light spot on the QPD surface. Typically, the distribution of the light spot irradiance on the QPD surface mostly depends on the used light source as well as on the object-coupling optics arrangement. However, most algorithms for light spot position estimation with respect to the QPD center are based on the position measurement of the light spot center of gravity irrespectively to the light spot irradiance distribution on the QPD surface. Usually, these algorithms enable high sensitivity, high speed, and high-resolution light spot position measurement. If the laser or the light coming out of the single-mode optical fiber is used as the light source, then the Gaussian irradiance distribution is usually obtained. Usually, it is assumed that the dimensions of the Gaussian distributed irradiance light spot are much smaller than the QPD dimensions. This implies that the total amount of the irradiated light flux impinges the QPD surface. The abovementioned assumption is often not satisfied. In the case where we have long beam paths due to the change of the beam propagation conditions, there can be significant changes of the light spot dimensions on the QPD surface. For example, in the case where one has servo system that should keep the system movements collinear with the laser beam along relatively long path, for example, in civil engineering or mining industry, than due to the laser beam divergence, we will have dimensions of the light spot fluctuating with respect to the system movement. Nevertheless, servo system should stay stable and still keep acceptable level of system parameters.

It is well known that the sensitivity of any sensor will greatly influence the overall sensing system working parameters. The dimensions of the light spot on the QPD surface greatly influence the QPD sensitivity irrespective of the irradiance distribution. To match the real systems exploitation conditions, the Gaussian distribution of the light spot irradiance on the QPD surface will be assumed. The QPD sensitivity depends on the ratio of the $1/e$ light spot radius at the QPD surface and the QPD. By increasing the light spot radius with respect to the QPD radius, there will be an increase of the interaxis cross talk, for example, the change of the position along one axis will influence the estimated position on the other axis. Therefore, the analysis of the influence of the interaxis cross talk on the measurement error and how to make it minimal while still keeping high linearity of the QPD sensor is very important task.

Direct measurements of the light spot position with respect to the QPD center by processing electrical current signals from the QPD are not possible. By measuring the current signals, one can determine only the ratios of positions of the light spot and QPD relevant dimensions. To determine these relations, one needs to know the irradiance distribution on the QPD surface. The exact measurements of these ratios are not possible, and only the estimations can be made with the following equations:

$$\chi = \frac{(i_{\mathrm{I}} + i_{\mathrm{IV}}) - (i_{\mathrm{II}} + i_{\mathrm{III}})}{(i_{\mathrm{I}} + i_{\mathrm{IV}}) + (i_{\mathrm{II}} + i_{\mathrm{III}})}, \tag{3.32}$$

$$\psi = \frac{(i_{\mathrm{I}} + i_{\mathrm{II}}) - (i_{\mathrm{III}} + i_{\mathrm{IV}})}{(i_{\mathrm{I}} + i_{\mathrm{II}}) + (i_{\mathrm{III}} + i_{\mathrm{IV}})}, \tag{3.33}$$

where χ and ψ are the estimated values of the ratios along X- and Y-axis, respectively, and i_Q is the electrical current acquired by the QPD's Qth quadrant, where Q = I, II, III, IV. For these currents, $i_Q = \Re P_Q$ is valid, where $\Re$ is the photodiode conversion factor, and P_Q is the optical power (flux) acquired by the QPD's Qth quadrant. Finally, from eq 3.33, the following is obtained for the Y-axis:

$$\psi = \frac{(P_{\mathrm{I}} + P_{\mathrm{II}}) - (P_{\mathrm{III}} + P_{\mathrm{IV}})}{(P_{\mathrm{I}} + P_{\mathrm{II}}) + (P_{\mathrm{III}} + P_{\mathrm{IV}})}, \tag{3.34}$$

where due to the symmetry only the Y-axis will be considered, for example, everything that is valid for one axis is valid for the other axis if the irradiance distribution is symmetrical with respect to light spot center.

Since the QPD is used in many critical applications such as laser-guided weapons for military purpose, for the beam alignment in the free-space optical communication links, as well as, in the civil and maritime engineering, where the system parameter changes can be significant, it is even more important to figure out in which way the sensor sensitivity will be changed. The QPD is usually used to sense the drift of the laser beam across its surface. To achieve the highest possible sensor linearity, usually single-mode lasers or the light coming out of the single-mode optical fiber are used where the irradiance distribution on the QPD surface is Gaussian. Due to the long light beam paths as well as due to the different propagation conditions of the light beam, the change of the light spot radius on the QPD surface can be significant. The relative change of the light spot radius with respect to the QPD radius will strongly influence the sensor sensitivity and hence strongly influence the overall sensing system characteristics.

To determine the sensitivity of the QPD sensor, the analysis will start by assuming the Gaussian distribution of the irradiance of the light spot falling onto the QPD surface where for the irradiance distribution $I(r)$, we have:

$$I(r) = \frac{P}{\pi w^2} \exp\left[-\left(\frac{r}{w}\right)^2\right], \tag{3.35}$$

where r is the radius, P is the overall light source optical power that reaches the QPD surface, and w is the light spot radius on the QPD surface for which the irradiance drops to $1/e$ value of its central value. Further, we will assume that the center of the light spot is shifted with respect to the QPD center at the position (x, y). The position of the light spot with respect to the QPD is shown in Figure 3.16, where all other relevant geometrical parameters are marked.

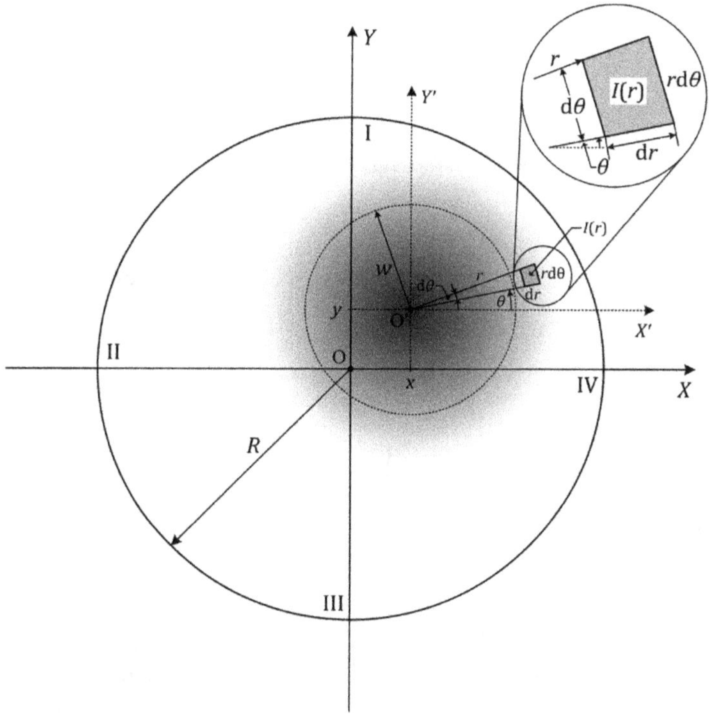

FIGURE 3.16 Geometry of the light spot onto the QPD surface with all relevant geometrical parameters.

First of all the overall optical power that is captured by the QPD will be found, for example, the signal obtained as the sum of all powers falling onto each quadrant of the QPD. For the overall optical power, we have the following:

$$P_{\Sigma}(x,y)=P_{\mathrm{I}}(x,y)+P_{\mathrm{II}}(x,y)+P_{\mathrm{III}}(x,y)+P_{\mathrm{IV}}(x,y), \backslash \tag{3.36}$$

where $P_Q(x, y)$ (Q = I, II, III, IV) is the optical power captured by Qth quadrant of the QPD when the light spot center is at the position (x, y) corresponding the QPD center. Instead of finding each optical power that falls on the corresponding quadrant, one will find the overall optical power by integrating the light spot irradiance that is captured by the active QPD surface. In the presented calculations, one will assume that the dimensions of the gaps between the QPD quadrants are much smaller than QPD radius and light spot radius. By careful analysis of the light spot geometry and neglecting the gaps influence, the following for the overall optical power is obtained:

$$P_{\Sigma}(x,y)=\int_{\theta=0}^{2\pi}\int_{r}^{l(x,y,\theta)} I(r)r\mathrm{d}r\mathrm{d}\theta, \tag{3.37}$$

where for the upper integration limit $l(x, y, \theta)$, the following is valid:

$$l(x,y,\theta)=\sqrt{R^2-(x\sin\theta-y\cos\theta)^2}-(x\cos\theta+y\sin\theta), \tag{3.38}$$

where R is the QPD radius. After partial solving of the integral given in eq 3.37, one has:

$$P_{\Sigma}(x,y)=P\left\{1-\frac{1}{2\pi}\int_0^{2\pi}\exp\left\{-\left[\frac{l(x,y,\theta)}{w}\right]^2\right\}\mathrm{d}\theta\right\}, \tag{3.39}$$

For the estimation of the position $\psi(x, y)$ of the light spot center with respect to the QPD center along the Y-axis according to eq 3.34, we have:

$$\psi(x,y)=\frac{\left[P_{\mathrm{I}}(x,y)+P_{\mathrm{II}}(x,y)\right]-\left[P_{\mathrm{III}}(x,y)+P_{\mathrm{IV}}(x,y)\right]}{\left[P_{\mathrm{I}}(x,y)+P_{\mathrm{II}}(x,y)\right]+\left[P_{\mathrm{III}}(x,y)+P_{\mathrm{IV}}(x,y)\right]}$$

$$=2\frac{P_{\mathrm{I}}(x,y)+P_{\mathrm{II}}(x,y)}{P_{\Sigma}(x,y)}-1. \tag{3.40}$$

To find the sum of the optical powers $P_{\mathrm{I}}(x, y) + P_{\mathrm{II}}(x, y)$ that fall on the first and the second quadrant, one will use the geometrical representation of the complete QPD geometry given in Figure 3.17. According to the geometry that one can see from Figure 3.17, for the optical power sum, one can write the following:

$$P_{\mathrm{I}}(x,y)+P_{\mathrm{II}}(x,y)=P_{y1}(x,y)+P_{y2}(x,y), \tag{3.41}$$

where $P_{y1}(x, y)$ is the optical power that is captured by the circle segment limited by the lines O′A, O′B and the upper part of the XOY coordinate system and $P_{y2}(x, y)$ is the optical power that is captured in the triangle AO′B. From eqs 3.40 and 3.41, we have the following:

$$\psi(x,y)=2\frac{P_{y1}(x,y)+P_{y2}(x,y)}{P_{\Sigma}(x,y)}-1=2\frac{P_{y+}(x,y)}{P_{\Sigma}(x,y)}-1, \tag{3.42}$$

where $P_{y+}(x, y)$ is the optical power that is captured by two upper quadrants. With respect to the geometry shown in Figure 3.17 and omitting the gaps influence, one has the following for the optical power $P_{y1}(x, y)$:

$$P_{y1}(x,y)=\frac{P}{\pi w^2}\int_{\theta=-\beta(x,y)}^{\pi+\alpha(x,y)}\int_{r=0}^{l(x,y,\theta)}\exp\left[-\left(\frac{r}{w}\right)^2\right]rdrd\theta, \tag{3.43}$$

where $\alpha(x, y)$ is the angle between the line O′A and the X-axis and $\beta(x, y)$ is the angle between the line O′B and the X-axis for which the following is valid:

$$\alpha(x,y)=\arctan\left(\frac{y}{R-x}\right). \tag{3.44}$$

$$\beta(x,y)=\arctan\left(\frac{y}{R+x}\right). \tag{3.45}$$

After combining and rearranging eqs 3.43–3.45, one has:

$$P_{y1}(x,y)=\frac{P}{2}\left\{\begin{matrix}1+\frac{1}{\pi}\left[\arctan\left(\frac{y}{R-x}\right)+\arctan\left(\frac{y}{R+x}\right)\right]\\ -\frac{1}{\pi}\int_{-\beta(x,y)}^{\pi+\alpha(x,y)}\exp\left\{-\left[\frac{l(x,y,\theta)}{w}\right]^2\right\}d\theta\end{matrix}\right\}. \tag{3.46}$$

Similarly for the optical power $P_{y2}(x, y)$, one has:

$$P_{y2}(x,y)=\frac{P}{\pi w^2}\int_{\theta=\pi+\alpha(x,y)}^{2\pi-\beta(x,y)}\int_{r=0}^{d(x,y,\theta)}\exp\left[-\left(\frac{r}{w}\right)^2\right]rdrd\theta, \tag{3.47}$$

where for $d\,(x, y\ \theta)$ is valid:

$$d(x,y,\theta)=-\frac{y}{\sin\theta}. \tag{3.48}$$

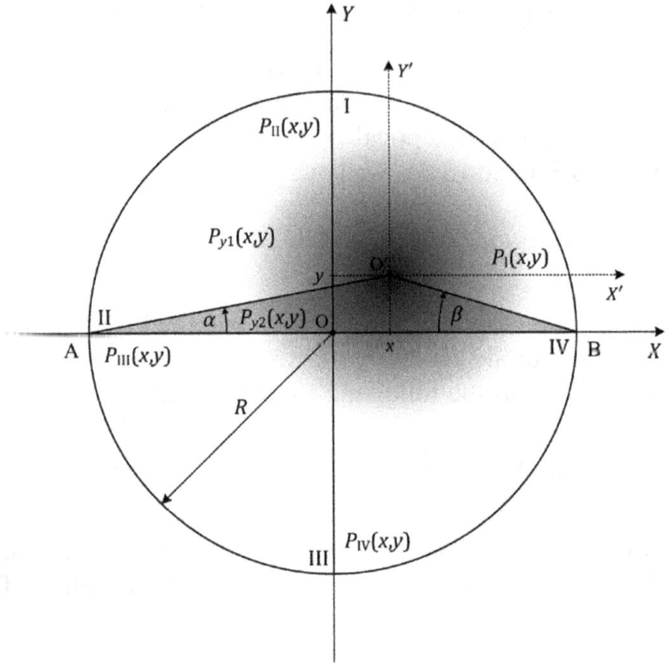

FIGURE 3.17 Geometry of the light spot onto the QPD surface that is used for the power sum calculation.

After combining and rearranging eqs 3.47 and 3.48, one has:

$$P_{y2}(x,y)=\frac{P}{2}\left\{\begin{array}{l}1-\dfrac{1}{\pi}\left[\arctan\left(\dfrac{y}{R-x}\right)+\arctan\left(\dfrac{y}{R+x}\right)\right]\\ -\dfrac{1}{\pi}\displaystyle\int_{\theta=\pi+\alpha(x,y)}^{2\pi-\beta(x,y)}\exp\left\{-\left[\dfrac{d(x,y,\theta)}{w}\right]^2\right\}\mathrm{d}\theta\end{array}\right\}. \tag{3.49}$$

Finally, for the optical power $P_{y+}(x, y)$ that is captured by two upper quadrants, we have:

$$P_{y+}(x,y)=P\left\{\begin{array}{l}1-\dfrac{1}{2\pi}\displaystyle\int_{-\beta(x,y)}^{\pi+\alpha(x,y)}\exp\left\{-\left[\dfrac{l(x,y,\theta)}{w}\right]^2\right\}\mathrm{d}\theta\\ -\dfrac{1}{2\pi}\displaystyle\int_{\theta=\pi+\alpha(x,y)}^{2\pi-\beta(x,y)}\exp\left\{-\left[\dfrac{d(x,y,\theta)}{w}\right]^2\right\}\mathrm{d}\theta\end{array}\right\}. \tag{3.50}$$

For small x and y in comparison with the light spot radius w and the QPD radius R, for example, $x, y \gg \min(w, R)$, we have:

$$\psi(x,y) \approx \left.\frac{\partial \psi(x,y)}{\partial(y/R)}\right|_{\substack{x=0\\y=0}} \cdot \left(\frac{y}{R}\right) + \left.\frac{\partial \psi(x,y)}{\partial(x/R)}\right|_{\substack{x=0\\y=0}} \cdot \left(\frac{x}{R}\right). \quad (3.51)$$

Due to the symmetry, the first partial derivative of function $\psi(x, y)$ with respect to the variable x is equal to zero for $x = y = 0$, so we have:

$$\psi(x,y) \approx \left.\frac{\partial \psi(x,y)}{\partial(y/R)}\right|_{\substack{x=0\\y=0}} \cdot \left(\frac{y}{R}\right) = S_y \frac{y}{R}, \quad (3.52)$$

where S_y is the QPD sensitivity. As can be noticed from the above-presented analysis, we defined the sensitivity of the QPD with respect to the relative displacement y/R instead to the absolute displacement y. The reason for this is to obtain the QPD sensitivity as a plain number irrespective of the QPD radius or light spot radius and because the QPD radius is the only parameter of the system that doesn't change with the change of the beam propagation conditions. As we will see later on, such defined QPD sensitivity only depends on the ratio of the light spot radius and the QPD radius. In this case, for the sensitivity, due to the symmetry at the point $x = y = 0$, the following is valid:

$$S_y = R\left.\frac{\partial \psi(x,y)}{\partial y}\right|_{\substack{x=0\\y=0}} = \frac{2R}{P_\Sigma(x=0, y=0)} \left.\frac{\partial P_{y+}(x,y)}{\partial y}\right|_{\substack{x=0\\y=0}}, \quad (3.53)$$

where the following is valid:

$$P_\Sigma(x=0, y=0) = P\left\{1 - \exp\left[-\left(\frac{R}{w}\right)^2\right]\right\}. \quad (3.54)$$

To find the QPD sensitivity, we must find the following:

$$\left.\frac{\partial P_{y+}(x,y)}{\partial y}\right|_{\substack{x=0\\y=0}} = -\frac{P}{2\pi}\frac{\partial}{\partial y}\left[A(x,y) + B(x,y)\right]\Big|_{\substack{x=0\\y=0}}. \quad (3.55)$$

where we have made the following substitutions:

$$A(x,y) = \int_{-\beta(x,y)}^{\pi+\alpha(x,y)} \exp\left\{-\left[\frac{l(x,y,\theta)}{w}\right]^2\right\} d\theta \text{ and} \quad (3.56)$$

$$B(x,y)=\int_{\theta=\pi+\alpha(x,y)}^{2\pi-\beta(x,y)}\exp\left\{-\left[\frac{d(x,y,\theta)}{w}\right]^2\right\}\mathrm{d}\theta. \tag{3.57}$$

To solve the equation given in eq 3.55, we will need to find the first derivatives of both functions given by eqs 3.56 and 3.57, for example, $a=\partial A(x,y)/\partial y\big|_{x=0,y=0}$ and $b=\partial B(x,y)/\partial y\big|_{x=0,y=0}$. For the parameter a, the following is valid:

$$\begin{aligned}a=&\int_{-\beta(x,y)}^{\pi+\alpha(x,y)}\frac{\partial}{\partial y}\exp\left\{-\left[\frac{l(x,y,\theta)}{w}\right]^2\right\}\Bigg|_{\substack{x=0\\y=0}}\mathrm{d}\theta\\
&+\exp\left\{-\left[\frac{l(0,0,\pi+\alpha(0,0))}{w}\right]^2\right\}\frac{\partial\alpha(x,y)}{\partial y}\Bigg|_{\substack{x=0\\y=0}}\\
&+\exp\left\{-\left[\frac{l(0,0,-\beta(0,0))}{w}\right]^2\right\}\frac{\partial\beta(x,y)}{\partial y}\Bigg|_{\substack{x=0\\y=0}}.\end{aligned} \tag{3.58}$$

By calculating the values given in eq 3.58 and by solving the simple integral given in the same equation, this equation becomes:

$$a=\frac{2}{R}\left[1+2\left(\frac{R}{w}\right)^2\right]\exp\left[-\left(\frac{R}{w}\right)^2\right]. \tag{3.59}$$

Further for parameter b, we have:

$$b=\lim_{y\to 0}\left\{\begin{array}{l}\displaystyle\int_{\theta=\pi+\alpha(x,y)}^{2\pi-\beta(x,y)}\frac{\partial}{\partial y}\exp\left\{-\left[\frac{d(x,y,\theta)}{w}\right]^2\right\}\Bigg|_{x=0}\mathrm{d}\theta\\
\displaystyle-\exp\left\{-\left[\frac{d(0,y,2\pi-\beta(0,y))}{w}\right]^2\right\}\frac{\partial\beta(x,y)}{\partial y}\Bigg|_{x=0}\\
\displaystyle-\exp\left\{-\left[\frac{d(0,y,\pi+\alpha(0,y))}{w}\right]^2\right\}\frac{\partial\alpha(x,y)}{\partial y}\Bigg|_{x=0}\end{array}\right\}, \tag{3.60}$$

where we took the limit of parameter b for $y\to 0$ instead of calculating b for $y=0$, because each of three elements in eq 3.60 has an undefined value in the case of $y = 0$. After careful solving of eq 3.60, we have the following:

$$b = -\frac{2}{R}\left\{c + \exp\left[-\left(\frac{R}{w}\right)^2\right]\right\}. \tag{3.61}$$

where the parameter c is given by:

$$c = \lim_{z \to 0} \frac{1}{z}\int_{z}^{\pi - z}\left(\frac{R}{w}\frac{z}{\sin\theta}\right)^2 \exp\left[-\left(\frac{R}{w}\frac{z}{\sin\theta}\right)^2\right] d\theta. \tag{3.62}$$

The closed-form solution for the parameter c exists but the calculation of this parameter is rather complicated and not straightforward, so only the final solution will be given:

$$c = \sqrt{\pi}\frac{R}{w}\operatorname{erf}\left(\frac{R}{w}\right), \tag{3.63}$$

where erf(x) is the error function of its argument, so finally, for the parameter b, we have:

$$b = -\frac{2}{R}\left\{\sqrt{\pi}\frac{R}{w}\operatorname{erf}\left(\frac{R}{w}\right) + \exp\left[-\left(\frac{R}{w}\right)^2\right]\right\}. \tag{3.64}$$

At the end, we have the following:

$$\left.\frac{\partial P_{y+}(x,y)}{\partial y}\right|_{\substack{x=0\\y=0}} = -\frac{P}{2\pi}(a+b) = \frac{P}{\pi R}\frac{R}{w}\left\{\sqrt{\pi}\operatorname{erf}\left(\frac{R}{w}\right) - \frac{2R}{w}\exp\left[-\left(\frac{R}{w}\right)^2\right]\right\}, \tag{3.65}$$

and finally, for the QPD sensitivity, we have:

$$S_y = \frac{2}{\sqrt{\pi}}\frac{R}{w}\frac{\operatorname{erf}(R/w)\exp\left[(R/w)^2\right] - \left(2/\sqrt{\pi}\right)(R/w)}{\exp\left[(R/w)^2\right] - 1}. \tag{3.66}$$

As it can be seen from eq 3.66, we obtained the QPD sensitivity in the case of a circular QPD and Gaussian distribution of the light spot irradiance onto the QPD surface in the closed form. To see how the QPD sensitivity is changed with respect of the light spot radius, we plot the diagram of the QPD sensitivity S_x and S_y (due to the light spot and QPD symmetry, the sensitivity along X-axis S_x and Y-axis S_y are identical, e.g., $S_x = S_y$) vs. the ratio of the light spot radius and QPD radius w/R. This diagram is presented in Figure 3.18. By analyzing Figure 3.18, one can notice two separate behaviors of the sensitivity function given in eq 3.66. For small values of ratio w/R ($w/R < 1$), the QPD sensitivity decreases with the increase of the light spot radius with the approximate rate of 20 dB/decade, where for large values of ratio w/R ($w/R < 1$), the QPD sensitivity decreases with the increase of the light spot radius with the approximate rate of 40 dB/decade.

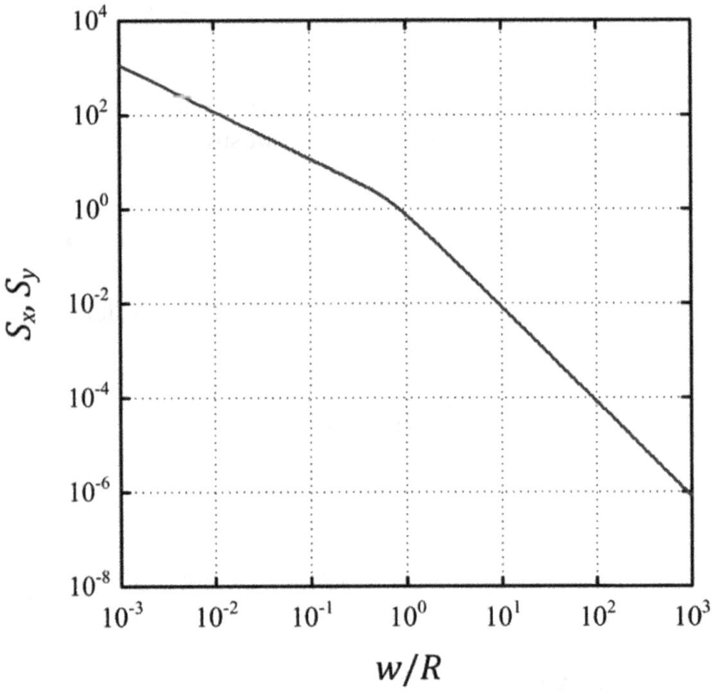

FIGURE 3.18 QPD sensitivity S_x and S_y vs. the ratio of light spot radius and QPD radius w/R.

Besides the significant change of the light spot radius onto the QPD surface due to the significant changes of the beam propagation conditions, which generally occur on a relatively long time scale, there, also, occur rather smaller but faster changes of the light spot radius due to some abrupt changes of the system configuration or externally induced vibration. In the former case, the small changes of the light spot radius will induce change of the QPD sensitivity, which on the other side can influence the overall sensing system characteristics. Therefore, it would be interesting to see how the relative change of the QPD sensitivity is changed with respect to the relative change of the light spot radius. Taking into consideration mentioned earlier, we will find the absolute change (ΔS_y of the QPD sensitivity vs. the absolute change of the light spot radius Δw. According to eq 3.66, we have:

$$\Delta S_y = \frac{\mathrm{d}S_y}{\mathrm{d}w}\Delta w = -\frac{R}{w^2}\frac{\mathrm{d}S_y}{\mathrm{d}(R/w)}\Delta w. \tag{3.67}$$

We are interested in the relative change $\Delta S_y/S_y$ of the QPD sensitivity vs. the relative change $\Delta w/w$ of the light spot radius, so according to eqs 3.66 and 3.67, we have:

$$\frac{\Delta S_y / S_y}{\Delta w / w} = -1 - 2\left(\frac{R}{w}\right)^2$$
$$\left\{\frac{1}{\exp\left[(R/w)^2\right]-1} + \frac{\left(2/\sqrt{\pi}\right)-(R/w)}{\mathrm{erf}(R/w)\exp\left[(R/w)^2\right]-\left(2/\sqrt{\pi}\right)(R/w)}\right\}, \tag{3.68}$$

The ratio of the QPD sensitivity relative change and the light spot radius relative change vs. the ratio of light spot radius and QPD radius is given in Figure 3.19. If we analyze the diagram given in Figure 3.19, we can notice an abrupt increase in absolute value of the ratio of the QPD sensitivity relative change and the light spot radius relative change with respect to the ratio of light spot radius and QPD radius. This abrupt change occurs when the light spot radius approaches the QPD radius.

Subnanometer resolution, relatively simple signal processing circuitry, high-speed measurements in a broad temperature range, makes the QPD probably the first choices for the position measurement in the nano world. Therefore, QPDs have been extensively used in the atomic force microscopes for microcantilever position sensing. When QPD has been illuminated with a laser source or with a light coming out of the single-mode optical fiber, a Gaussian irradiance distribution develops on the QPD surface. As the above-presented analysis shows the smaller the light spot, the higher the QPD sensitivity in light spot position measurement as well as the higher measurement resolution. However, the small insensitive gaps between the quadrants (photodetectors) significantly influence the measurement for a very small light spot. The gap typically represents the dead area of the QPD that has zero responsivity.

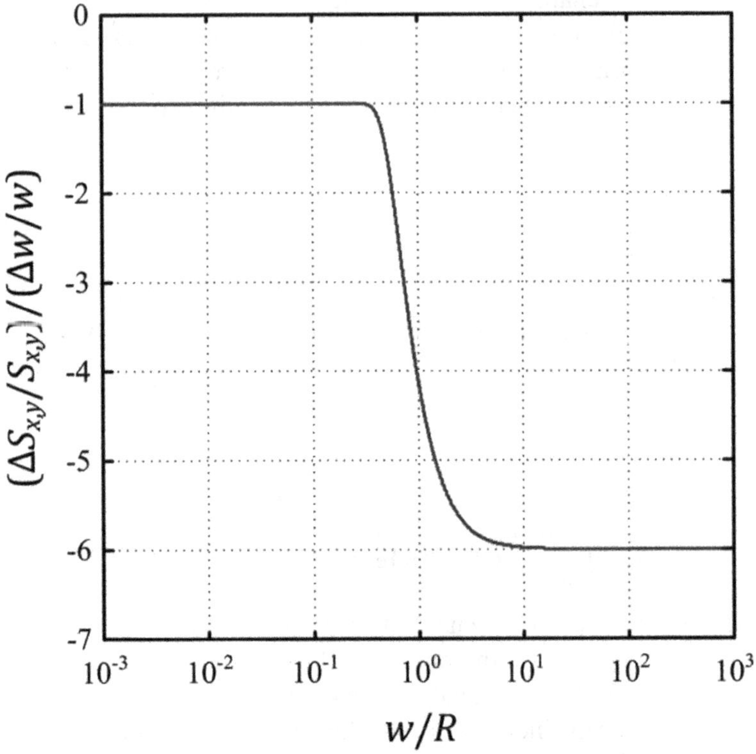

FIGURE 3.19 Ratio of the QPD sensitivity relative change and the light spot radius relative change vs. the ratio of light spot radius and QPD radius.

To minimize the cross talk between QPD elements and thus to reduce the measurement error, there is a narrow gap between each element as is depicted in Figure 3.20. Although very narrow these gaps influence the QPD sensor sensitivity, this will be found under the assumption of Gaussian irradiance distribution on the QPD surface. Therefore, for the irradiance distribution $I(x, y)$, we have:

$$I(x,y)=\frac{P}{\pi\rho^2}\exp\left[-\frac{(x-\chi)^2+(y-\psi)^2}{\rho^2}\right]. \tag{3.69}$$

where x and y are the coordinates in the coordinate system of the QPD, χ and ψ are the coordinates of the light spot center, P is the optical power of the light source that arrive at the QPD surface, and ρ is the radius of the light spot for which the irradiance falls down to the $1/e$ value of its maximal value. According to eq 3.69, the corresponding optical powers P_Q(Q = I, II, III, IV) that have been captured by the corresponding quadrants are given by:

$$P_{\mathrm{I}}(\chi,\psi)=\frac{P}{4}\mathrm{erfc}\left(\frac{(w/2)-\chi}{\rho}\right)\mathrm{erfc}\left(\frac{(w/2)-\psi}{\rho}\right), \tag{3.70}$$

$$P_{\mathrm{II}}(\chi,\psi)=P_{\mathrm{I}}(-\chi,\psi), \tag{3.71}$$

$$P_{\mathrm{III}}(\chi,\psi)=P_{\mathrm{I}}(-\chi,-\psi), \tag{3.72}$$

$$P_{\mathrm{IV}}(\chi,\psi)=P_{\mathrm{I}}(\chi,-\psi), \tag{3.73}$$

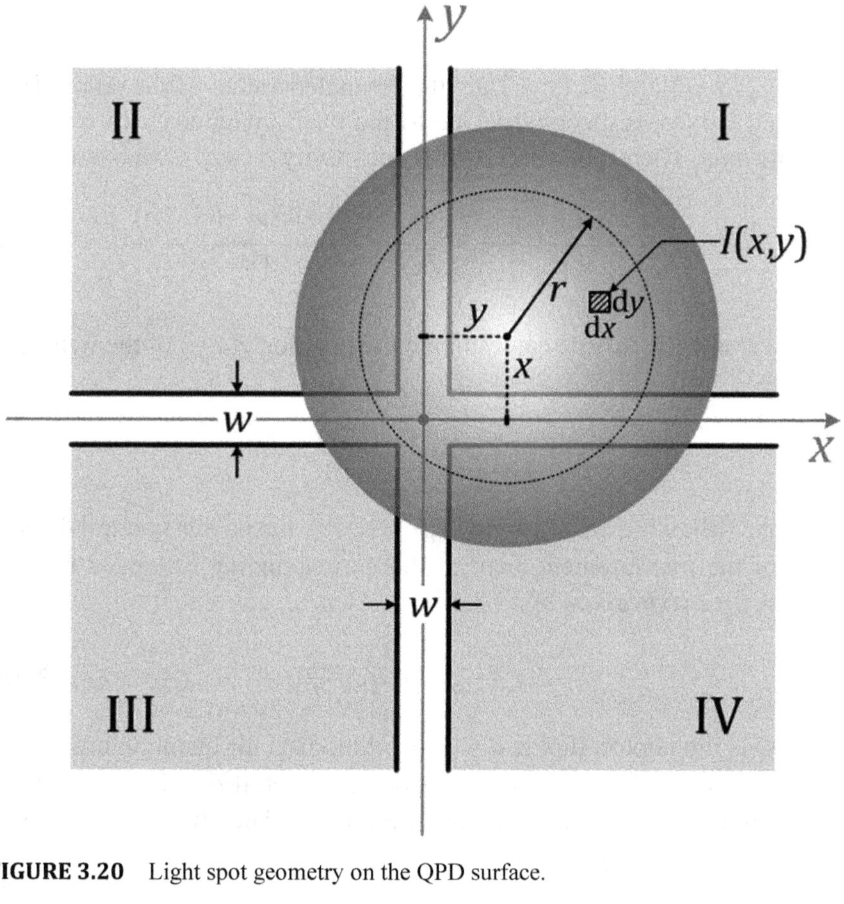

FIGURE 3.20 Light spot geometry on the QPD surface.

where w is the interquadrant gap width and erfc(x) is the complementary error function of its argument. Due to the QPD geometry, only indirect measurements of the light spot position with respect to the QPD center are possible. By simple processing of the QPD signals that are further proportional to the corresponding optical powers, it is only possible to determine the ratios of the light spot positions and the QPD's relevant geometrical parameters. The accurate measurements of these ratios are not possible but only their estimations can be made by using the following equations, which was also shown in the above-presented analysis:

$$\hat{\chi}_{\mathrm{R}}(\chi,\psi)=\frac{P_{\mathrm{I}}(\chi,\psi)+P_{\mathrm{IV}}(\chi,\psi)-P_{\mathrm{II}}(\chi,\psi)-P_{\mathrm{III}}(\chi,\psi)}{P_{\mathrm{I}}(\chi,\psi)+P_{\mathrm{II}}(\chi,\psi)+P_{\mathrm{III}}(\chi,\psi)+P_{\mathrm{IV}}(\chi,\psi)}, \tag{3.74}$$

$$\hat{\psi}_{\mathrm{R}}(\chi,\psi)=\frac{P_{\mathrm{I}}(\chi,\psi)+P_{\mathrm{II}}(\chi,\psi)-P_{\mathrm{III}}(\chi,\psi)-P_{\mathrm{IV}}(\chi,\psi)}{P_{\mathrm{I}}(\chi,\psi)+P_{\mathrm{II}}(\chi,\psi)+P_{\mathrm{III}}(\chi,\psi)+P_{\mathrm{IV}}(\chi,\psi)}, \tag{3.75}$$

where $\hat{\chi}_{\mathrm{R}}(\chi,\psi)$ and $\hat{\psi}_{\mathrm{R}}(\chi,\psi)$ are the estimated values of the ratios along the x and y axes, respectively. Due to the QPD symmetry, we will treat only one axis. Therefore, for the QPD sensitivity $S(w, \rho)$, we have:

$$S(w,\rho)=\left.\frac{\partial\hat{\chi}_{\mathrm{R}}(\chi,\psi)}{\partial\chi}\right|_{\substack{\chi=0\\ \psi=0}}=\frac{4}{\sqrt{\pi}w}\frac{(w/2\rho)\exp\left[-(w/2\rho)^{2}\right]}{\mathrm{erfc}(w/2\rho)}. \tag{3.76}$$

Based on eq 3.76, one can estimate the position $\hat{\chi}(\chi,\psi)$ of the light spot center as:

$$\hat{\chi}(\chi,\psi)\approx\frac{\hat{\chi}_{\mathrm{R}}(\chi,\psi)}{S(w,\rho)}, \tag{3.77}$$

where the following was assumed $\chi,\psi \ll \rho$. For the power spectral density $\langle\hat{\chi}_{\mathrm{Rn}}^{2}\rangle$ of the measurement error of the corresponding estimated relative position, $\hat{\chi}_{\mathrm{R}}(\chi,\psi)$ is given by:

$$\langle\hat{\chi}_{\mathrm{Rn}}^{2}\rangle\approx\frac{1}{2\eta\Phi\,\mathrm{erfc}^{2}(w/2\rho)}, \tag{3.78}$$

where Φ is the photon flux and η is the photodetector quantum efficiency. According to eqs 3.76 and 3.78, the power spectral density $\langle\hat{\chi}_{n}^{2}\rangle$ of the measurement error of the corresponding estimated position $\hat{\chi}(\chi,\psi)$ is given by:

$$\langle \hat{\chi}_n^2 \rangle \approx \frac{\pi w^2}{32\eta\Phi} \frac{1}{(w/2\rho)^2 \exp^2\left[-(w/2\rho)^2\right]}. \tag{3.79}$$

The power spectral density $\langle \hat{\chi}_{Rn}^2 \rangle$, given by eq 3.79, takes the minimum value for the optimal value ρ_{opt} of radius of the light spot for which the irradiance falls down to the $1/e$ value of its maximal value $\rho_{opt} = w/\sqrt{2}$, which is given by:

$$\langle \hat{\chi}_n^2 \rangle_{min} \approx \frac{e\pi w^2}{16\Phi}, \tag{3.80}$$

where for a high-efficiency QPD, $\eta \approx$ has been taken into account. If an integration time of τ has been assumed in the light spot position measurement, for the minimal variance $\Delta\chi_{min}^2$ of the position measurement, we have:

$$\Delta\chi_{min}^2 = \frac{2}{\tau}\int_0^\tau R_\chi(t)\left(1-\frac{t}{\tau}\right)dt, \tag{3.81}$$

where $R_\chi(t)$ represents the autocorrelation function of the light spot measurement error, which is given by:

$$R_\chi(\mathrm{t}) \approx \frac{e\pi w^2}{16\Phi}\delta(t), \tag{3.82}$$

where $\delta(t)$ is the Dirac delta function. Finally, according to eqs 3.81 and 3.82 for the minimal standard deviation $\Delta\chi_{min}$ of the position measurement, we have:

$$\Delta\chi_{min} \approx \sqrt{\frac{e\pi}{8N}}w, \tag{3.83}$$

where $N = \tau$ represents the average number of photons that reach the QPD during the measurement. In Figure 3.21, it is presented the dependence of the minimal standard deviation of the light spot position measurement on the QPD captured photons number for three typical values of the gap width (1) w = 100 μm, (2) w = 10 μm, and (3) w = μm. In the same diagram, it is also depicted the standard quantum limited (SQL) resolution in the optical path difference measurement with the He–Ne laser-based interferometer, given by:

$$\Delta l = \frac{1}{2\pi\sqrt{N}}\lambda, \tag{3.84}$$

where λ is the laser wavelength. Based on the above-presented analysis, one can notice that the QPD resolution is severely limited by the interquadrant gap width. Nevertheless, by using very narrow gap, with the width of only 1 μm, the resolution that can be achieved is just an order of magnitude lower than the SQL optical path difference measurement resolution with the He–Ne laser-based interferometer.

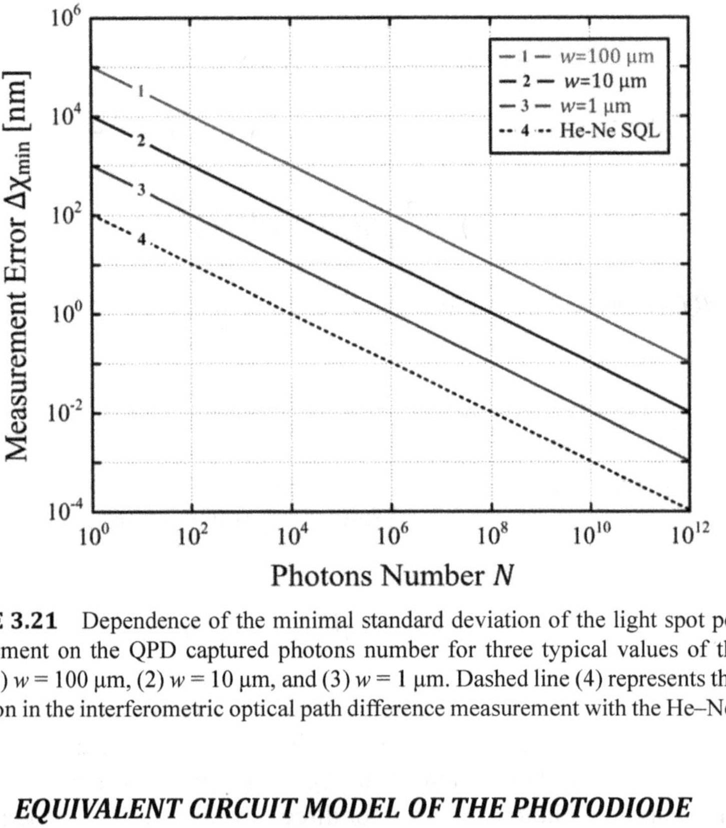

FIGURE 3.21 Dependence of the minimal standard deviation of the light spot position measurement on the QPD captured photons number for three typical values of the gap width (1) $w = 100$ μm, (2) $w = 10$ μm, and (3) $w = 1$ μm. Dashed line (4) represents the SQL resolution in the interferometric optical path difference measurement with the He–Ne laser.

3.3.7 EQUIVALENT CIRCUIT MODEL OF THE PHOTODIODE

The equivalent circuit model of a typical photodiode can be represented by a current source in parallel with an ideal diode as is presented in Figure 3.22. The photocurrent that is generated by the incident optical radiation is represented as the current generator. The ideal diode represents the p-n junction.

In addition, a junction capacitance C_J and a shunt resistance R_{SH} are placed in parallel with the other components. Finally, the series resistance R_S is connected in series with all components in this model.

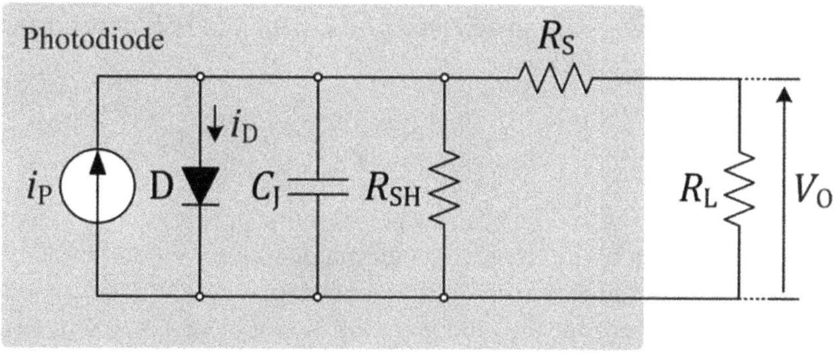

FIGURE 3.22 Equivalent circuit model for a typical photodiode.

Shunt resistance R_{SH} represents the slope of the current–voltage curve of the photodiode for the zero bias voltage, that is, $R_{SH} = \partial V_D / \partial i_D |_{V_D=0}$, with V_D being the bias voltage of the photodiode and i_D its current. Although an ideal photodiode should have an infinite shunt resistance, the real value of the shunt resistance ranges from several tenths of MΩ to several thousands of MΩ. Experimentally, it can be measured by applying the voltage of approximately $V_D = \pm 10$ mV and measuring the current and calculating the resistance. Shunt resistance is important parameter to determine the equivalent power spectral density of the noise current in the photodiode with no bias (photovoltaic mode). For best photodiode performance, that is, the lowest noise, the possible highest shunt resistance is required.

Series resistance R_S of a photodiode represents the resistance of the contacts and the resistance of the undepleted silicon. It is given by:

$$R_S = \frac{\rho(w_S - w_D)}{A} + R_C, \tag{3.85}$$

with w_S being the thickness of the substrate, w_D being the width of the depleted region, A being the diffused area of the junction, ρ being the resistivity of the substrate, and R_C being the contact resistance. Series resistance is typically used to determine the linearity of the photodiode in the photovoltaic mode ($V_D = 0$). Although an ideal photodiode should have no series resistance, typical values ranging from several tenths to several thousands of Ω.

The boundaries of the photodiode depletion region form the plates of a parallel plate capacitor. The photodiode junction capacitance is directly proportional to the diffused area and inversely proportional to the width of the depletion region. In addition, higher resistivity substrates have lower junction capacitance. The capacitance of the photodiode is dependent on the reverse bias in the following way:

$$C_J = \frac{\varepsilon_R \varepsilon_0 A}{\sqrt{2\varepsilon_R \varepsilon_0 \mu \rho \left(V_D + V_B\right)}}, \tag{3.86}$$

where $\varepsilon_0 = 8.854 \times 10^{-12}$ F/m is the permittivity of free space, ε_R is the relative permittivity of the semiconductor material the photodiode has been built and where in the case of silicon, we have $\varepsilon_R = 11.9$, μ is the mobility of the electrons where for the silicon at 300 K, we have $\mu = 0.14$ m^2/Vs, ρ is the resistivity of the semiconductor material, V_B is the built-in voltage of semiconductor, and V_D is the applied bias voltage. Photodiode junction capacitance is used to determine the speed of the response of the photodiode.

Neglecting the charge collection time of the carriers in the depleted region of the photodiode and the charge collection time of the carriers in the undepleted region of the photodiode, and by taking into consideration only the RC time constant of the diode-circuit combination, as presented in Figure 3.22, the frequency response of the diode circuit is given by:

$$\frac{V_O(j\omega)}{i_P(j\omega)} = \frac{R_L}{R_L + R_S} \frac{\left(R_{SH}\left(R_L + R_S\right)\right) / \left(R_{SH} + R_L + R_S\right)}{1 + j\omega C_J \left(R_{SH}\left(R_L + R_S\right) / R_{SH} + R_L + R_S\right)}, \tag{3.87}$$

where $i_P(j\omega)$ is the photocurrent and $V_O(j\omega)$ is output voltage measured at the load resistor R_L. Typically, the following conditions are fulfilled $R_S \ll R_L \ll R_{SH}$, thus giving the frequency response of the photodiode equivalent circuit in the following way:

$$\frac{V_O(j\omega)}{i_P(j\omega)} \approx \frac{R_L}{1 + j\omega C_J R_L}. \tag{3.88}$$

According to eq 3.88, the 3 dB bandwidth of the equivalent photodiode circuit is given by:

$$B_{3\mathrm{dB}} = \frac{1}{2\pi C_J R_L}, \tag{3.89}$$

where the transimpedance gain is equal to the load resistance R_L. Taking into consideration the available transimpedance gain and bandwidth, one can notice that the gain is severely influenced by the load resistance. To increase the speed of the circuit, one needs to reduce the resistance of the load resistor. However, the reduction of the load resistance leads to the increase of the overall thermal noise in the circuit and thus lowering the measurement precision of the circuit. Therefore, one needs to introduce active electronic components in the signal processing chain to increase the bandwidth of the transimpedance amplifier. Typical constructions of the transimpedance amplifiers together with their corresponding parameters will be given in the next chapter.

3.3.8 PHOTODIODE AMPLIFIER CIRCUIT

According to eq 3.55, capacitance of the large diffused area photodiode can easily reach nanofarad range. Such photodiode in combination with the simple high sensitive transimpedance amplifier, consisted of a single high resistance load resistor, gives a transimpedance amplifier which bandwidth may be in the range of several hertzs. Being inherently high-speed detectors, photodiode in this way cannot take the advantage of its basic characteristics. Therefore, an inverting amplifier with very high input impedance must be employed in the signal processing chain, as is presented in Figure 3.23. The voltage amplifier with the gain of $-A$, $A > 1$, has the resistor R_F placed in the feedback arm. Due to the Miller effect, the impedance that loads the photodiode represents equivalently the input impedance of the amplifier, which is further given by:

$$R_{\text{in}} = \frac{R_F}{1+A}. \tag{3.90}$$

This load resistance limits the bandwidth of the transimpedance amplifier to the value of:

$$B_{3\text{dB}} = \frac{1}{2\pi C_J \left(R_F / (1+A)\right)}, \tag{3.91}$$

while keeping the transimpedance gain equal to R_F. This simple schematic increases the gain–bandwidth product of the transimpednce amplifier by a factor of $1 + A$.

Usually, there is also a parasitic input capacitance C_A of the used high impedance amplifier as well as the parasitic capacitance C_F of the feedback resistor R_F, as is presented in Figure 3.24.

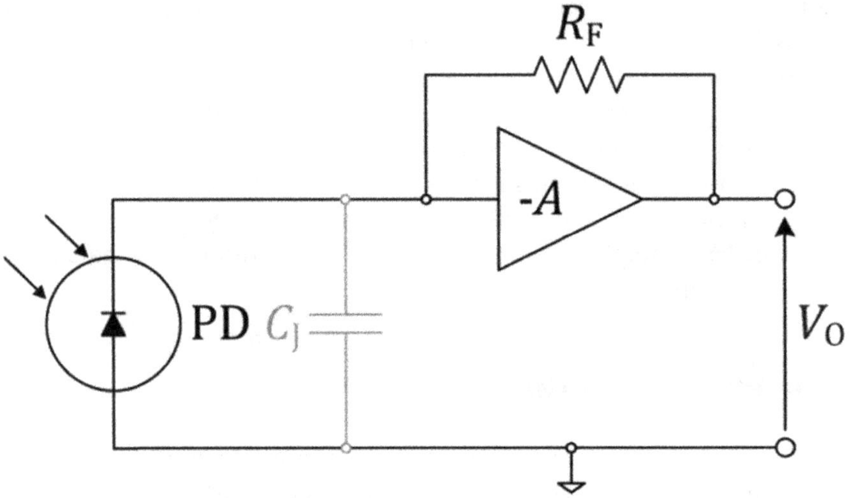

FIGURE 3.23 A simple high-speed transimpedance amplifier.

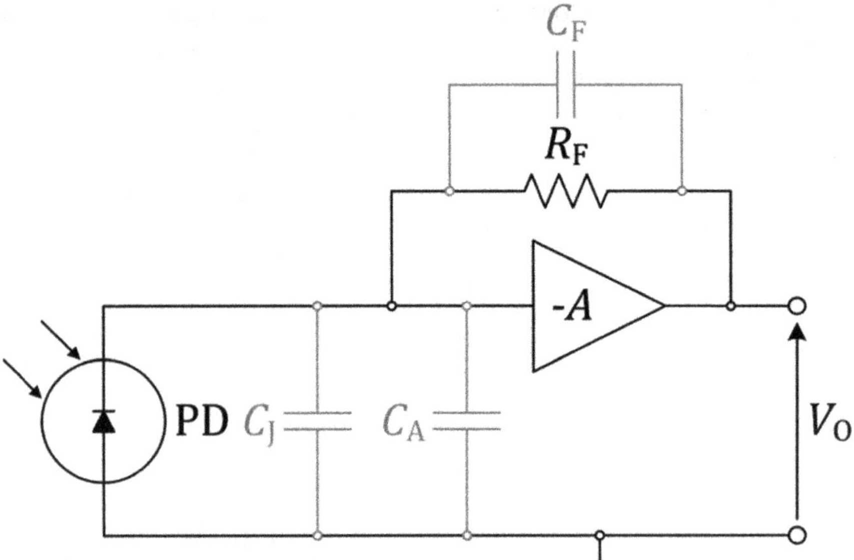

FIGURE 3.24 Simple high-speed transimpedance amplifier with parasitic elements that influence the bandwidth.

The output voltage signal is in this case given by:

$$V_O(j\omega) = \frac{R_F}{1 + j\omega\left(\left((C_J + C_A)/A\right) + C_F\right)R_F} i_P(j\omega), \tag{3.92}$$

with $V_O(j\omega)$ being the output voltage and $i_P(j\omega)$ being the photocurrent. To maximize the bandwidth of the transimpedance amplifier, one needs to choose high gain amplifier so that $C_F \gg (C_J + C_A)/A$ is fulfilled. In this case, eq 3.92 becomes:

$$V_O(j\omega) \approx \frac{R_F}{1 + j\omega C_F R_F} i_P(j\omega), \tag{3.93}$$

where the 3dB bandwidth is equal to:

$$B_{3\text{dB}} = \frac{1}{2\pi C_F R_F}. \tag{3.94}$$

Typical parasitic capacitance of the feedback resistor is in the range form $C_F = 0.02$ pF for the 0201 SMD (surface mount device) case size up to $C_F = 0.2$ pF for a through-hole component. For a typical resistor value of $R_F = 1$ MΩ, the bandwidth of the transimpedance amplifier is $B_{3\text{dB}} \approx 800$ kHz – 8 MHz.

Typically instead of an inverting amplifier, an operational amplifier is used. Instead of the above-supposed flat frequency response of the amplifier an operational amplifier gain is frequency dependent, typically in the following way:

$$A(j\omega) \approx \frac{A_0}{1 + j(\omega/\omega_0)}, \tag{3.95}$$

where $A_0 \gg 1$ is the low-frequency gain and $\omega_0/2\pi$ is the bandwidth of the operational amplifier. The analysis shows that the output voltage in this case is given by:

$$V_O(j\omega) = \frac{A(j\omega)Z_F}{1 + A(j\omega) + (Z_F/Z_{\text{in}})} i_P(j\omega), \tag{3.96}$$

where we have $Z_{\text{in}} = 1/j\omega(C_J + C_A)$ and $Z_F = R_F/(1 + j\omega C_F R_F)$. After rearranging eq 3.96, one has:

$$V_O(j\omega) = \frac{R_F}{1+(1/A_0)+j\omega\{(1/A_0\omega_0)+R_F[C_F(1+(1/A_0))+((C_J+C_A)/A_0)]\}-\omega^2((R_F(C_J+C_A+C_F))/\omega_0)} i_P(j\omega). \tag{3.97}$$

Taking into consideration the following conditions that are typically satisfied for the transimpedance amplifier: $A_0 \gg 1$, $C_F \gg (C_J + C_A)/A_0$, and $C_J + C_A \gg C_F$, eq 3.97 becomes:

$$V_O(j\omega) \approx \frac{R_F}{1+j\omega((1/A_0\omega_0)+R_F C_F)-\omega^2((R_F(C_J+C_A))/\omega_0)} i_P(j\omega), \tag{3.98}$$

or equivalently:

$$|V_O(j\omega)| \approx \frac{R_F}{\sqrt{1+\omega^2\left[((1/A_0\omega_0)+R_F C_F)^2-(2R_F(C_J+C_A))/\omega_0\right]+\omega^4\left[(R_F(C_J+C_A))/\omega_0\right]^2}} |i_P(j\omega)|. \tag{3.99}$$

To avoid gain peaking of the transimpedance amplifier and thus oscillation of the output voltage signal, one needs to fulfill the following condition:

$$\left(\frac{1}{A_0\omega_0}+R_F C_F\right)^2 - \frac{2R_F(C_J+C_A)}{\omega_0} \geq 0. \tag{3.100}$$

Since one can choose the feedback capacitance, we have the following condition to avoid gain peaking:

$$C_F \geq \sqrt{\frac{2(C_J+C_A)}{\omega_0 R_F}} - \frac{1}{A_0\omega_0 R_F}, \tag{3.101}$$

where in the case when the left side of eq 3.101 is equal to the right site of eq 3.101, we obtain the maximally flat frequency response, that is, the transimpedance amplifier behaves as the Butterworth filter.

In the occasion of low light level measurements, one of the limiting factors in achieving high-resolution measurements is noise. Noise model of the transimpedance amplifier is presented in Figure 3.25 where are included all relevant equivalent noise sources such as i_{PDn} representing the shot noise of the photodiode, e_{An} representing the equivalent input voltage noise of the inverting amplifier, i_{An} representing the equivalent input current noise of the inverting amplifier, and e_{R_F} representing the thermal voltage noise of the feedback resistor R_F.

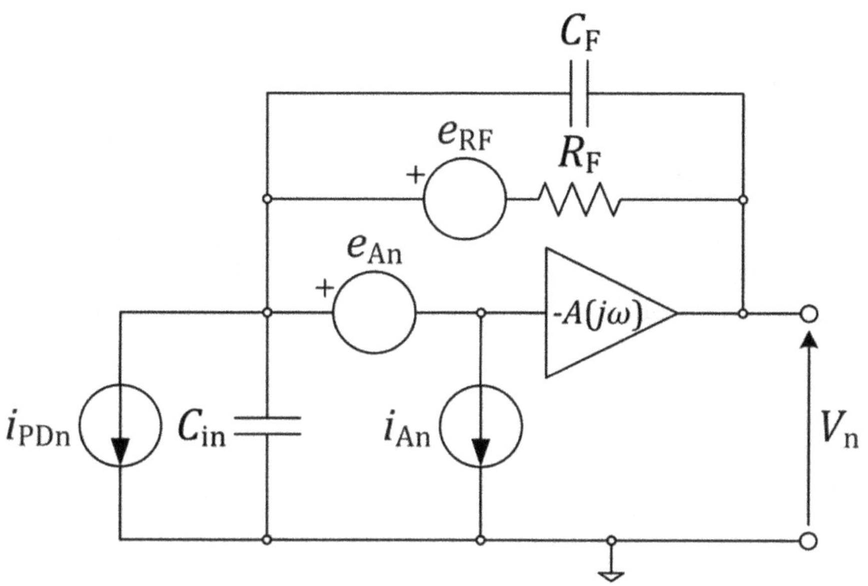

FIGURE 3.25 Noise model of the transimpedance amplifier.

The power spectral density of the photodiode shot noise is given as:

$$i_{\mathrm{PDn}}^2 = 2qI_{\mathrm{PD}}, \qquad (3102)$$

where q is the elementary charge and I_{PD} is the current flowing through the photodiode. The power spectral densities of the equivalent input voltage e_{An}^2 and current i_{An}^2 noise sources have the values specified from the manufacturers, that is, one can find the corresponding data in the operational amplifiers datasheet. The power spectral density of the thermal voltage noise of the feedback resistor R_{F} is given by:

$$e_{R_F}^2 = 4k_B T R_F, \qquad (3.103)$$

with k_B being the Boltzmann constant and T being the absolute resistor temperature. Based on the analysis of the circuit presented in Figure 3.25, the power spectral density of the output voltage is given by:

$$V_n^2 = R_T^2(\omega)\left(i_{\mathrm{PDn}}^2 + i_{\mathrm{An}}^2 + \frac{e_{R_F}^2}{R_F^2}\right) + \frac{1+\left[\omega R_F\left(C_{\mathrm{in}}+C_F\right)\right]^2}{1+\left(\omega R_F C_F\right)^2} e_{\mathrm{An}}^2, \qquad (3.104)$$

where all the noise sources are assumed to be independent and where for the transimpedance amplifier gain $R_T(\omega)$ the following is valid:

$$R_T(\omega) = \frac{|V_O(j\omega)|}{|i_P(j\omega)|} \approx$$

$$\frac{R_F}{\sqrt{1+\omega^2\left[\left((1/A_0\omega_0)+R_F C_F\right)^2-\left(2R_F C_{\text{in}}/\omega_0\right)\right]+\omega^4\left(R_F C_{\text{in}}/\omega_0\right)^2}}, \quad (3.105)$$

where we have $C_{\text{in}} = C_J + C_A$ is the overall input capacitance. In the case of low-frequency optical signal detection and amplification, the power spectral density of the overall noise voltage at the amplifier output the following is obtained:

$$V_n^2 = R_F^2\left(i_{\text{PDn}}^2 + i_{\text{An}}^2 + \frac{e_{R_F}^2}{R_F^2}\right) + e_{\text{An}}^2. \quad (3.106)$$

The equivalent power spectral density of the equivalent input current noise source i_n^2 is given by:

$$i_n^2 = \frac{V_n^2}{R_F^2} = i_{\text{PDn}}^2 + i_{\text{An}}^2 + \frac{e_{R_F}^2 + e_{\text{An}}^2}{R_F^2}. \quad (3.107)$$

In terms of the best possible signal-to-noise ratio at the output of the transimpedance amplifier, one needs to add additional resistor between the photodiode and the noninverting input of the operational amplifier as is presented in Figure 3.26 where the optimal topology of the transimpedance amplifier is presented. This topology is especially convenient in the case of a single supply operational amplifier since its noninverting input isn't grounded. The output voltage of the presented optimal topology transimpedance amplifier is given by:

$$V_O = (R_F + R_L)\,i, \quad (3.108)$$

with R_L being the load resistor.

In Figure 3.27, it is presented the schematic of the transimpedance amplifier together with all noise relevant sources. According to the equivalent noise generators presented in Figure 3.27, the equivalent input noise current i_n^2 is given by:

$$i_n^2 = \frac{V_n^2}{R_F^2} = i_{\text{PDn}}^2 + \frac{i_{\text{An+}}^2}{\left(1+(R_F/R_L)\right)^2} + \frac{i_{\text{An-}}^2}{\left(1+(R_L/R_F)\right)^2} + \frac{e_{R_F}^2 + e_{R_L}^2 + e_{\text{An}}^2}{\left(R_F + R_L\right)^2}. \quad (3.109)$$

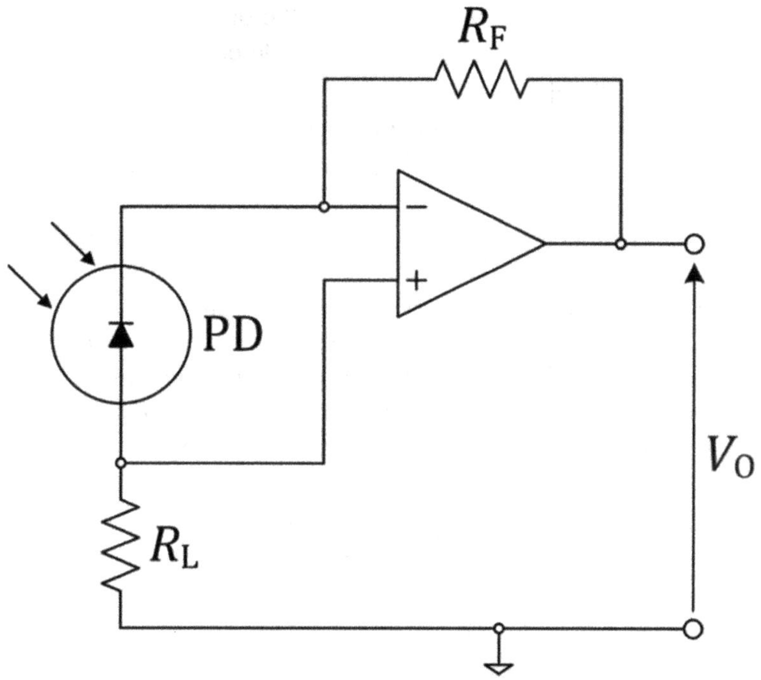

FIGURE 3.26 Optimal topology transimpedance amplifier.

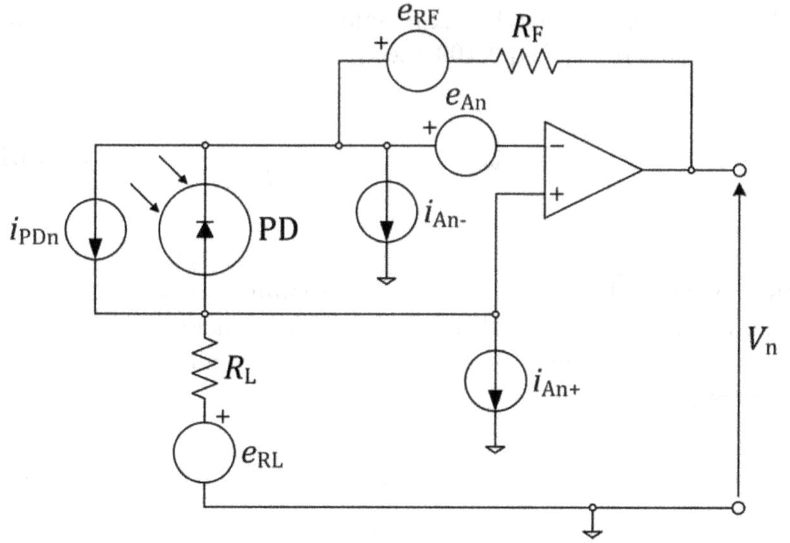

FIGURE 3.27 Noise model of the optimal topology transimpedance amplifier.

where V_n^2 is the power spectral density of the output voltage noise, i_{PDn}^2 is the power spectral density of the photodiode equivalent noise current, $i_{\mathrm{An+}}^2$ and $i_{\mathrm{An-}}^2$ are the operational amplifier equivalent input noise currents power spectral densities, e_{An}^2 is the operational amplifier equivalent input voltage noise power spectral density, and $e_{R_F}^2$ and $e_{R_L}^2$ are voltage noises power spectral densities of the resistors R_F and R_L, respectively (where for the transimpedance gain R_T we have $R_T = R_F + R_L$). In eq 3.109, we neglect the influence of the operational amplifier input capacitances and photodiode capacitance on the equivalent input current noise power spectral densities because it is assumed that transimpedance amplifier has relatively low bandwidth and because the equivalent operational amplifier input voltage noise is much smaller than the resistor voltage noises. For the corresponding noise signals, we have the following:

$$i_{\mathrm{PDn}}^2 = 2qI_{\mathrm{PD}} + \frac{4k_BT}{R_{\mathrm{PD}}} \approx 2qI_{\mathrm{PD}}, \tag{3.110}$$

$$e_{R_F}^2 = 4k_BTR_F, \text{ and} \tag{3.111}$$

$$e_{R_L}^2 = 4k_BTR_L, \tag{3.112}$$

where R_{PD} is the photodiode resistance at zero bias voltage for which is valid $R_{\mathrm{PD}} \gg R_F, R_L$. Further, by taking into account that typically following is fulfilled: $e_{\mathrm{An}}^2 \ll e_{R_F}^2, e_{R_L}^2$, eq 3.109 becomes:

$$i_n^2 \approx i_{\mathrm{PDn}}^2 + \frac{4k_BT}{R_T} + i_{\mathrm{An}}^2\left[\frac{1}{\left(1+\left(R_F/R_L\right)\right)^2} + \frac{1}{\left(1+\left(R_L/R_F\right)\right)^2}\right], \tag{3.113}$$

where we have $i_{\mathrm{An}}^2 \approx i_{\mathrm{An+}}^2 \approx i_{\mathrm{An-}}^2$. The equivalent input noise current has the minimum value when $R_L = R_F$ is satisfied. To minimize the overall equivalent input noise current both of the transimpedance amplifier resistors must have the same value. In this case, the equivalent input current noise is given by:

$$i_n^2 \approx i_{\mathrm{PDn}}^2 + \frac{4k_BT}{R_T} + \frac{1}{2}i_{\mathrm{An}}^2. \tag{3.114}$$

KEYWORDS

- **avalanche photodiode**
- **optical detector**
- **photoconductor**
- **photodiode**
- **photomultiplier**
- **photon counting**

REFERENCES

Alexander, S. B. *Optical Communications Receiver Design*; SPIE – The International Society for Optical Engineering: Bellingham, 1997.

Cox, III C. H. *Analog Optical Links: Theory and Practice*; Cambridge University Press: Cambridge, 2004.

Graeme, J. *Photodiode Amplifiers: Op Amp Solutions*; McGraw-Hill Companies Inc.: New York, NY, 1996.

Henini, M.; Razeghi, M., Eds. *Handbook of Infrared Detection Technologies*; Elsevier Science Ltd.: Oxford, 2002.

Hermans, C.; Steyaert, M. *Broadband Opto-Electrical Receivers in Standard CMOS*; Springer: Dordrecht, 2007.

Johnson, M. *Photodetection and Measurement: Maximizing Performance in Optical Systems*; The McGraw-Hill Companies Inc.: New York, NY, 2003.

Manojlović, L. M. *Optics and Applications*; Arcler Press: Oakville, Canada, 2018.

Manojlović, L. M. Quadrant Photodetector Sensitivity. *Appl. Opt.* July **2011,** *50* (20), 3461–3469.

Manojlović, L. M. Resolution Limit of the Quadrant Photodetector. *Optik – Int. J. Light Electron Opt.* Oct **2016,** *127* (19), 7631–7634.

Manojlović, L. M.; Živanov, M. B.; Slankamenac, M. P.; Bajić, J. S.; Stupar, D. Z. High-Speed and High-Sensitivity Displacement Measurement with Phase-Locked Low-Coherence Interferometry. *Appl. Opt.* July **2012,** *51* (9), 4333–4342.

Piotrowski, J.; Rogalski, A. *High-Operating-Temperature Infrared Photodetectors*; The Society of Photo-Optical Instrumentation Engineers: Bellingham, 2007.

Piprek, J. *Semiconductor Optoelectronic Devices: Introduction to Physics and Simulation*; Elsevier Science: Amsterdam, 2003.

Reed, G. T.; Knights, A. P. *Silicon Photonics: An Introduction*; John Wiley & Sons Ltd: Chichester, 2004.

Righini, G. C.; Tajani, A.; Cutolo, A. *An Introduction to Optoelectronic Sensors*; World Scientific Publishing Co. Pte. Ltd.: Singapore, 2009.

Rossi, L.; Fischer, P.; Rohe, T.; Wermes, N. *Pixel Detectors: From Fundamentals to Applications*; Springer-Verlag: Berlin Heidelberg, 2006.

Säckinger, E. *Broadband Circuits for Optical Fiber Communication*; John Wiley & Sons, Inc.: Hoboken, 2003.

Schneider, K.; Zimmermann, H. *Highly Sensitive Optical Receivers*; Springer: Berlin Heidelberg, 2006.

Stocker, A. A. *Analog VLSI Circuits for the Perception of Visual Motion*; John Wiley & Sons Ltd: Chichester, 2006.

CHAPTER 4

Coherence and Interference of Light

ABSTRACT

Finite size of the spectral bandwidth of a light source requires more powerful theory to process the optical phenomena accompanying such sources. The coherence theory, which is basically a statistical theory that describes the properties of the radiation field in terms of the correlation between the vibrations at different points in the field, offers the possibility to describe such phenomena. The waves originating from different points of a finite-size light source or even from the same point of a thermal source, exhibit random amplitude and phase fluctuations. If such wave fields illuminate, two points in space or the same point at different instants of time exhibit only partial correlation.

4.1 TWO-BEAM INTERFERENCE

A light ray, which concept was used in describing the geometrical optics phenomena, fails trying to explain optical phenomena like diffraction and interference. Therefore, instead of a light ray, the wave nature of electromagnetic radiation will be taken into consideration in this chapter. For the purpose of simplicity, we will consider only a linearly polarized plane wave propagating in a vacuum in the z-axis direction. In this case, the electric field at any point along the direction of propagation and any time moment can be represented by a harmonic function of distance and time as:

$$E(z,t) = E_0\cos(\omega t - kz), \tag{4.1}$$

where E_0 is the electrical field amplitude, k is the wavenumber, and ω is the angular frequency. Sometimes, it is more convenient to represent an electrical field of a plane wave by using Euler representation of complex number as:

$$E(z,t) = \mathrm{Re}\{E_0 \exp[j(\omega t - kz)]\}, \tag{4.2}$$

where Re{•} represents the real part of a complex number. If the mathematical operations on electrical field are linear, then, without losing the meaning of the analysis, we can represent the electrical field as a complex number in the following form:

$$E(z,t) = E_0 \exp\,[j(\omega t - kz)], \tag{4.3}$$

and take the real part of the final result at the end of the calculation. Equation 4.3 can be reorganized as:

$$E(z,t) = E_0 \exp(-jkz)\, \exp(j\omega t) = A\, \exp(j\omega t), \tag{4.4}$$

where A is the complex amplitude given by:

$$A = E_0 \exp(-jkz). \tag{4.5}$$

Due to the extremely high frequencies of optical waves (in the visible range, we typically have $\nu \approx 6 \times 10^{14}$ Hz for $\lambda = 500$ nm), direct measurement of the electric field is not possible. It is only possible to measure directly the radiation parameters that are related to the optical wave's energetic content. Therefore, one of the most important quantities is the irradiance, which represents the time average of the amount of energy which, in unit time, crosses a unit area normal to the direction of the energy flow. The irradiance is thus proportional to the time average of the square of the electric field, that is:

$$I = \langle E^2(t) \rangle = \lim_{T \to \infty} \frac{1}{T} \int_0^T E^2(t)\, \mathrm{d}t. \tag{4.6}$$

The combination of eqs 4.1 and 4.6 is:

$$I = \lim_{T \to \infty} \frac{1}{T} \int_0^T E_0^2 \cos^2(\omega t - kz)\, \mathrm{d}t = \frac{E_0^2}{2} = \frac{|A|^2}{2}. \tag{4.7}$$

Ignoring the scale factor 1/2, the important conclusion that can be extracted is that the irradiance is directly proportional to the electrical field amplitude squared.

In the case of two monochromatic waves, which propagate in the same direction and are polarized in the same plane, a superposition of the waves occurs at a certain point in space, where the total electric field at this point is given by:

$$E = E_1 + E_2, \tag{4.8}$$

where E_1 and E_2 are the electric fields of two waves. In the case where both waves have the same frequency, the irradiance at this point is:

$$I = |A_1 + A_2|^2, \tag{4.9}$$

where $A_1 = E_{01} \exp(-jkz_1) = E_{01} \exp(-j\varphi_1)$ and $A_2 = E_{02} \exp(-jkz_2) = E_{02} \exp(-j\varphi_2)$ are the complex amplitudes of the two waves. In this case, eq 4.9 becomes:

$$I = |A_1|^2 + |A_2|^2 + |A_1 A_2^*| + |A_1^* A_2|, \tag{4.10}$$

$$I = I_1 + I_2 + 2\sqrt{I_1 I_2}\cos\Delta\varphi, \tag{4.11}$$

where the asterisk sign (∗) marks the complex conjugate, I_1 and I_2 are the intensities of the waves and $\Delta\varphi = \varphi_1 - \varphi_2$ is the phase difference between them. Depending on the phase difference, that is, on the optical path difference between the waves, the irradiance at the point of observation can take any value between the maximum and minimum values that are given by:

$$I_{\min} = I_1 + I_2 - 2\sqrt{I_1 I_2}, \tag{4.12}$$

$$I_{\max} = I_1 + I_2 + 2\sqrt{I_1 I_2}. \tag{4.13}$$

An appropriate measure of the interference phenomenon contrast is the visibility of the interference fringes, which is defined as:

$$V = \frac{I_{\max} - I_{\min}}{I_{\max} + I_{\min}} = \frac{2\sqrt{I_1 I_2}}{I_1 + I_2}, \tag{4.14}$$

which can take any value between 0 and 1 ($0 \le V \le 1$).

If we take into consideration a light from a thermal source, even if it consists of a single spectral line, this light source is not strictly monochromatic. The reason for this is the random rapid fluctuations of the electric field amplitude and phase. So, if we observe two such independent light sources, which fluctuations are uncorrelated, interference effects cannot be observed. At this stage, we will consider that to obtain interference phenomena, we must obtain two light beams from the same light source. Therefore, we shall consider that the observed interference phenomena between two beams can be classified according to the method used to obtain these beams.

4.1.1 WAVEFRONT DIVISION METHOD

To obtain two beams from a single light source, one can use two portions of the original wavefront. Then, both of the beams are superimposed to obtain interference pattern. This method is known as wavefront division. A simple optical system based on Fresnel's mirrors arrangement is presented in Figure 4.1. The light from a point-like source S has been used to illuminate two mirrors M_1 and M_2 that are mutually inclined to a very small angle α ($\alpha \ll 1$). The reflected light is directed toward the screen where the beams are overlapped and thus the interference fringes occur on the screen. Equivalently, we can consider that the interference takes place between the light from the two virtual sources S_1 and S_2, that is, the images of the source S in the mirrors M_1 and M_2. According to the presented mirrors and source geometry, we have:

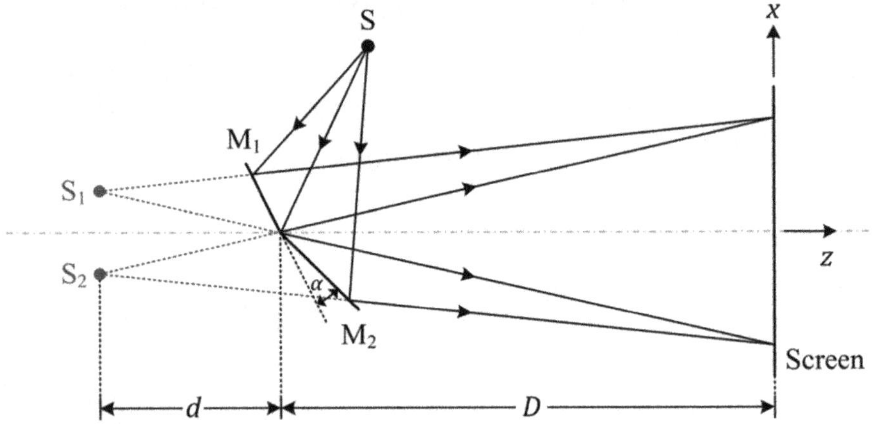

FIGURE 4.1 Optical interference produced by wavefront division (Fresnel's mirrors).

$$\overline{S_1S_2} \approx 2\alpha d, \tag{4.15}$$

where $\overline{S_1S_2}$ is the distance between the virtual sources and d is the distance between the mirrors and the virtual sources. The optical path difference between the beams at the screen point positioned at the distance x from the optical is, to a first approximation, given by:

$$\text{OPD} \approx \frac{2\alpha x d}{d + D}, \tag{4.16}$$

where D is the distance between the mirrors and the screen. As the successive maxima or minima in the interference pattern correspond to a change in the optical path difference equals to the light source wavelength λ, the distance between the fringes is given by:

$$\Delta x = \frac{\lambda\left(d+D\right)}{2\alpha d}. \tag{4.17}$$

There are also other arrangements such as Young's double-slit interferometer, presented in Figure 4.2, which uses the diffracted beams from a pair of slits S_1 and S_2, and Lloyd's mirror, presented in Figure 4.3, which uses one beam from the source S and another from a mirror illuminated at near-grazing incidence.

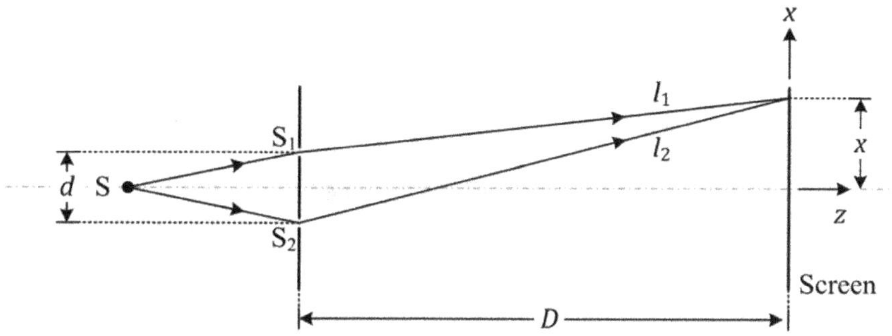

FIGURE 4.2 Young's double-slit interferometer.

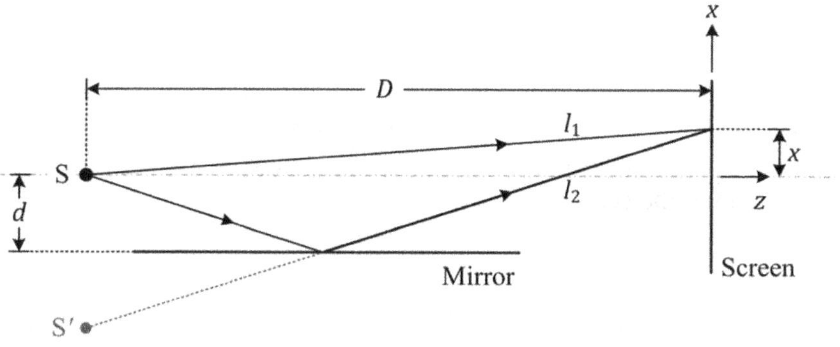

FIGURE 4.3 Lloyd's mirror.

In the case of Young's double-slit interferometer, the optical path difference between the beams that impinge the screen at the point, which is positioned at the distance x from the optical axis (axis of symmetry), is given by:

$$\text{OPD} \approx l_1 - l_2, \tag{4.18}$$

where l_1 and l_2 are the beam lengths from the slits S_1 and S_2, respectively. These distances are given with:

$$l_1 = \sqrt{D^2 + \left(x - \frac{d}{2}\right)^2}, \tag{4.19}$$

$$l_2 = \sqrt{D^2 + \left(x + \frac{d}{2}\right)^2}, \tag{4.20}$$

where d is the distance between the slits and D is the distance between the slits and the screen. Equations 4.18–4.20 in the case when $D \gg d, x$ give:

$$\text{OPD} \approx \sqrt{D^2 + \left(x - \frac{d}{2}\right)^2} - \sqrt{D^2 + \left(x + \frac{d}{2}\right)^2} \approx -\frac{d}{D}x. \tag{4.21}$$

As the successive maxima or minima in the interference pattern correspond to a change in the optical path difference equals to the light source wavelength λ, the distance between the fringes is given by:

$$\Delta x = \frac{D\lambda}{d}. \tag{4.22}$$

In the case of Lloyd's interferometer, the optical path difference between the beams that impinge the screen at the point which is positioned at the distance x from the optical axis (axis of symmetry) is given by:

$$\text{OPD} \approx l_1 - l_2, \tag{4.23}$$

where l_1 is the beam length from the source S to the point at the screen located at the position x from the optical axis, and l_2 is the reflected beam length, which is equal to the length from the virtual source S′ to the observation point at the screen. These distances are given with:

$$l_1 = \sqrt{D^2 + (x - d)^2}, \tag{4.24}$$

$$l_2 = \sqrt{D^2 + (x + d)^2}, \tag{4.25}$$

where d is the distance between the source S and the horizontal mirror and D is the distance between the source and the screen. Equations 4.23–4.22 in the case when $D \gg d, x$ give:

$$\text{OPD} \approx \sqrt{D^2 + (x - d)^2} - \sqrt{D^2 + (x + d)^2} \approx -\frac{2d}{D}x \tag{4.26}$$

As the successive maxima or minima in the interference pattern correspond to a change in the optical path difference equals to the light source wavelength λ, the distance between the fringes is given by:

$$\Delta x = \frac{D\lambda}{2d}. \tag{4.27}$$

4.1.2 AMPLITUDE DIVISION METHOD

The other method, where two beams can be obtained from a single source, is by division of the amplitude over the same section of the wavefront. To perform amplitude division of the wavefront, there is one way where we can use a surface that partially reflects the incident light and partially transmits it. If a transparent plane-parallel plate is illuminated by the monochromatic point-source S, as presented in Figure 4.4, then at a given point at the observation plane two beams, reflected from the upper and lower surface of the plane-parallel plate, arrive approximately with the same amplitude. Therefore, from the setup symmetry, the observed interference fringes have circular shape. In the case when the observation plane is positioned at infinity, the interference fringes can be obtained at the back focal plane of the lens, as shown in Figure 4.5. The optical path difference between the beams is given by:

$$\text{OPD} \approx n\left(\overline{\text{AB}} + \overline{\text{BC}}\right) - \overline{\text{AD}}, \tag{4.28}$$

where n is the plan-parallel plate index of refraction, and lengths $\overline{\text{AB}}$, $\overline{\text{BC}}$, and $\overline{\text{AD}}$ are presented in Figure 4.5. The geometry from Figure 4.5 also gives the following relations:

$$\overline{\text{AB}} = \overline{\text{BC}} = \frac{d}{\cos\theta_2}, \tag{4.29}$$

$$\overline{\text{AD}} = \overline{\text{AC}}\sin\theta_1 = 2d\tan\theta_2\sin\theta_1, \tag{4.30}$$

where d is the plate thickness. Taking into consideration Snell's law, we have $\sin\theta_1 = n\sin\theta_2$, we have:

$$\text{OPD} \approx 2nd\cos\theta_2. \tag{4.31}$$

However, we also have to take into account a phase shift of π introduced by reflection at one of the plate surfaces. Therefore, the effective optical path difference between the interfering wavefronts is:

$$\mathrm{OPD} \approx 2nd\cos\theta_2 \pm \frac{\lambda}{2}. \tag{4.32}$$

The conditions for the bright and dark fringes are given:

$$\mathrm{OPD} \approx 2nd\cos\theta_2 \pm \frac{\lambda}{2} = k\lambda, \tag{4.33}$$

$$\mathrm{OPD} \approx 2nd\cos\theta_2 \pm \frac{\lambda}{2} = (2k+1)\frac{\lambda}{2}, \tag{4.33}$$

respectively, where k is the integer.

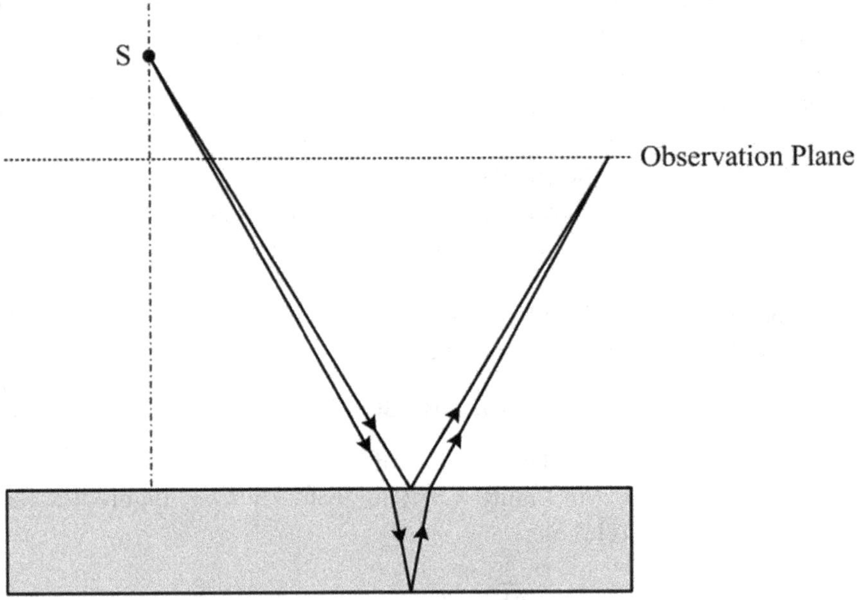

FIGURE 4.4 Interference fringes formation by the reflection in a plane-parallel plate.

Many interferometer applications require an optical arrangement where two interfering beams travel along separate paths before they are recombined. Therefore, a number of interferometers for specific purposes have been developed. Most of them are based on the division of amplitude, with the exception of the Rayleigh interferometer and its variations. Here, it will be presented three typical interferometers with separate beam paths and division of amplitude. To perform amplitude division, one of them uses partially reflecting film made of metal or dielectric commonly known as a beam splitter. Other uses a birefringent element to produce two orthogonally polarized beams. In this case, both beams must be derived from a single polarized beam and then brought into the same polarization state. The last one uses a scatter plate, which transmittance varies in a random manner, to produce one or more diffracted beams, as well as a reflected or transmitted beam.

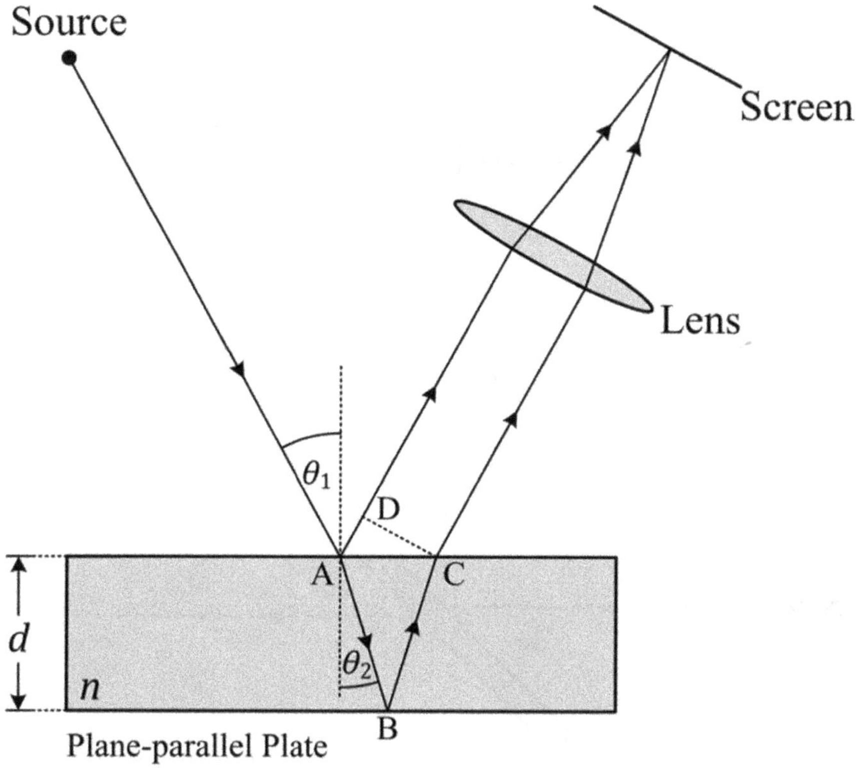

FIGURE 4.5 Interference fringes formation of equal inclination by reflection in a plane-parallel plate.

There are also two beam interferometers that can have different topologies where the most prominent are the Michelson, Mach–Zehnder, and Sagnac interferometers, which will be analyzed in more detail in the following chapters.

4.1.3 THE MICHELSON INTERFEROMETER

The Michelson interferometer, as shown in Figure 4.6, uses a light source S that is divided by a semireflecting coating deposited on one surface of a plane-parallel glass plate BS (known as a beam splitter) into two beams with approximately equal amplitudes. The beams, which are transmitted and reflected from the beam splitter, are back-reflected at two plane mirrors M_1 and M_2 and returned to the beam splitter. After the recombination in the beam splitter, they impinge the optical detector D.

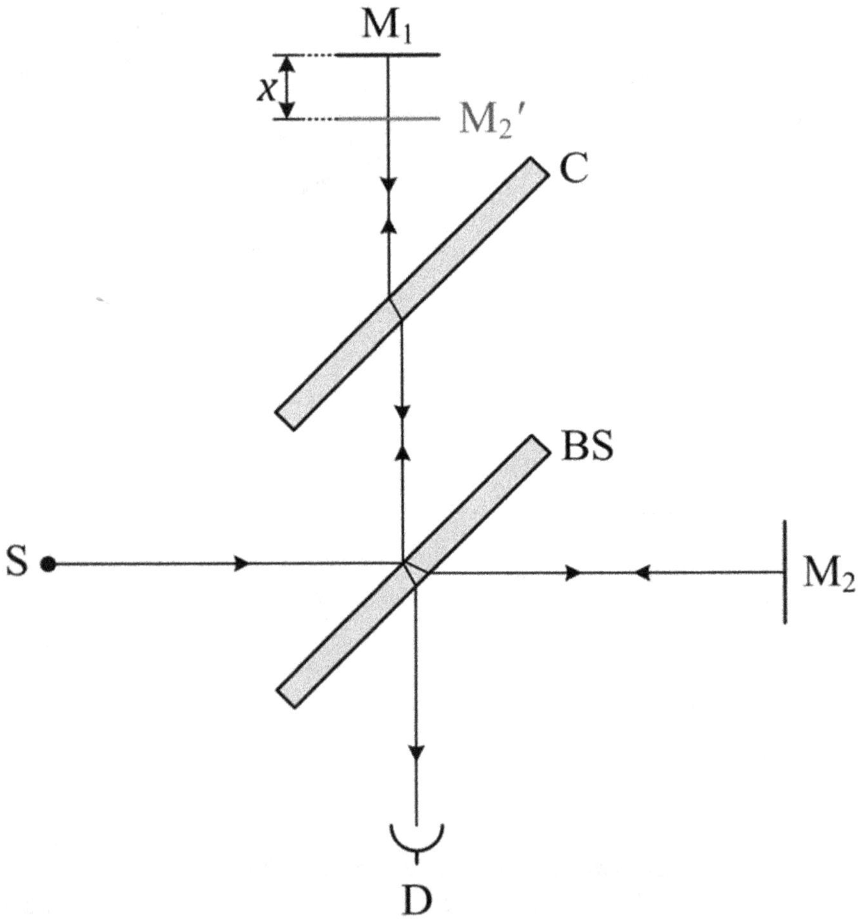

FIGURE 4.6 The Michelson interferometer.

The optical path difference between interferometer arms is given by the difference of the optical paths in each arm where the optical path of each arm is given by the sum of the products of the thickness of each medium traversed and its refractive index. So, for the optical path difference, we can write:

$$\text{OPD} = \sum_i n_i t_i - \sum_j n_j t_j, \tag{4.34}$$

where n_i is the index of refraction of the *i*th component and t_i its thickness in one arm and n_j is the index of refraction of the *j*th component and t_j its thickness in the other arm. The interferometer components are made of a material which index of refraction is wavelength dependent. Therefore, the optical path difference will be wavelength dependent. To eliminate the influence of the wavelength on the optical path difference, it is necessary that both arms contain the same thickness of glass having the same dispersion. The compensating plate C has the role to compensate the influence of the beam splitter BS and has the same thickness and it is made of the same material as the beam splitter. In this way, each beam passes three times through the glass material of the beam splitter and the compensating plate.

Reflection at the beam splitter forms an image of mirror M_2 at the position of the virtual mirror M_2'. Therefore, the interference pattern is the same as that formed by a layer of air bounded by the mirror M_1 and M_2', that is, the optical path difference is equal to twice the distance between them, OPD = $2x$.

Two virtual point sources (images) S_1 and S_2, which are the images of the monochromatic point-source S reflected from mirrors M_1 and M_2, form the nonlocalized fringes on a screen placed at O, as shown in Figure 4.7. The case when the beams impinge the mirrors at normal incidence, the line that connects virtual sources S_1 and S_2 is collinear with the beam directed toward the mirror M_2 is shown in Figure 4.7(a). The observed fringes have circular form. If the mirrors are inclined, as presented in Figure 4.7(b), the line that connects virtual sources S_1 and S_2 is normal to the beam directed toward the mirror M_2 and the observed fringes are straight lines.

4.1.4 THE MACH–ZEHNDER INTERFEROMETER

Figure 4.8 represents the typical setup of the Mach–Zehnder interferometer. Instead of one beam splitter, which has been used in tandem with the compensation plate in the Michelson interferometer, the Mach–Zehnder interferometer uses two beam splitters BS_1 and BS_2 and two mirrors M_1

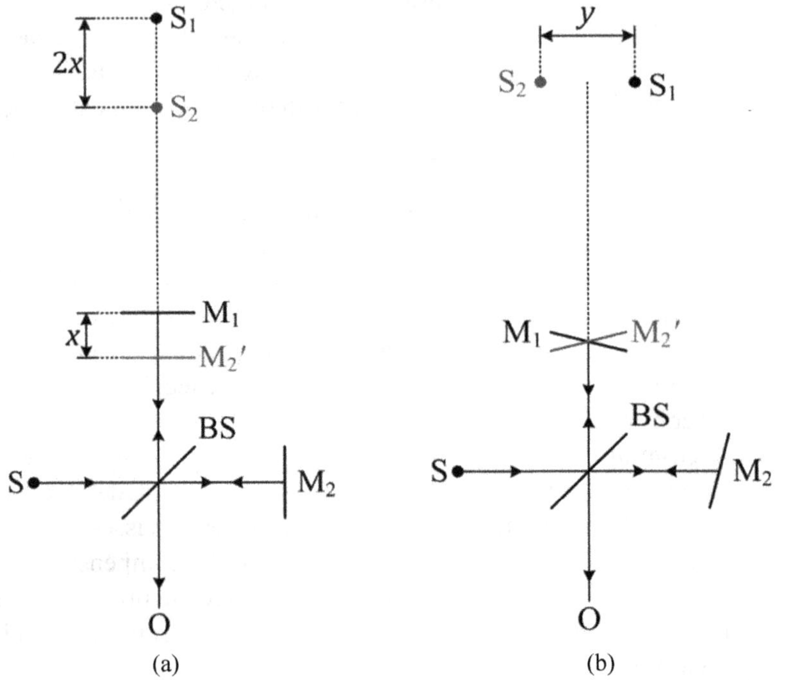

FIGURE 4.7 Formation of nonlocalized fringes in a Michelson interferometer.

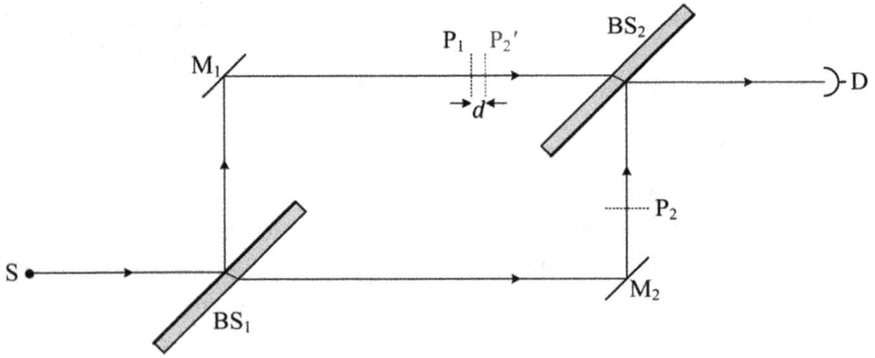

FIGURE 4.8 The Mach–Zehnder interferometer.

and M_2. At the first place, the light from the source S is divided at the semireflecting surface of beam splitter BS_1 into two beams, which, after reflection at the mirrors M_1, M_2, are recombined at the semireflecting surface of the beam splitter BS_2. Typically, beam splitters and mirrors are positioned so that they are approximately parallel, and the paths traversed by the beams form a rectangle or a parallelogram. If it is assumed that the interferometer is illuminated with a collimated beam, then we have two plane waves in the interferometer arms. If the wavefront plane P_2' is the image of the wavefront plane P_2 in the beam splitter BS_2, then the phase difference between P_1 and P_2' is given by:

$$\Delta\varphi = \frac{2\pi}{\lambda} nd, \tag{4.35}$$

where d is the distance between planes P_1 and P_2' and n is the refractive index of the medium between them.

Thanks to its topology, the Mach–Zehnder interferometer is a more versatile optical instrument than the Michelson interferometer because each beam arm is passed over only once, and the fringes can be localized in any plane. Therefore, the Mach–Zehnder interferometer has been widely used in measurement systems where the measurements of the refractive index changes that occur in one arm are related to the changes in pressure, temperature, etc. One of the drawbacks of the Mach–Zehnder interferometer is its adjustment to get good fringe visibility with an extended broadband light source.

4.1.5 THE SAGNAC INTERFEROMETER

The Sagnac interferometer is designed in such a way that the two beams travel around the same closed circuit in opposite directions, as presented in Figure 4.9. As the beams travel along the same path, the interferometer is extremely stable. Moreover, as the optical paths of the two beams are always very nearly equal, it is very easy to align even with a broadband source. Due to the path lengths balance, the optical path difference between two beams is equal to zero.

In the case when the interferometer setup rotates along the axis perpendicular to the plane of the interferometer, a phase shift between the beams occurs, which is given by:

$$\Delta\varphi = \frac{4nA\Omega}{\lambda c}, \tag{4.36}$$

where n is the index of refraction, A is the area of the loop, and Ω is the angular frequency of rotation.

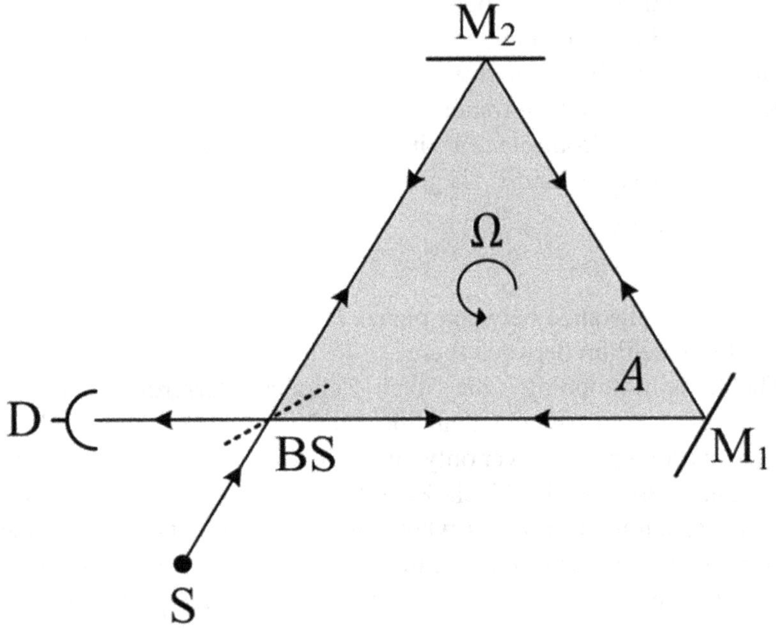

FIGURE 4.9 The Sagnac interferometer.

4.2 COHERENCE

Finite size of the spectral bandwidth of a light source requires more powerful theory to process the optical phenomena accompanying such sources. The coherence theory, which is basically a statistical theory that describes the properties of the radiation field in terms of the correlation between the vibrations at different points in the field, offers the possibility to describe such phenomena. In the case of a broadband light source, the overall electric field can be expressed as:

$$U_r(t) = \int_0^{+\infty} u(\omega)\cos[\omega t + \theta(\omega)]\,d\omega, \tag{4.37}$$

where the electric field $U_r(t)$ has been presented as a sum of the individual monochromatic electric field components, and $u(\omega)$ and $\theta(\omega)$ are the amplitude and phase, respectively, of a monochromatic component of angular frequency ω. To develop a complex representation of the electric field of the broadband light, we will introduce the following function:

$$U_i(t) = \int_0^{+\infty} u(\omega)\sin[\omega t + \theta(\omega)]\mathrm{d}\omega. \quad (4.38)$$

Based on the last two equations, the complex electric field can be represented as:

$$U(t) = U_r(t) + U_i(t) = \int_0^{+\infty} u(\omega)\exp\{j[\omega t + \theta(\omega)]\}\mathrm{d}\omega, \quad (4.39)$$

or, similarly, if $e(\omega)$ is the Fourier transform of the electric field $U_r(t)$, we have:

$$U_r(t) = \frac{1}{2\pi}\int_{-\infty}^{+\infty} e(\omega)\exp(-j\omega t)\mathrm{d}\omega, \quad (4.40)$$

and consequently:

$$U(t) = \frac{1}{\pi}\int_0^{+\infty} e(\omega)\exp(-j\omega t)\mathrm{d}\omega. \quad (4.41)$$

It can be easily shown that the following is valid:

$$\langle U(t)U^*(t)\rangle = \langle 2U_r^2(t)\rangle. \quad (4.42)$$

and if the factor 1/2 is ignored, the irradiance is given by:

$$I = \langle U(t)U^*(t)\rangle. \quad (4.43)$$

Consequently, if the operations on $U_r(t)$ are linear, it is possible to replace it by $U(t)$ and take the real part at the end of the calculation.

4.2.1 THE MUTUAL COHERENCE FUNCTION

The waves originating from different points of a finite-size light source or even from the same point of a thermal source exhibit random amplitude and phase fluctuations. If such wave fields illuminate, two points in space or the same point at different instants of time exhibit only partial correlation. To evaluate such correlation, we will consider an optical setup as presented in Figure 4.10, which is similar to Young's double-slit interference experiment.

A quasimonochromatic source of finite size illuminates a measurement plane with two pinholes S_1 and S_2, and the light passing through these pinholes forms an interference pattern at a screen. The electric fields caused by the source at the pinholes S_1 and S_2 are $U_1(t)$ and $U_2(t)$, respectively.

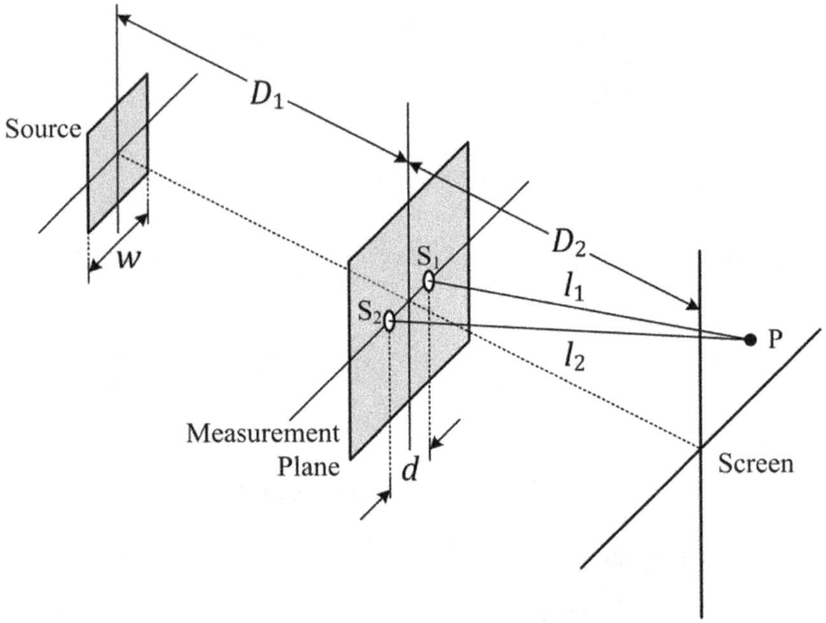

FIGURE 4.10 Measurement of the coherence of the light produced by a finite size source.

Therefore, pinholes can be considered as two secondary sources, so that the electric field at a point P on the screen is given by:

$$U_P(t) = \alpha_1 U_1(t - t_1) + \alpha_2 U_2(t - t_2), \tag{4.44}$$

where $t_1 = l_1/c$ and $t_2 = l_2/c$ are the times taken for the waves to travel from the pinholes S_1 and S_2 to travel to the point P on the screen, respectively, and α_1 and α_2 are the scaling factors, the magnitudes of which are determined by the geometry of the whole setup, such as the size of the pinholes and the distance from the pinholes to the point P. As the interference pattern is stationary, it doesn't depend on the selected time origin, so we can in general case also write:

$$U_P(t) = \alpha_1 U_1(t) + \alpha_2 U_2(t - \tau_2), \tag{4.45}$$

where $\tau = t_2 - t_1$ is the time delay between the beams. For the corresponding irradiance at the point P on the screen, we can write:

$$I_P = \langle U_P(t) U_P^*(t) \rangle = |\alpha_1|^2 I_1 + |\alpha_2|^2 I_2 + 2|\alpha_1\alpha_2| \mathrm{Re}\{\Gamma_{12}(\tau)\}, \tag{4.46}$$

where I_1 and I_2 are the irradiances at the pinholes S_1 and S_2, respectively, and

$$\Gamma_{12}(\tau) = \langle U_1(t) U_2^*(t-\tau) \rangle, \tag{4.47}$$

is the mutual coherence function of the wave fields at S_1 and S_2. The mutual coherence function has the same dimensions as intensity. If we normalize it, we obtain the dimensionless quantity:

$$\gamma_{12}(\tau) = \frac{\Gamma_{12}(\tau)}{\sqrt{I_1 I_2}}, \tag{4.48}$$

known as the complex degree of coherence of the wave fields at S_1 and S_2. The parameters $|\alpha_1|^2 I_1$ and $|\alpha_2|^2 I_2$ are the irradiances at the point P due to the secondary sources S_1 and S_2 acting separately. Taking the following substitutions $I_{P1} = |\alpha_1|^2 I_1$ and $I_{P2} = |\alpha_2|^2 I_2$ in eqs 4.46 and 4.48, we have:

$$I_P = I_{P1} + I_{P2} + 2\sqrt{I_{P1} I_{P2}} \operatorname{Re}\{\gamma_{12}(\tau)\}, \tag{4.49}$$

which represents the general law of interference for partially coherent light. The complex degree of coherence can be represented as the product of its modulus and its phase factor, so we have:

$$\gamma_{12}(\tau) = |\gamma_{12}(\tau)| \exp\{j[\varphi_{12}(\tau) - \langle\omega\rangle\tau]\}, \tag{4.50}$$

where $\varphi_{12}(\tau)$ is the phase difference between the waves incident at S_1 and S_2, and χs the mean angular frequency. The parameter $|\gamma_{12}(\tau)|$ is known as the degree of coherence. According to eq 4.46 and the Cauchy–Schwarz–Bunyakovsky inequality, we have the following condition fulfilled:

$$0 \leq |\gamma_{12}(\tau)| \leq 1. \tag{4.51}$$

If it is satisfied $|\gamma_{12}(\tau)| = 1$, the wave fields at S_1 and S_2 are coherent, and if it is satisfied $|\gamma_{12}(\tau)| = 0$ the wave fields at S_1 and S_2 are incoherent. In all other cases, they are partially coherent. The combination of eqs 4.49 and 4.50 gives:

$$I_P = I_{P1} + I_{P2} + 2\sqrt{I_{P1} I_{P2}} |\gamma_{12}(\tau)| \cos[\varphi_{12}(\tau) - \langle\omega\rangle\tau], \tag{4.52}$$

Typically, in the case of the quasimonochromatic light, the parameters $|\gamma_{12}(\tau)|$ and $\varphi_{12}(\tau)$ slowly vary with respect to the time delay τ variation when compared with $\langle\omega\rangle\tau$, that is, $\langle\omega\rangle \gg d|\gamma_{12}(\tau)|/d\tau, d\varphi_{12}(\tau)/d\tau$ is fulfilled. According to eq 4.52, the fringe visibility is given by:

$$V = \frac{2\sqrt{I_{P1} I_{P2}}}{I_{P1} + I_{P2}} |\gamma_{12}(\tau)|, \tag{4.53}$$

which in the case of two equally intense beams reduces to:

$$V = |\gamma_{12}(\tau)|. \tag{4.54}$$

4.2.2 SPATIAL COHERENCE

To evaluate the spatial coherence of the wave field, we will consider two points P_1 and P_2 located at the screen, as shown in Figure 4.11, illuminated by a small source *S*, the dimensions of which are negligible if compared to its distance from the points P_1 and P_2. Further, we will assume that $\Delta U_1(t)$ and $\Delta U_2(t)$ are the elementary electric fields at the points P_1 and P_2 caused by the small radiating element ΔS positioned at the surface of small source S at the position (x,y), the overall electric fields originating from the whole source S at the points P_1 and P_2 are given by:

$$\Delta U_1(t) = \sum_S \Delta U_1(t), \tag{4.55}$$

$$\Delta U_2(t) = \sum_S \Delta U_1(t). \tag{4.56}$$

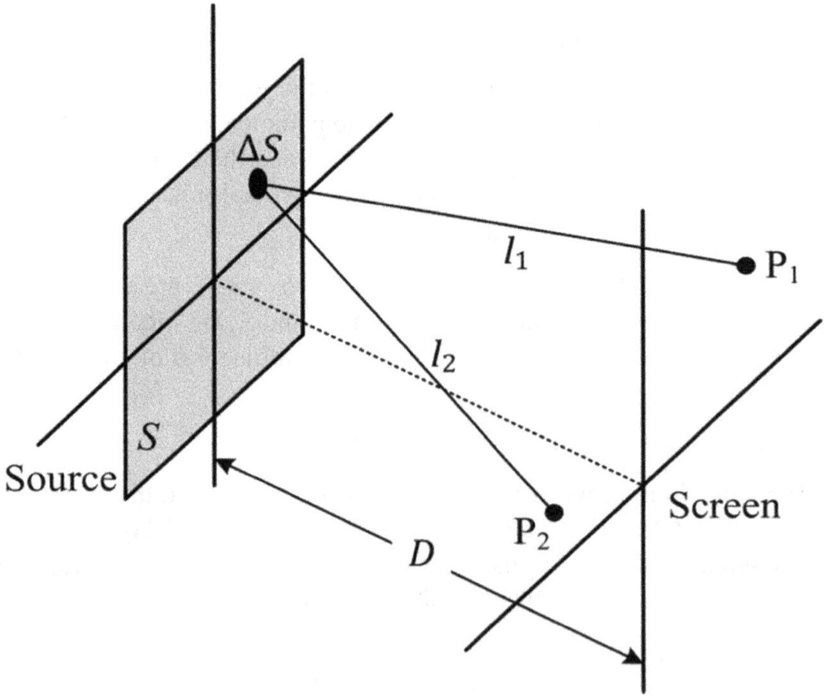

FIGURE 4.11 Spatial coherence of the fields at two points P_1 and P_2 illuminated by an extended source.

The mutual irradiance is given by:

$$I_{12} = \langle U_1(t)U_2^*(t)\rangle = \sum\nolimits_S \langle \Delta U_1(t)\Delta U_2^*(t)\rangle, \tag{4.57}$$

where due to the independent fluctuations of the wave fields, originating from different elementary areas of the source, the sum elements in eqs 4.58 and 4.59 that represent the time average of such fields are omitted. The elementary electric fields at the points P_1 and P_2 can be represented as:

$$\Delta U_1(t) = \frac{\Delta u(t-(l_1/c))}{l_1}\exp\left[-j\langle\omega\rangle\left(t-\frac{l_1}{c}\right)\right], \tag{4.58}$$

$$\Delta U_2(t) = \frac{\Delta u(t-(l_2/c))}{l_2}\exp\left[-j\langle\omega\rangle\left(t-\frac{l_2}{c}\right)\right], \tag{4.59}$$

where $\Delta u(t)$ is the complex amplitude of the wave field at the element ΔS, and l_1 and l_2 are the distances from ΔS to the points to P_1 and P_2, respectively. Consequently, we have:

$$\Delta U_1(t)\Delta U_2^*(t) = \frac{\Delta u(t-(l_1/c))\Delta u^*(t-(l_2/c))}{l_1 l_2}\exp\left[-j\langle\omega\rangle\left(\frac{l_2-l_1}{c}\right)\right]. \tag{4.60}$$

With a quasimonochromatic source, which emits only in a narrow range of frequencies of $\langle \nu\rangle \pm \Delta\nu/2$, the complex amplitude $\Delta u(t)$ varies slowly enough, thus, in the case where time interval $(l_2 - l_1)/c$ is much smaller if compared with the time interval $1/\Delta\nu$ we have $\Delta u(t - l_1/c) \approx \Delta u(t - l_2/c) = \Delta u(t)$ and consequently:

$$\Delta U_1(t)\Delta U_2^*(t) = \frac{|\Delta u(t)|^2}{l_1 l_2}\exp\left[-j\langle\omega\rangle\left(\frac{l_2-l_1}{c}\right)\right]. \tag{4.61}$$

To obtain the overall mutual irradiance emerging from the whole source, the last equation must be integrated over the whole area of the source, so we have:

$$I_{12} = \int_S \frac{I(x,y)}{l_1 l_2}\exp\left[-j\langle k\rangle(l_2-l_1)\right]\mathrm{d}S, \tag{4.62}$$

where it was taken into account the following relation:

$$I(x,y)\,\mathrm{d}S = |\Delta u(t)|^2, \tag{4.63}$$

and $\langle k\rangle = \langle\omega\rangle/c = 2\pi/\langle\lambda\rangle$. If I_1 and I_2 are the irradiances at the points P_1 and P_2 due to the whole source S, respectively, we similarly have:

$$I_1 = \int_S \frac{I(x,y)}{l_1^2} \mathrm{d}S, \tag{4.64}$$

$$I_2 = \int_S \frac{I(x,y)}{l_2^2} \mathrm{d}S. \tag{4.65}$$

Finally, for the complex degree of coherence of the fields at the points P_1 and P_2, we obtain:

$$\mu_{12} = \frac{1}{\sqrt{I_1 I_2}} \int_S \frac{I(x,y)}{l_1 l_2} \exp\left[-j\langle k\rangle(l_2 - l_1)\right] \mathrm{d}S. \tag{4.66}$$

If the dimensions of the source, and the separation of P_1 and P_2, are much smaller than the distances of P_1 and P_2 from the source, then the last equation can be simplified.

4.2.3 COHERENCE TIME AND COHERENCE LENGTH

In the case of a source having very small dimensions, that is, in the case of a point-like source, which radiates over a range of wavelengths, we have to deal with the temporal coherence of the field. Therefore, the complex degree of coherence depends only on the difference in the transit times from the source to P_1 and P_2, and the mutual coherence function, as a matter of fact, represents the autocorrelation function:

$$\Gamma_{11}(\tau) = \langle U(t)\, U^*(t-\tau)\rangle, \tag{4.67}$$

and the degree of temporal coherence of the field is then given by:

$$\gamma_{11}(\tau) = \frac{\langle U(t) U^*(t-\tau)\rangle}{\langle U(t) U^*(t)\rangle}. \tag{4.68}$$

The degree of temporal coherence can be obtained from the visibility of the interference fringes as the optical path difference varies.

According to the Wiener–Khinchin theorem, the radiation power spectrum is given by the Fourier transform of the mutual coherence function (autocorrelation function), so that the mutual coherence function and the radiation spectrum represent the Fourier transform pair:

$$S(v) = \mathrm{F}\{\Gamma_{11}(\tau)\}, \tag{4.69}$$

$$\Gamma_{11}(\tau) = \mathrm{F}^{-1}\{S(v)\}, \tag{4.70}$$

where $S(\nu)$ is the radiation power spectrum. Consequently, the complex degree of coherence is given by:

$$\gamma_{11}(\tau) = \frac{\mathrm{F}^{-1}\{S(\nu)\}}{\int_{-\infty}^{+\infty} S(\nu)\,\mathrm{d}\nu}. \tag{4.71}$$

If we now consider the radiation source that emits the radiation evenly distributed over the frequency range of $\langle \nu \rangle - \Delta\nu/2 \le \nu \le \nu + \Delta\nu/2,$ that is, centered at the mean frequency of $\langle \nu \rangle$ and with the spectrum width equal to $\Delta\nu$, then according to eq 4.71, the complex degree of coherence of the radiation is given by:

$$\gamma_{11}(\tau) = \frac{\sin(\Delta\nu\tau)}{\Delta\nu\tau}. \tag{4.72}$$

The sinc(x) function is a damped oscillating function, the first zero of which occurs at a time difference τ_C given by:

$$\Delta\nu\tau_C = 1. \tag{4.73}$$

The time interval τ_C is also known as the coherence time of the radiation. Consequently, the coherence length is defined as:

$$l_C = c\tau_C = \frac{c}{\Delta\nu} \approx \frac{\lambda^2}{\Delta\lambda}. \tag{4.74}$$

On a clear, dark night, our eyes can see about 6000 or so stars in the sky. One of the most interesting phenomena of the clear night sky is why some stars are twinkling and some not. To determine which kind of celestial bodies twinkle and which not, we will observe a star with the radius R located at a very large distance D from the observer. Two rays, which emerge from the point P located at the star surface, travel along two paths with the distance l_1 and l_2 until reach the observer eye. After the refraction at the eye lens, the rays impinge the retina at the point R. Due to the different ray paths, the turbulence of the atmosphere can cause random phase modulation of the wave fields that further lead to the random fluctuations at the observer retina known as the star twinkling. According to the geometry from Figure 4.12, we have:

$$l_1 = \sqrt{D^2 + \left(R - \frac{d}{2}\right)^2} \approx D\left(1 - \frac{Rd}{2D^2}\right), \tag{4.75}$$

$$l_2 = \sqrt{D^2 + \left(R + \frac{d}{2}\right)^2} \approx D\left(1 + \frac{Rd}{2D^2}\right), \tag{4.76}$$

where d is the pupil diameter and where the following condition $d \ll R \ll D$ were taken into consideration.

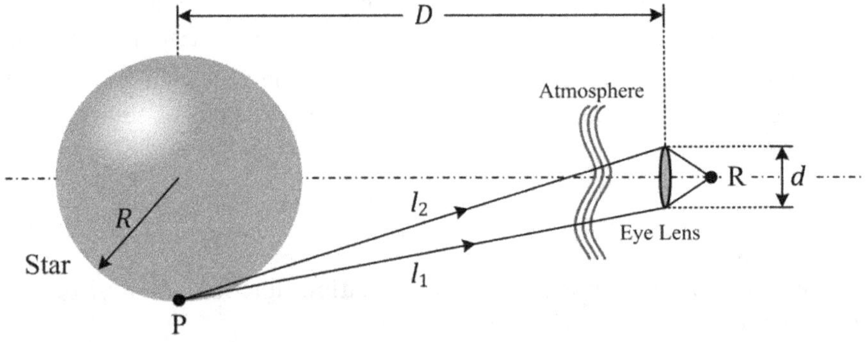

FIGURE 4.12 Star twinkle.

The optical path difference of these two rays is given by:

$$\Delta l = l_2 - l_1 \approx \frac{Rd}{D}. \tag{4.77}$$

Roughly speaking the star will twinkle if the optical path difference is smaller than the coherence length ($\Delta l < l_C$) and will not if the optical path difference is larger than the coherence length ($\Delta l < l_C$). Therefore, if the star apparent size, defined as:

$$\varphi \approx \frac{2R}{D}, \tag{4.78}$$

is smaller than:

$$\varphi < \frac{2l_C}{d}, \tag{4.79}$$

the star will twinkle and in opposite will not. The coherence length of the light emitted by the star depends on the star temperature (black body radiation) and the atmosphere absorption. Therefore, the coherence length has not the fixed value but depends on the particular star characteristics. However, regardless the star light spectrum, our eyes are sensitive only in a certain wavelength range. According to the spectral luminous efficiency function for photopic vision, an average human eye has a spectral bandwidth ranging from $\lambda_L \approx 380$ nm to $\lambda_H \approx 770$ nm, thus having a spectral width of $\Delta\lambda = \lambda_H - \lambda_L \approx 390$ nm, and the peak sensitivity at $\langle\lambda\rangle \approx 555$ nm. Therefore, the effective coherence length, given by eq 4.74, is equal to $l_C \approx 0.8$ μm. As the observer looks at the sky during the night, the pupil has the maximal openings of approximately $d \approx 8$ mm. So, according to eq 4.79, if the apparent size is smaller then 200 μrad or 40″, the star will twinkle and in opposite will not or its twinkling will be weaker. If we compare this approximate limit value for the star apparent size and the apparent sizes of the particular celestial bodies, shown in Table 4.1, one can see why stars twinkle distinctively from the planets which in some cases may also twinkle but not so noticeable.

TABLE 4.1 Apparent Sizes of Some Celestial Bodies.

	Planets			Stars		
Celestial body	Venus	Mars	Jupiter	Sirius	Alpha Centauri	Betelgeuse
Apparent size	9.6″–66″	3.5″–25″	30″–50″	0.006″	0.007″	0.05″–0.06″

4.3 WHITE-LIGHT INTERFEROMETRY

If the interferometer has been illuminated with a monochromatic light, one can notice a distinctive interferometric pattern on the observation screen or if an optical detector has been used to capture the fringes when the optical path difference in the interferometer changes, one can notice a very large number of fringes, which in the case of a linear optical path difference sweep produces a periodic fringe pattern. In theory, we will have an unlimited number of fringes. However, typical light source has a finite spectrum width thus limiting the number of observed/detected fringes. According to the Wiener–Khinchin theorem, the broader the spectrum, the shorter the interferometric pattern. One of the most common light sources used in the white-light interferometer or also called low-coherence interferometer is the superluminescent diode. Typical superluminescent diode emits infrared light, which has Gaussian spectrum shape given as:

$$S(\nu) = \frac{2}{\Delta\nu}\sqrt{\frac{\ln 2}{\pi}}\exp\left[-4\ln 2\left(\frac{\nu - \bar{\nu}}{\Delta\nu}\right)^2\right], \tag{4.80}$$

where $\langle \nu \rangle$ is the central frequency and $\Delta\nu$ is the full width at half maximum (FWHM) spectrum width. The complex degree of coherence of the radiation is given by:

$$|\gamma_{11}(\tau)| = \exp\left[-\left(\frac{\pi\Delta\nu}{2\sqrt{\ln 2}}\tau\right)^2\right]. \tag{4.81}$$

Consequently, the FWHM coherence time and the coherence length are:

$$\tau_C = \frac{2\sqrt{\ln 2}}{\pi}\frac{1}{\Delta\nu}, \tag{4.82}$$

$$l_C = \frac{2\sqrt{\ln 2}}{\pi}\frac{\lambda^2}{\Delta\lambda}. \tag{4.83}$$

Due to the very short wavelength of the light, interferometry offers the possibility of measuring very small distances. However, the periodicity of the interferometric pattern, when the monochromatic source has been used, limits the dynamic range of the measurement. The 2π ambiguity in interferometer phase unwrapping can limit the maximal distance measurement to the fraction of the wavelength. Fringe counting is one of the possible solutions in measurement larger distances with high-coherence interferometry, which in general case requires moving optical elements that can influence measurement precision and accuracy. Where the classical high-coherence interferometry fails, the white-light interferometry manages to measure larger distances even without moving parts. As the interferometer optical path difference changes with the changes of a distance of some interferometer components, the center of the complex degree of coherence of the radiation also proportionally changes. Therefore, by extracting the position of the maximum value of the complex degree of coherence, one can absolutely measure the interferometer optical path difference and thus the measuring object position, as presented in Figure 4.13 where the white-light Twyman–Green interferometer is presented.

When the sensor mirror SM, which is fixed to an object, which position is to be measured, moves along one of the interferometer arm to measure its position, the reference mirror RM must perform scanning along the other interferometer arm. To be able to measure the absolute position of SM the scanning mechanism of the reference mirror must be equipped with an internal RM position measurement during the scanning process. The scanning of RM will give the interferometric patterns as presented in Figure 4.14.

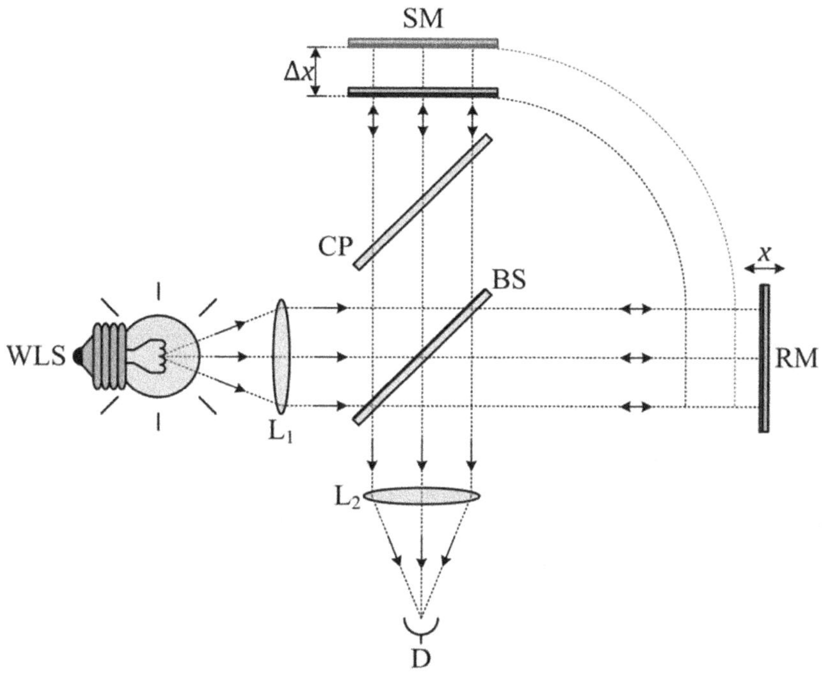

FIGURE 4.13 White-light Twyman–Green interferometer.

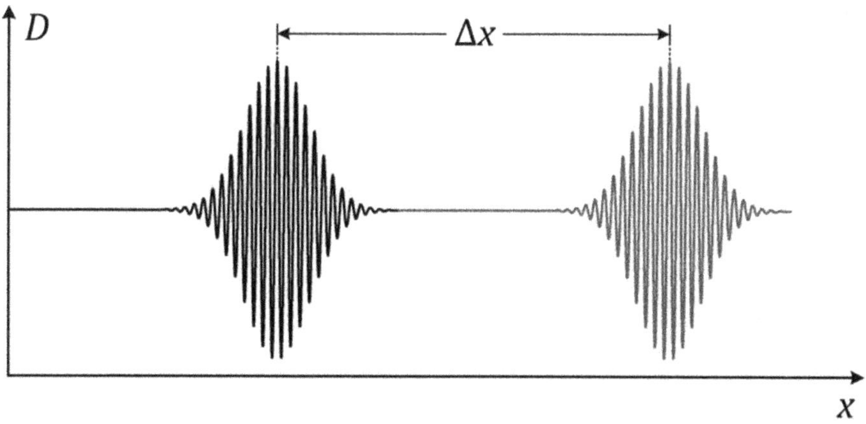

FIGURE 4.14 White-light interferometric patterns for two different optical path differences.

According to Figure 4.14, if SM is displaced for the distance Δx, the maximum of the complex degree of coherence will be also shifted along the x axis for the same distance. The drawback of such a setup is the movable parts, which can influence the measurement. The movable parts can be omitted if the electronic scanning of the white-light interferometric pattern is employed as presented in Figure 4.15. Instead of moving the RM, an optical wedge in tandem with CCD linear sensor can be used. The optical wedge with a small apex angle θ is made of a glass with the refractive index n, which planes are partially reflective, as presented in Figure 4.16. Due to the linear increase of the optical path difference in the wedge along the CCD linear array a low-coherence interferometric pattern emerges. The corresponding optical path difference at the distance x from the wedge apex is given by:

$$\text{OPD} + 2n\theta x. \tag{4.84}$$

The maximum or the center of gravity of the complex degree of coherence will be positioned at the CCD linear array pixel, the position x_C of which is determined according to the optical path difference balance in the sensing interferometer and the receiving interferometer made of the wedge, so we have:

$$\text{OPD} = 2n\theta x_C = 2\Delta x \Rightarrow x_C = \frac{\Delta x}{n\theta}, \tag{4.85}$$

thus giving the linear relation between the interferogram center position and the absolute position of the sensing mirror.

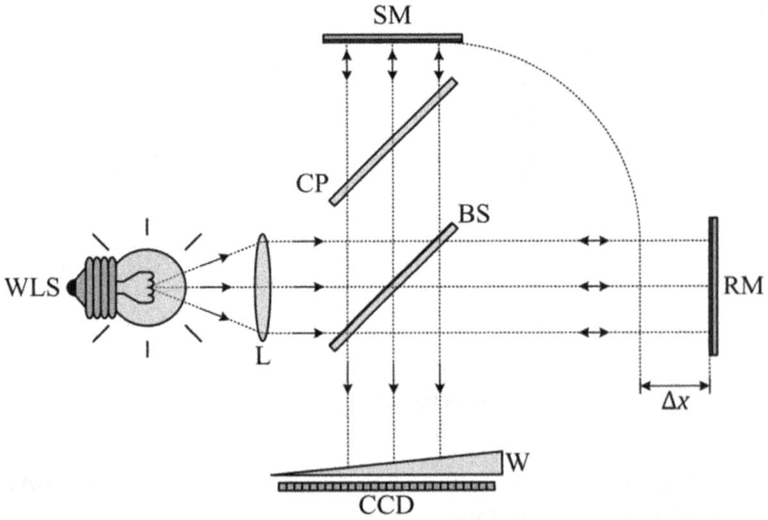

FIGURE 4.15 White-light Twyman–Green interferometer with electronic scanning.

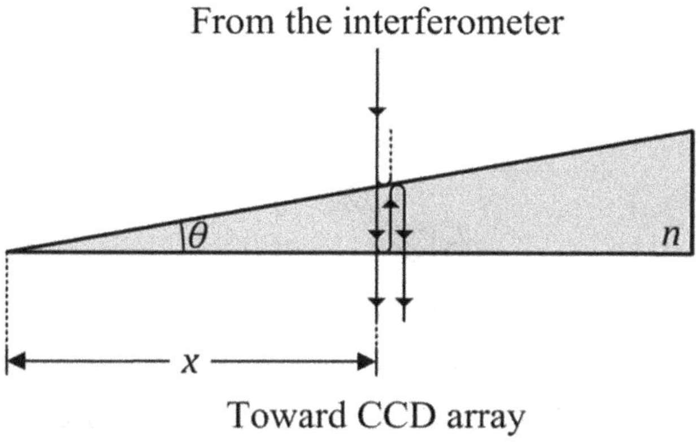

FIGURE 4.16 Optical wedge for electronic scanning.

If instead of an optical detector such as photodiode or CCD linear array, a dispersive element has been employed, one can obtain the spectral fringes or the so-called channeled spectrum. The optical setup for obtaining the channeled spectrum is presented in Figure 4.17. The light, obtained from a low-coherence interferometer, illuminates the slit S where after passing through the slit the broadband light illuminates an optical spectrometer OS. The radiation that illuminates the spectrometer is consisted of two wave fields:

$$U_{\mathrm{os}}(t) = aU(t) + bU(t - \Delta t), \tag{4.86}$$

that is, the superposition of the reference beam field $U(t)$ and the same field delayed by propagating the path length difference $2\Delta x$ $U(t - \Delta t)$ where $\Delta t = 2\Delta x/c$ is fulfilled and where the parameters a and b are inserted in the equation to account the possible differences in relative amplitudes between the reference and delayed fields. The spectrum of the wave field captured by the optical spectrometer is found from the Fourier transform of eq 4.86 as:

$$S(\nu) = (a^2 + b^2)WLS(\nu)\left[1 + V\cos\left(\frac{4\pi\Delta x}{c}\nu\right)\right], \tag{4.87}$$

where $WLS(\nu)$ is the spectrum of the white-light source WLS and V is the spectral fringe visibility given by:

$$V = \frac{2ab}{a^2 + b^2}. \tag{4.88}$$

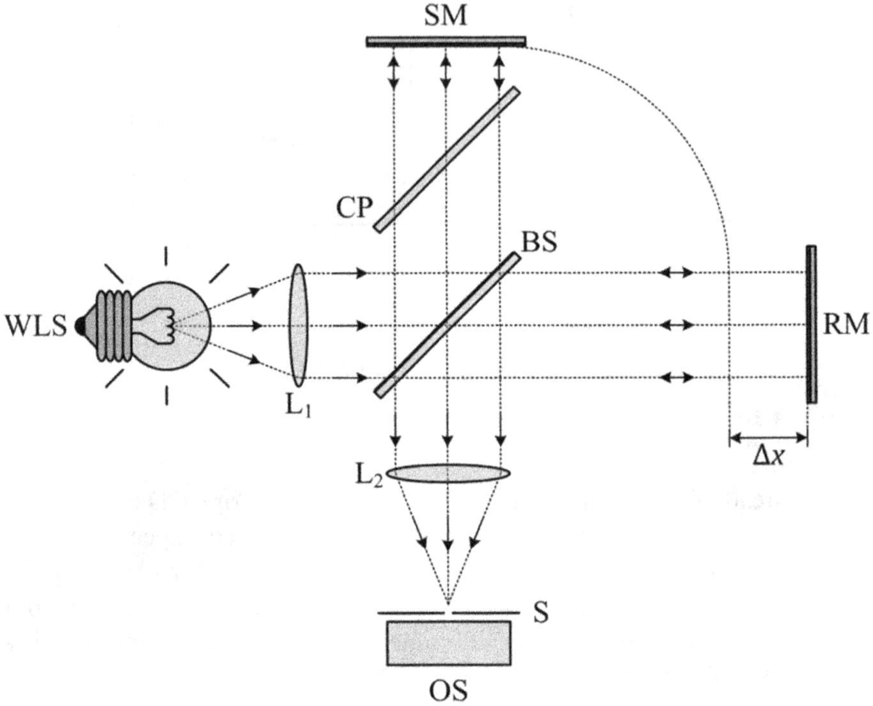

FIGURE 4.17 Channeled spectrum.

The typical shape of the channeled spectrum is presented in Figure 4.18 where are shown two spectral fringes for $\Delta x = 90$ μm and $\Delta x = 110$ μm where as a broadband light source a superluminescent diode is used. The diode has a central wavelength of 850 nm and spectrum width of approximately 20 nm, which further results in the coherence length of the diode of about 36 μm.

Spectral interferometry is a very powerful interrogation tool especially in the case of a large number of quasidistributed extrinsic low-finesse Fabry–Pérot interferometric sensors aimed, for example, for distributed measurements of temperature, pressure, force, and strain. This technique allows us multiplexing a large number of sensors placed along a single optical fiber. As shown in Figure 4.18, a white-light source spectrum is modulated, where the modulation "frequency" depends on the optical path difference. So, in the case of several sensing interferometers, we will have more complex modulation scheme with the same number as the sensor number of modulation "frequencies". Therefore, by locking on a single "frequency" of the captured channeled spectrum, one can track the changes in "frequency" of the particular sensor and perform the necessary measurements on this particular sensor.

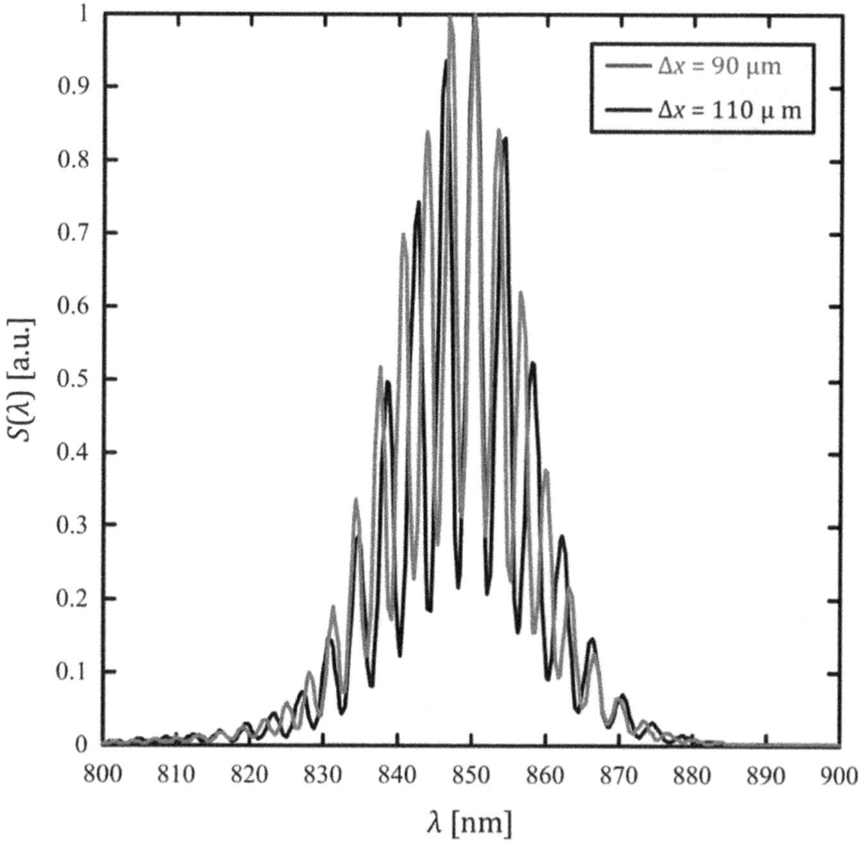

FIGURE 4.18 Measured channeled spectrum.

Multiplexing capabilities of the white-light interferometry, that is, the simultaneous measurement of several optical path differences, are used in the optical coherence-domain reflectometry that is suitable for determining the positions and magnitudes of reflection sites within a miniature multilayered optical assembly, as presented in Figure 4.19. Typical optical coherence reflectometer/tomography measurement system captures the internal reflections from the structured miniature object. The internal reflections are caused by the scattering of the incident optical radiation on a nonhomogeneous internal structure of the object. To investigate the positions, sizes, and partial characteristics of the layers of such internal structured object, the reference mirror performs scanning where at the detector D are captured the corresponding low-coherence interferometric patterns. These patterns are presented in Figure 4.20 where each pattern corresponds to a single reflection. Position of each layer corresponds to the interferogram center position and where the reflection coefficient can be determined from the interferogram amplitude. One can notice that the resolution of the optical coherence tomography is approximately equal to the coherence length of the WLS. Typical optical coherence tomography systems have the resolution of about 10 μm.

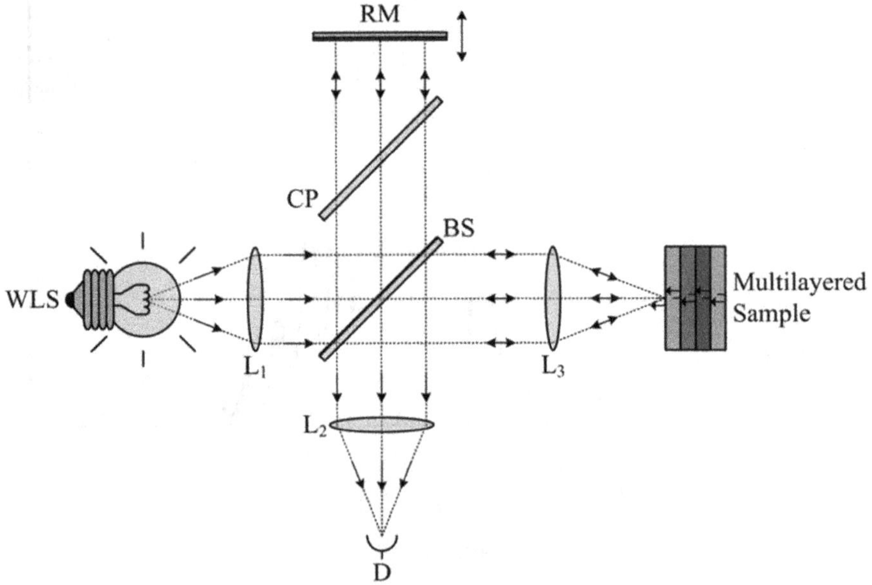

FIGURE 4.19 Optical coherence-domain reflectometry.

Instead of mechanical scanning and a simple optical detector, one can use the fixed reference mirror in tandem with the optical spectrometer. However, such a system suffers from lower signal-to-noise ratio, then one with the movable reference mirror and single optical detector, which can be the limiting factor in the case of weak scattering.

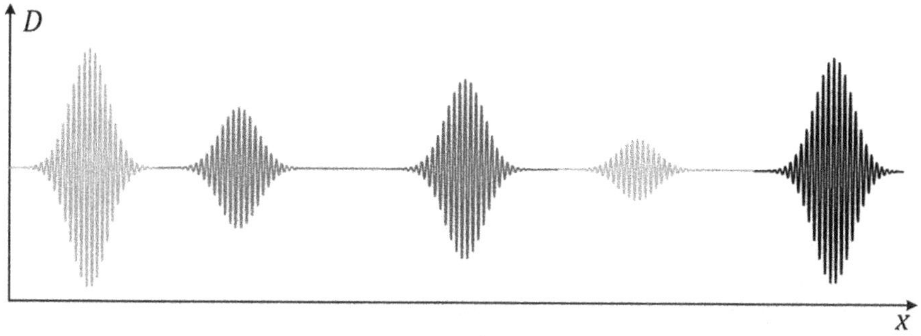

FIGURE 4.20 Captured optical coherence tomography interferometric patterns where each pattern corresponds to a single reflection.

One of the most prominent applications of the optical coherence tomography is in medical imaging to obtain micrometer-resolution, three-dimensional images of biological tissues. Depending on the properties of the used white-light source such as superluminescent diodes, ultrashort pulse lasers, and supercontinuum lasers, optical coherence tomography has achieved submicrometer resolution typically with a very wide-spectrum source.

White-light interferometry represents also a powerful tool for measuring a surface's profile, to quantify its roughness as presented in Figure 4.21. A light that is back-reflected from the rough surface is captured by a fast CCD two-dimensional array (camera). To perform surface profiling, the reference mirror is moved typically with the help of a piezo actuator. The camera captures images where the signal at each camera pixel represents the low-coherence interferometric pattern obtained from a particular point at the rough surface. By finding the maximum of the interferometric pattern, one can obtain the position of this particular point with respect to some reference plane. By analyzing in such way each interferogram obtained from all the camera pixels, one can obtain the three-dimensional image of the surface.

4.4 MULTIPLE-BEAM INTERFERENCE

In the previous discussion in this chapter regarding the interferometric phenomena occurring in the plane-parallel plate, we have taken into account only two reflected beams and ignored any multiple reflections.

If the slab has low refection coefficients at its planes, we can consider this interferometer to be low-finesse, that is, we can consider this interferometer as a two-beam interferometer. However, if special measures are taken to increase the reflection coefficients so that they are enhanced, allowing more beams to be combined, the fringes change their character and become very much sharper. If a thin metallic film is deposited onto the slab surfaces it is possible to increase the reflection coefficient. In this way, the slab front face reflects a large fraction of the light incident upon it. The small amount of light that passes through to the inside of the slab is reflected back and forth many times. Therefore, we have the interference of these many emerging rays, which gives multiple interferences its special character.

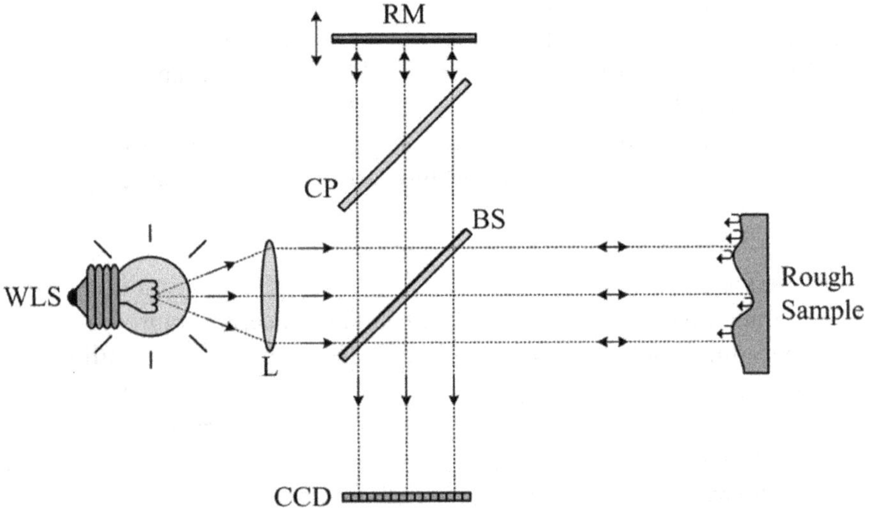

FIGURE 4.21 White-light profilometry.

The case of simple multiple interferences is presented in Figure 4.22. Light source S emits the light toward the plane-parallel plate of thickness d and index of refraction n. Due to the multiple reflections, we have reflected rays R_1, R_2, R_3, … as well as transmitted rays T_1, T_2, T_3, …, where each ray with the larger index is weaker than one with the smaller index due to the larger number of internal reflections and thus larger losses. If the amplitude of the incident ray is U and r and t the reflection and transmission coefficients from the surrounding medium to the slab, respectively, and r'

and t' the corresponding quantities from slab to the surrounding medium, we have the following for the amplitude of the reflected rays:

$$U_{R_1} = rU, \tag{4.89}$$

$$U_{R_i} = tt'r'^{2i-1}U,\ i > 1, \tag{4.90}$$

and of the transmitted rays:

$$U_{T_i} = tt'r'^{2(i-1)}U. \tag{4.91}$$

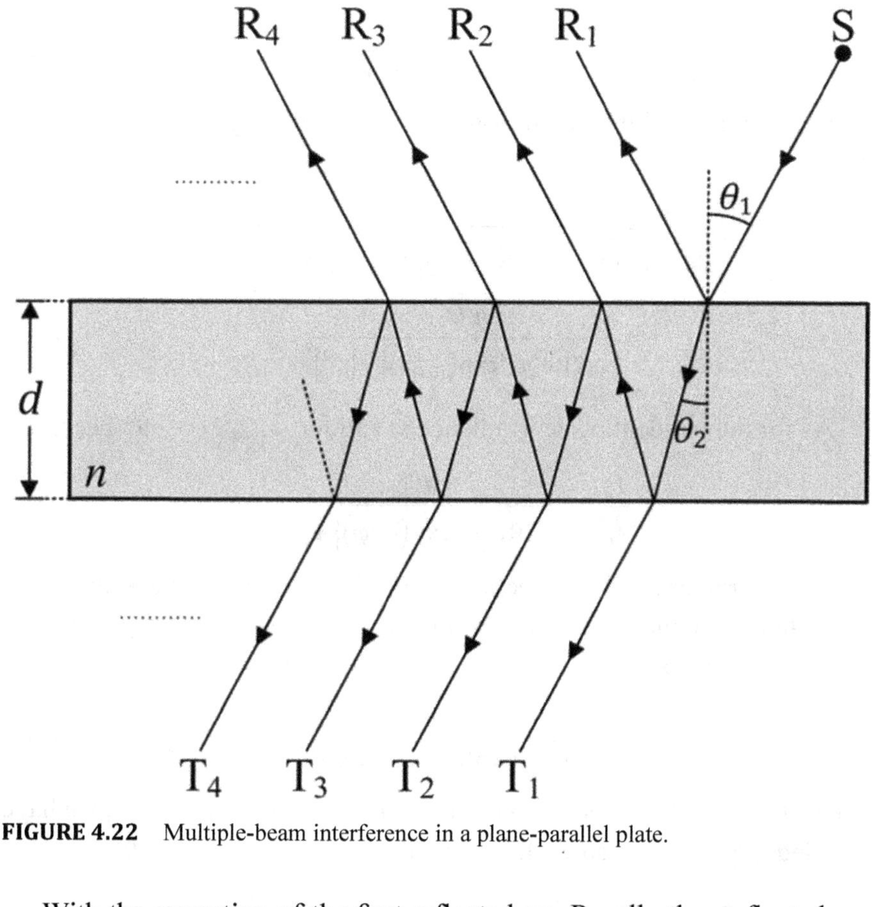

FIGURE 4.22 Multiple-beam interference in a plane-parallel plate.

With the exception of the first reflected ray R_1, all other reflected ray amplitude follows the geometric progression. Each ray phase depends on the angle of refraction inside the plate, its thickness, and its refraction index. The relative phase of the rays on a plane outside the slab perpendicular to the ray depends as:

$$\varphi = \frac{4\pi n d\cos\theta_2}{\lambda}. \tag{4.92}$$

In the case of the overall transmitted beam amplitude, we have the following:

$$U_{\mathrm{T}} = \sum_{i=1}^{+\infty} U_{\mathrm{T}_i} = \sum_{i=1}^{+\infty} tt'r'^{2(i-1)}U\exp\left[-j(i-1)\varphi\right], \tag{4.93}$$

where after infinite summation the following is obtained:

$$U_{\mathrm{T}} = tt'U\frac{1}{1-r'^2\exp(-j\varphi)}. \tag{4.94}$$

Bearing in mind that the transmitted irradiance is given by $I_{\mathrm{T}} = U_{\mathrm{T}}U_{\mathrm{T}}^*$, we have:

$$I_{\mathrm{T}} = \frac{|tt'U|^2}{\left[1-r'^2\exp(-j\varphi)\right]\left[1-\left(r'^2\right)^*\exp(j\varphi)\right]}$$
$$= \frac{|tt'U|^2}{1-2\mathrm{Re}\left\{r'^2\exp(-j\varphi)\right\}+\left|r'^2\right|^2}. \tag{4.95}$$

As for the incident wave irradiance is valid $I_{\mathrm{I}} = |U|^2$, eq 4.95 becomes:

$$\frac{I_{\mathrm{T}}}{I_{\mathrm{I}}} = \frac{|tt'|^2}{1-2\mathrm{Re}\left\{r'^2\exp(-j\varphi)\right\}+\left|r'^2\right|^2}. \tag{4.96}$$

In general case, the reflection coefficients from to the slab to the surrounding medium is a complex quantity where is valid $r' = |r'|\exp(j\phi)$, where ϕ is the phase change due to the reflection so, eq 4.96 becomes:

$$\frac{I_{\mathrm{T}}}{I_{\mathrm{I}}} = \frac{|tt'|^2}{1-2|r'|^2\cos\psi+|r'|^4}, \tag{4.97}$$

where $\psi = 2\phi - \varphi$ is valid. Based on the energy conservation low, we have fulfilled $|tt'|^2 = 1-|r'|^2$, so eq 4.97 becomes:

$$\frac{I_{\mathrm{T}}}{I_{\mathrm{I}}} = \frac{1-|r'|^2}{1-2|r'|^2\cos\psi+|r'|^4}, \tag{4.98}$$

where after the rearrangement of eq 4.98, we have:

$$\frac{I_{\mathrm{T}}}{I_{\mathrm{I}}} = \frac{1}{1 + F\sin^2\left(\frac{\psi}{2}\right)}, \tag{4.99}$$

and:

$$F = \frac{2|r'|^2}{\left(1 - |r'|^2\right)^2}. \tag{4.100}$$

One can notice from the last equation that if parameter $|r'|^2$ approaches unity the parameter F becomes very large. For example, if $|r'|^2 = 0.9$, $F = 180$. The consequence of this is to keep the transmitted irradiance very small for a broad ranges of wavelengths, except when ψ is close enough to $2k\pi$, where k is integer, and where $\sin^2(\psi/2)$ becomes less than $1/F$. The transmission coefficient I_T/I_I plots for five different values of the parameter F are given in Figure 4.23. The sharp fringes of multiple-beam interference are known as Fabry–Pérot fringes.

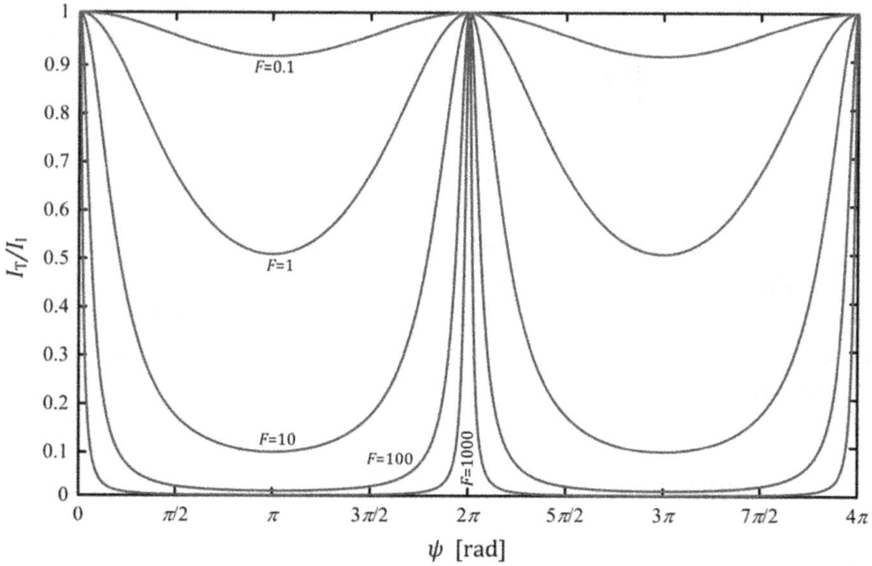

FIGURE 4.23 Fabry–Pérot fringes for five values of the parameter F.

The Fabry–Pérot fringe's sharpness is often itemized as the finesse Φ, which is determined as the ratio of the separation of adjacent maximums to the fringe half-width. Equation 4.99 gives the phase shift $\psi_{½}$ for the fringe irradiance to be halved as:

$$1 + F\sin^2\left(\frac{\psi_{½}}{2}\right) = 2, \tag{4.101}$$

or:

$$\psi_{½} = 2\arcsin\left(\frac{1}{\sqrt{F}}\right). \tag{4.102}$$

Typically, parameter F is large, and therefore, we have arcsin $\left(1/\sqrt{F}\right) \approx 1/\sqrt{F}$. Since the phase difference between two adjacent fringes is 2π, the finesse Φ is the ratio $2\pi/2\psi_{½}$:

$$\ddot{O} = \frac{2\pi}{2\psi_{½}} \approx \frac{\pi\sqrt{F}}{2}. \tag{4.103}$$

The Fabry–Pérot interferometer, presented in Figure 4.24, uses the previously discussed characteristic of such a multiple-beam interferometer to obtain sharp interference fringes from the source S. The interference pattern, observed on the plate P, is in the shape of concentric rings, which are the images of those points on the source produced with the help of the first lens. Usually, the cavity is made of two glass plates. These plates have their inner surfaces coated with the films of high reflectivity. These parallel plates are held at a certain distance by an optical spacer made of invar or silica. This device is called a Fabry–Pérot etalon.

When the Fabry–Pérot etalon is illuminated with the help of a light containing two spectral components, one can observe double ring system. With the etalon it is possible to optically distinguish hyperfine structure of spectral lines directly. Therefore, the Fabry–Pérot etalon since its introduction dominates the field of high-resolution spectrometry.

4.5 MULTILAYER THIN FILMS

Optical surfaces characterized by an arbitrary reflectance and transmittance coefficients can be produced by means of thin film coatings. These films are usually deposited onto a glass or metal substrates. It is well known that the antireflection coatings are deposited onto camera lenses and other optical instruments components. There are also other applications of thin films such as heat-reflecting and heat-transmitting mirrors ("hot" and "cold" mirrors), one-way mirrors, and optical filters.

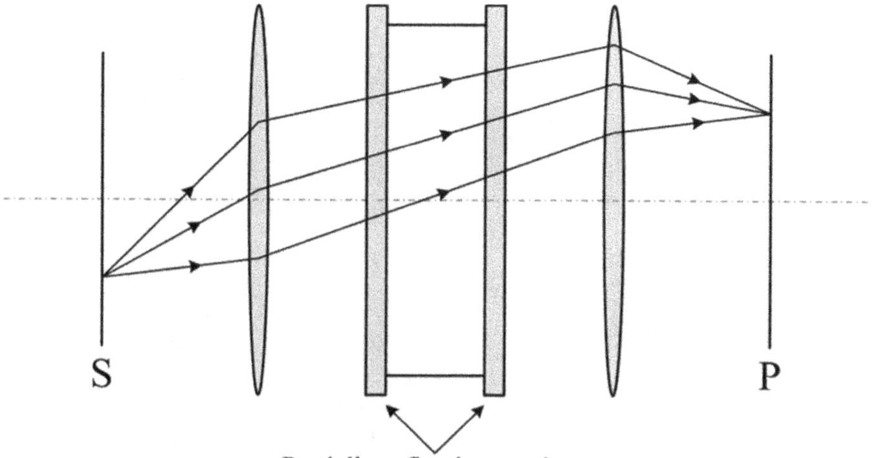

FIGURE 4.24 A simple Fabry–Pérot interferometer.

First of all, we will consider the case of a single thin layer of dielectric material of index of refraction n and thickness t deposited on a semi-infinite substrate with the refractive index n_S as it is presented in Figure 4.25. For the sake of simplicity, we will consider an incident wave that propagates perpendicularly to the thin film. The modification for an arbitrary incident wave direction can be easily made. The electric field vector of the incident beam is $\mathbf{E}_\mathbf{I}$ where the back-reflected electric field vector is $\mathbf{E}_\mathbf{R}$, and the transmitted electric field vector is equal to $\mathbf{E}_\mathbf{T}$. The electric field vectors inside the thin film are $\mathbf{E}'_\mathbf{T}$ and $\mathbf{E}'_\mathbf{R}$, for the forward and backward propagating waves, respectively.

According to the boundary conditions, which require that, the electric and magnetic fields to be continuous at each interface, for the first interface, we have:

$$E_\mathrm{I} + E_\mathrm{R} = E'_\mathrm{T} + E'_\mathrm{R}, \tag{4.104}$$

for electric fields amplitudes and:

$$H_\mathrm{I} - H_\mathrm{R} = H'_\mathrm{T} - H'_\mathrm{R}, \tag{4.105}$$

for magnetic fields amplitudes, or equivalently:

$$E_{\mathrm{I}} - E_{\mathrm{R}} = n\left(E'_{\mathrm{T}} - E'_{\mathrm{R}}\right). \tag{4.106}$$

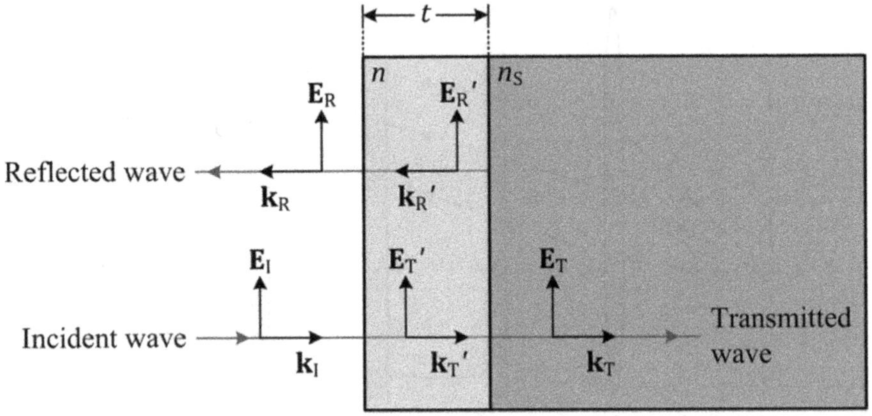

FIGURE 4.25 Wave vectors and their associated electric fields for the case of normal incidence on a single dielectric layer.

Consequently, for the second interface, we have:

$$E'_{\mathrm{T}}\exp(jkt) + E'_{\mathrm{R}}\exp(-jkt) = E_{\mathrm{T}}, \tag{4.107}$$

for electric fields amplitudes and:

$$H'_{\mathrm{T}}\exp(jkt) - H'_{\mathrm{R}}\exp(-jkt) = H_{\mathrm{T}}, \tag{4.108}$$

for magnetic fields amplitudes, or equivalently:

$$n\left[E'_{\mathrm{T}}\exp(jkt) - E'_{\mathrm{R}}\exp(-jkt)\right] = n_{\mathrm{S}}E_{\mathrm{T}}. \tag{4.109}$$

By eliminating amplitudes inside the thin film, we obtain:

$$1 + \frac{E_{\mathrm{R}}}{E_{\mathrm{I}}} = \frac{E_{\mathrm{T}}}{E_{\mathrm{I}}}\left[\cos(kt) - j\frac{n_{\mathrm{S}}}{n}\sin(kt)\right], \tag{4.110}$$

$$1 - \frac{E_{\mathrm{R}}}{E_{\mathrm{I}}} = \frac{E_{\mathrm{T}}}{E_{\mathrm{I}}}\left[n_{\mathrm{S}}\cos(kt) - jn\sin(kt)\right], \tag{4.111}$$

or, in matrix form:

$$\begin{bmatrix}1\\1\end{bmatrix} + \begin{bmatrix}1\\-1\end{bmatrix}\frac{E_{\mathrm{R}}}{E_{\mathrm{I}}} = \begin{bmatrix}\cos(kt) & -j\dfrac{\sin(kt)}{n}\\ -jn\sin(kt) & \cos(kt)\end{bmatrix}\begin{bmatrix}1\\n_{\mathrm{S}}\end{bmatrix}\frac{E_{\mathrm{T}}}{E_{\mathrm{I}}}, \tag{4.112}$$

which can be further reduced to:

$$\begin{bmatrix}1\\1\end{bmatrix}+\begin{bmatrix}1\\-1\end{bmatrix}\rho=\mathbf{M}\begin{bmatrix}1\\n_{\mathrm{S}}\end{bmatrix}\tau, \tag{4.113}$$

where we have introduced the reflection coefficient as:

$$\rho=\frac{E_{\mathrm{R}}}{E_{\mathrm{I}}}, \tag{4.114}$$

and the transmission coefficient as:

$$\tau=\frac{E_{\mathrm{T}}}{E_{\mathrm{I}}}. \tag{4.115}$$

The transfer matrix is also given by:

$$\mathbf{M}=\begin{bmatrix}\cos(kt) & -j\dfrac{\sin(kt)}{n}\\ -jn\sin(kt) & \cos(kt)\end{bmatrix}. \tag{4.116}$$

If we now assume that we have N layers of dielectric medium each having index of refraction n_i where $i = 1, 2, \ldots, N$, and thickness t_i we will derive the corresponding reflection and transmission coefficients in the same way as in the case of a single layer. Based on eq 4.113, one can show that the reflection and transmission coefficients are given by the following matrix equation:

$$\begin{bmatrix}1\\1\end{bmatrix}+\begin{bmatrix}1\\-1\end{bmatrix}\rho=\mathbf{M}_1\mathbf{M}_2\cdots\mathbf{M}_N\begin{bmatrix}1\\n_{\mathrm{S}}\end{bmatrix}\tau=\mathbf{M}\begin{bmatrix}1\\n_{\mathrm{S}}\end{bmatrix}\tau, \tag{4.117}$$

where $\mathbf{M}_i$ corresponds to the transfer matrix of the i th layer which have the same structure as given with eq 4.116 but with the corresponding index of refraction and thickness. For the overall transfer matrix **M**, we can write the following:

$$\mathbf{M}=\begin{bmatrix}A & B\\ C & D\end{bmatrix}. \tag{4.118}$$

By slowing eqs 4.117 and 4.118 in terms of the overall transmission matrix elements, we have:

$$\rho=\frac{A-C+n_{\mathrm{S}}(B-D)}{A+C+n_{\mathrm{S}}(B+D)}, \tag{4.119}$$

$$\tau = \frac{2}{A + C + n_S(B + D)}. \tag{4.120}$$

The reflectance and the transmittance in terms of irradiances are gives as: $R = |\rho|^2$ and $T = |\tau|^2$. In the case of a single layer and according to eqs 4.112 and 4.113, we have the following for the reflection coefficient:

$$\rho = \frac{n(1 - n_S)\cos(kt) - j(n_S - n^2)\sin(kt)}{n(1 + n_S)\cos(kt) - j(n_S + n^2)\sin(kt)}. \tag{4.121}$$

If the thickness of the film is 1/4 of the wavelength, we have $kt = \pi/2$, so the reflectance in this case is given by:

$$R = |\rho|^2 = \frac{(n_S - n^2)^2}{(n_S + n^2)^2}, \tag{4.122}$$

which has zero value, that is, there is no reflection (antireflective coating) if it is fulfilled:

$$n = \sqrt{n_S}, \tag{4.123}$$

Magnesium fluoride is usually used for coating the lenses. As the lens has been built typically from a glass material having index of refraction of about $n_S \approx 1.5$, the requirement given by eq 4.123 isn't fulfilled as the magnesium fluoride has the index of refraction of n = 1.35. Nevertheless, with a quarter-wave thickness of the magnesium fluoride coating the reflectance is reduced to about 1%. If a two-layer coating has been used, one can obtain zero reflectance for particular wavelength for a given coating material. More layers provide more extensive possibilities. Therefore, by using three suitably chosen layers, one can reduce the reflectance below 0.25% over almost entire visible spectrum.

To obtain a high reflectance value in a multilayer stack, a stack of alternating layers of high n_H and low n_L index of refraction has been built where each layer has the thickness equal to a quarter of wavelength, as shown in Figure 4.26. The transfer matrices are all of the same form where the product of two adjacent is given by:

$$\mathbf{m} = \begin{bmatrix} 0 & -j\frac{1}{n_H} \\ -jn_L & 0 \end{bmatrix} \begin{bmatrix} 0 & -j\frac{1}{n_H} \\ -jn_L & 0 \end{bmatrix} = \begin{bmatrix} -\frac{n_H}{n_L} & 0 \\ 0 & -\frac{n_L}{n_H} \end{bmatrix}. \tag{4.124}$$

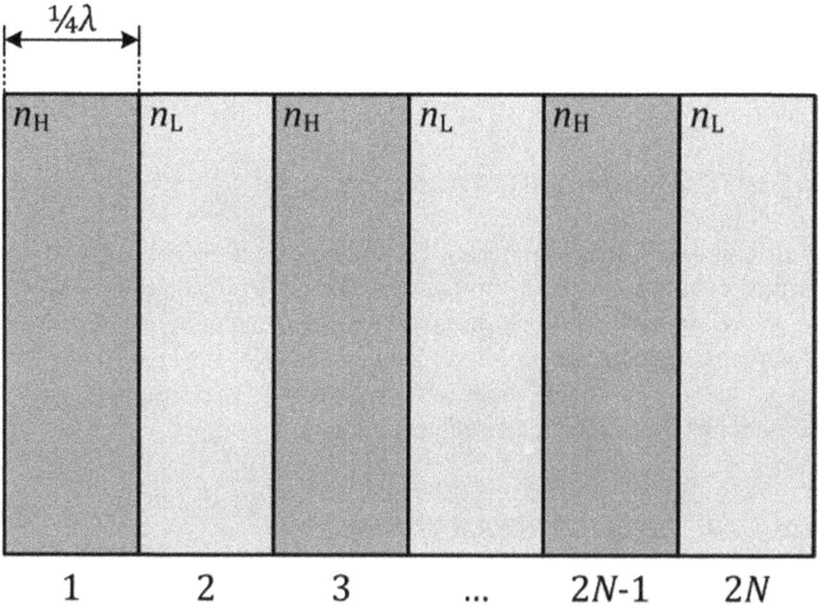

FIGURE 4.26 Multilayer stack for producing high reflectance.

If the stack consists of $2N$ layers, then the overall transfer matrix of the complete multilayer film is given by:

$$\mathbf{M} = \mathbf{m}^N = \begin{bmatrix} \left(-\frac{n_H}{n_L}\right)^N & 0 \\ 0 & \left(-\frac{n_L}{n_H}\right)^N \end{bmatrix}. \tag{4.125}$$

According to eq 4.119, the reflectance of a multilayer film is given by:

$$R = |\rho|^2 = \left[\frac{\left(-(n_H/n_L)\right)^N - \left(-(n_L/n_H)\right)^N}{\left(-(n_H/n_L)\right)^N + \left(-(n_L/n_H)\right)^N}\right]^2 = \left[\frac{(n_H/n_L)^{2N} - 1}{(n_H/n_L)^{2N} + 1}\right]^2. \tag{4.126}$$

One can notice from the last equation that the reflectance approaches unity for a large N. For example, for $N = 4$ (eight-layer stack) of zinc sulfide ($n_H = 2.3$) and magnesium fluoride ($n_L = 1.35$), we have the reflectance of about $R = 97\%$ which is higher than the reflectance of a pure silver layer in the visible spectrum. Further, in the case of a 30-layer stack, the reflectance is better than 99.9. The maximum reflectance is obviously achieved for a single wavelength but it is possible to broad the range by combining different layer thicknesses.

4.6 INTERFEROMETRIC SENSORS

Being highly sensitive, interferometers can be used to measure different physical quantities with an ultimate precision. Typically, interferometers perform vibration amplitude measurement, flow velocity as well as flow velocity distribution measurement, rotation angles as well rotation velocity measurement. Moreover, today's state-of-the-art measurement such as gravitational wave detectors are interferometer based.

4.6.1 THE RAYLEIGH REFRACTOMETER

The Rayleigh refractometer is based on Young's double-slit interferometer, although any other two-beam interferometer could be used. To measure the refractive index of materials having index of refraction close to unity such as gases, Rayleigh put in one of the arms, named the sensing arm, the tube T_1 field with the gas which index of refraction is to be measured and in the other arm, named the reference arm, the second tube T_2 filed with the gas with the known index of refraction or simply evacuated to form a vacuum, as presented in Figure 4.27. The tubes T_1 and T_2 are in the separate light paths from the slits S_1 and S_2, illuminated coherently from a single source.

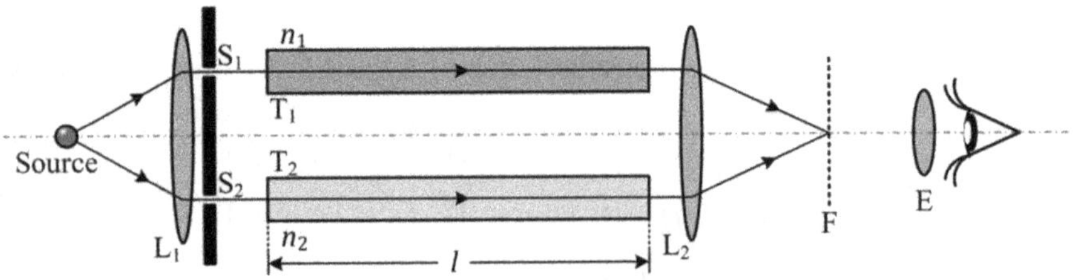

FIGURE 4.27 The Rayleigh refractometer.

If the gas pressure is changed in one of the tubes, the fringe system, viewed by an eyepiece E, at the focus of a long-focus lens L_2, moves across the field of view. When moving, the fringe count N provides a direct measurement of the change in the optical path through the tube caused by the change in refractive index δn as the amount of gas pressure changes. If the tube lengths are l and wavelength of the light of the used source in vacuum λ_0, we have the following relation:

$$N = \frac{l\delta n}{\lambda_0}. \tag{4.127}$$

In the case of a dilute gas the refractive index n differs from unity by an amount proportional to density, so that for a fixed temperature $n - 1$ is proportional to pressure. To measure the gas pressure, one can obtain the refractive index from a single measurement, which then can be used to calculate the value for any other pressure by a simple proportion.

4.6.2 LASER-DOPPLER INTERFEROMETRY

The flow velocity can be easily measured by the laser-Doppler interferometry. This technique uses the fact that light scattered by a moving particle has its frequency shifted according to the Doppler effect. The frequency shift can be measured by the beats produced between the scattered light and a reference beam. A typical measurement configuration is based on two intersecting laser beams, making mutual angle θ, which illuminate the test field where the flow takes a part, as shown in Figure 4.28. Light, which is scattered by a particle, passing through the beam overlapping region is focused on a photodetector. Assuming that particles have the same velocity v in the plane of the beams, at right angles to the direction of observation, the frequency of the beat signal is given by:

$$\Delta \nu = \frac{2v\sin(\theta/2)}{\lambda}. \tag{4.128}$$

To resolve the flow direction, one can introduce a small frequency shift between two intersecting beams by a pair of acousto-optic modulators operated at slightly different frequencies. Simultaneous measurements of the velocity components along two orthogonal directions can be arranged by using two orthogonal illuminating beams. To avoid the beams, interaction lasers with different wavelengths can be employed.

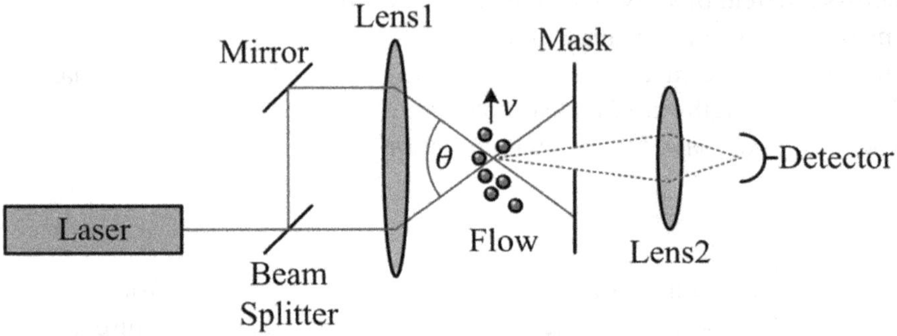

FIGURE 4.28 Laser-Doppler interferometer measures flow velocity.

4.6.3 VIBRATION AMPLITUDES MEASUREMENTS

Measurements of very small vibration amplitudes can be performed with the help of interferometry. Typically, one of the beams in an interferometer is back-reflected from a mirror attached to the object, in which vibration amplitude is to be measured. Due to the Doppler effect, there is a frequency shift of the back-reflected beam. To measure the frequency shift, the reflected beam is made to interfere with a reference beam with a known offset frequency. The photodetector time-varying output signal consists of a component at the offset frequency, that is, the carrier and two sidebands. The vibration amplitude A can then be obtained from the following relation:

$$\frac{2\pi}{\lambda}A = \frac{I_{\mathrm{SB}}}{I_0}, \tag{4.129}$$

where I_0 is the power at the offset frequency, and I_{SB} is the power in each of the sidebands.

4.6.4 MICHELSON'S STELLAR INTERFEROMETER

The principle of Michelson's stellar interferometer is shown in Figure 4.29, which is generally based on Young's double-slit interferometer. However, there are some modifications of the simple Young interferometric system to make an interferometer capable of producing fringes by the light of an observed star. Instead of two narrow slits in the Young interferometer

Michelson's stellar, interferometer uses two large plane mirrors M_1 and M_2 positioned at a distance D from each other. Each mirror reflects light toward two mirrors M_3 and M_4, which further reflect two parallel beams of light into a telescope objective. Both beams form images of the star in the focal plane of the lens where typically a CCD camera can be placed. As a result, fringes can be seen, two apertures, positioned in front of the telescope objective, S_1 and S_2, limit the size of the beams, acting like Young's double-slit interferometer with the spacing d.

If the observed star has a very small angular diameter, then the wavefronts falling onto mirrors M_1 and M_2 are coherent thus providing a coherent illumination of the slits S_1 and S_2. In the telescope objective focal plane, where the camera has been positioned, the interference pattern of the slits is observed, which consists of the fringes with the spacing $\lambda f/d$, where f is the focal length of the telescope.

To measure the angular diameter of the star, we will consider star as a light source, the brightness distribution of which depends on the incoming angle ψ. Let this be $B(\psi)$ and let $B(\psi)$ be symmetrical about the center of the source. In this case, each elementary strip of the source $d\psi$ wide and at a small angle ψ from the center of the source forms a fringe pattern with the intensity proportional to $B(\psi)$ and displaced by angle ψ from center of the fringe system. Therefore, the fringe intensity maximum is given by:

$$I_{\max} = C\int_{-\psi_0/2}^{\psi_0/2} B(\psi)\cos^2\left(\frac{\pi D}{\lambda}\psi\right)\mathrm{d}\psi, \tag{4.130}$$

where C is a constant and ψ_0 is the star angular diameter. Similarly, fringe intensity minimum is given by:

$$I_{\min} = C\int_{-\psi_0/2}^{\psi_0/2} B(\psi)\sin^2\left(\frac{\pi D}{\lambda}\psi\right)\mathrm{d}\psi. \tag{4.131}$$

The fringe visibility of the observed interferometric pattern is given by:

$$V = \frac{I_{\max} - I_{\min}}{I_{\max} + I_{\min}} = \frac{\int_{-\psi_0/2}^{\psi_0/2} B(\psi)\cos\left((2\pi D/\lambda)\psi\right)\mathrm{d}\psi}{\int_{-\psi_0/2}^{\psi_0/2} B(\psi)\,\mathrm{d}\psi}. \tag{4.132}$$

For a star with uniform brightness, we have:

$$V = \frac{\sin\left((\pi D/\lambda)\psi_0\right)}{(\pi D/\lambda)\psi_0}. \tag{4.133}$$

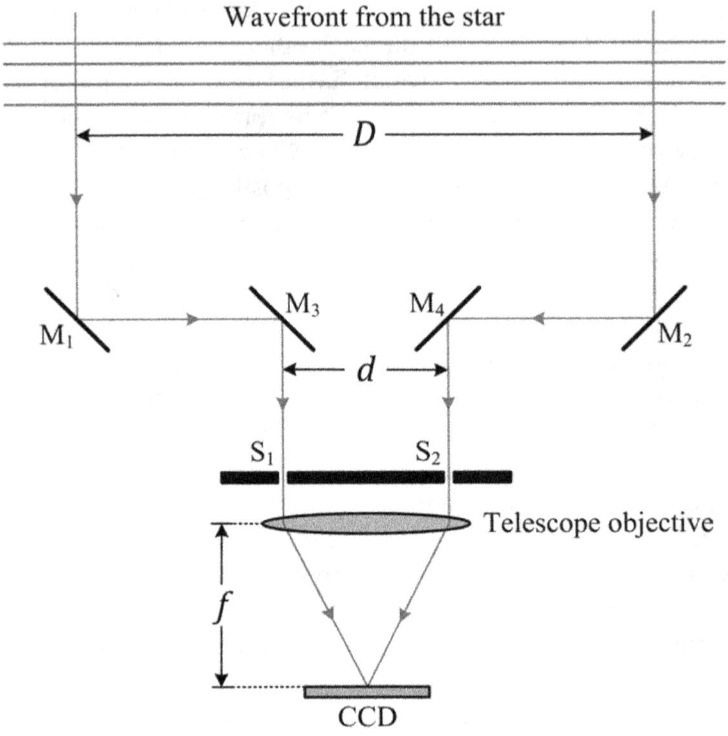

FIGURE 4.29 Michelson's stellar interferometer.

Equation 4.133 represents the variation of fringe visibility with ψ_0 at a fixed spacing D or alternatively if D varies, the change of fringe visibility can be used to measure angular diameter for a given star.

KEYWORDS

- **amplitude division**
- **coherence of light**
- **interference of light**
- **wavefront division**
- **white-light interferometry**

REFERENCES

Glindemann, A. *Principles of Stellar Interferometry*; Springer-Verlag: Berlin Heidelberg, 2011.

Grangeat, P., Ed. *Tomography*; ISTE Ltd: London, 2009.

Hariharan, P. *Basics of Interferometry*, 2nd ed.; Elsevier Inc.: Amsterdam, 2007.

Hariharan, P. *Optical Interferometry*, 2nd ed.; Elsevier Science, USA: Amsterdam, 2003.

Kreis, T. *Handbook of Holographic Interferometry: Optical and Digital Methods*; WILEY-VCH Verlag GmbH & Co.KgaA: Weinheim, 2005.

Labeyrie, A.; Lipson, S. G.; Nisenson, P. *An Introduction to Optical Stellar Interferometry*; Cambridge University Press: Cambridge, 2006.

Malacara, D.; Servín, M.; Malacara, Z. *Interferogram Analysis for Optical Testing*, 2nd ed.; Taylor & Francis Group, LLC: Boca Raton, FL, 2005.

Manojlović, L. M. *Optics and Applications*; Arcler Press: Oakville, Canada, 2018.

Razumovsky, I. A. *Interference-Optical Methods of Solid Mechanics*; Springer-Verlag: Berlin Heidelberg, 2011.

Richichi, A.; Delplancke, F.; Paresce, F.; Chelli, A., Eds.. *The Power of Optical/IR Interferometry: Recent Scientific Results and 2nd Generation Instrumentation - Proceedings of the ESO Workshop Held in Garching, Germany, 4- 2005 8 April*. Springer-Verlag: Berlin Heidelberg, 2008.

Smith, F. G.; King, T. A.; Wilkins, D. *Optics and Photonics: An Introduction*, 2nd ed.; John Wiley & Sons Ltd: Chichester, 2007.

Tuchin, V. V., Ed. *Coherent-Domain Optical Methods: Biomedical Diagnostics, Environmental and Material Science*; Kluwer Academic Publishers: New York, NY, 2004.

CHAPTER 5

Fiber Optics

ABSTRACT

The accessibility of generating and transferring the multimedia content over Internet to an average customer on a daily basis pushes hardly the telecom companies to provide telecom networks that are capable of transmitting the data with very high capacity at relatively long distances. The technology that is capable to fulfill such stringent requirements is based on a network technology that comprises the use of an optical fiber as the signal propagating medium. Nowadays, the standard telecom optical fibers are capable of transferring the data with the capacity that exceeds 400 Gb/s per fiber, which are based on the DWDM (Dense Wavelength Division Multiplexing) network technology. Although optical fibers are mostly used in the telecommunication networks for the high-capacity data transfer, there is also another specific use of the optical fibers. For example, having in mind that optical interferometry is one of the most sensitive measuring technique that is capable of high-precision and high-speed measurements of different physical quantities, a large number of different sensors have been developed which are based on the optical fibers. The fiber optic–based sensors, which are capable of measuring different physical as well as chemical parameters, such as temperature, strain, force, pressure, pH, concentration, and index of refraction can be organized in such a way that they can be positioned within the fiber, that is, to be the integral part of the optical fiber, or to be positioned on the tip of the fiber end.

In the last two decades, we are the witnesses of almost exponential growth of Internet traffic. The accessibility of generating and transferring the multimedia content over Internet to an average customer on a daily basis pushes hardly the telecom companies to provide telecom networks that are capable of transmitting the data with very high capacity at relatively long distances. The technology that is capable to fulfill such stringent requirements is based on a network technology that comprises the use of an optical fiber as the signal propagating medium. Nowadays, the standard telecom optical fibers are capable of transferring the data with the capacity that exceeds 400 Gb/s per fiber, which are based on the DWDM (***D**ense **W**avelength **D**ivision **M**ultiplexing*) network technology. Bearing in mind, that the standard single-mode fiber-optical cable, which is commonly manufactured, consists of maximum 864 fibers where the fibers are placed in 36 ribbons each containing 24 strands of fibers, the capacity per optical cable can reach a few hundreds of Tb/s. Moreover, the distances of 100 km can be reached between two optical nodes with an unamplified and uncompensated system with a bit rate of around 10 Gb/s. Therefore, the presented capabilities of the fiber optic–based telecommunication networks will further push the use of optical fibers to be the backbone of the future data transfer systems.

Although optical fibers are mostly used in the telecommunication networks for the high-capacity data transfer, there is also another specific use of the optical fibers. For example, having in mind that optical interferometry is one of the most sensitive measuring techniques that is capable of high-precision and high-speed measurements of different physical quantities, a large number of different sensors have been developed which are based on the optical fibers. The fiber optic–based sensors, which are capable of measuring different physical as well as chemical parameters, such as temperature, strain, force, pressure, pH, concentration, and index of refraction can be organized in such a way that they can be positioned within the fiber, that is, to be the integral part of the optical fiber, or to be positioned on the tip of the fiber end. It is common, that based on the channeled spectrum analysis of the sensor back-reflected light, we can arrange even a lab on a fiber tip where simultaneously can be interrogated several different physical and chemical quantities. Bearing in mind that optical fibers are passive device, which do not require electrical power supply, that are immune to the environmental electromagnetic radiation, and that they have zero electromagnetic radiation emission, they represent an ideal solution for different sensing applications in the hazard environments as well as in potentially explosive areas.

Sensors based on the optical fibers offer the possibility of simple multiplexing a large number of sensors even along a single optical fiber. In this way, one can arrange quasidistributed or even distributed measurements of different physical and chemical parameters. Therefore, the fiber-optical sensors are one of the most frequent choices when the structural integrity of different objects such as bridges, tunnels, and large buildings must be permanently monitored.

All the mentioned earlier shows that the optical fiber is one of the most important optical components, which significantly changed our lives in the last couple of decades enabling high capacity data transfer to the every person whether it uses a standalone or a mobile device with the Internet access. This is the reason why the whole chapter is dedicated to the optical fibers.

5.1 OPTICAL FIBERS

At the very beginning of the development of the optical waveguides, which were aimed for the light transmission, it was very well known that the microwaves can be easily transmitted through the very simple waveguides. However, the problem with optical waveguides was the very high absorption of the optical materials that were used at the time thus severely limiting the transmission range. The researchers tested different but among the best suited was the glass known as fused silica consisting of silica in amorphous noncrystalline form. Even fused silica attenuated the light by at least one third after a distance no longer than 1 m that severely limited the transmission over the long distances.

The significant breakthrough in developing low-loss optical fibers occurred in 1966 when the researchers became aware that the losses within the optical fibers are caused by the material impurities and not by the material itself. At that time, it was predicted that a glass-based optical fibers can be produced with the attenuation instead of typical 1000 dB/km of "only" 20dB/km dB/km. Since that time the optical losses of the optical fibers were constantly reducing reaching nowadays 0.2 dB/km.

The optical fibers can be roughly divided into two main groups, that is, in the multimode and single-mode fibers. The multimode fibers are characterized by a relatively large diameter of the core through which the light is guided. The diameter of the multimode core is usually much larger than the wavelength of the light as presented in Figure 5.1. Typically, the core diameter is 50 μm embedded in a fiber cladding with the outside diameter of 125 μm. In contrast to the multimode fibers, the single-mode fibers have a core diameter that is slightly larger than the wavelength with the typical values ranging between 7 and 10 μm. However, the cladding diameter of the single-mode fibers is of the same value of 125 μm as in the case of the multimode fibers, which is also presented in Figure 5.1.

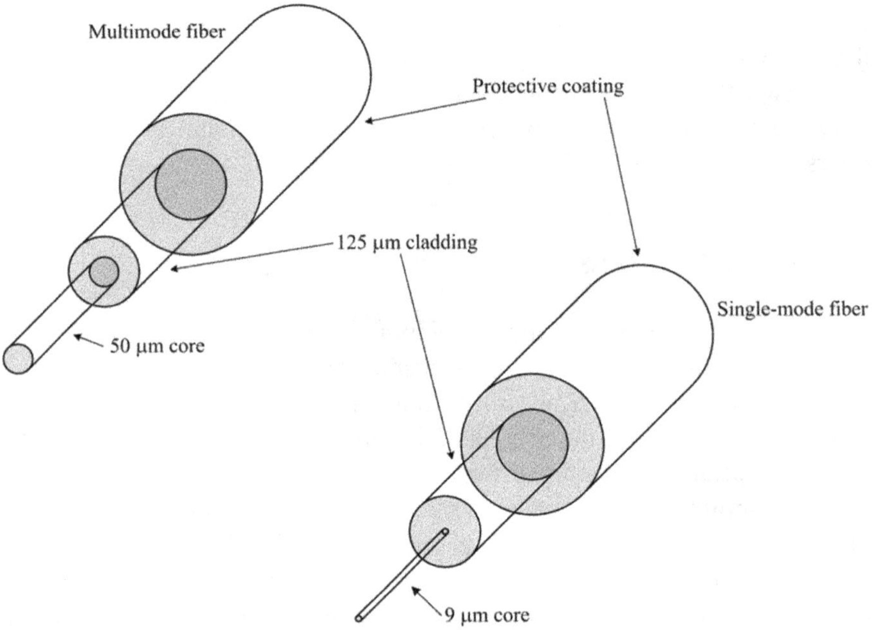

FIGURE 5.1 Multimode and single-mode fiber dimensions.

The first applications of the optical fibers were reserved for the multimode type. This type of optical fibers allows better mutual coupling efficiency of the fibers as there were fewer losses when connecting fibers together. In comparison with the multimode fibers, the single-mode fibers allow higher data transmission rates over longer distances. Due to the small core diameter of the single-mode fibers, the connector tolerance must be very tight to keep coupling losses low. At the beginning, this was the limiting factor in using single-mode fibers. As the problem with the connector tolerance was solved satisfactorily, single-mode fibers became the favored choice, especially for the long hauls. Nowadays, the multimode fibers are used only for the short distances in particular in the local area networks.

5.1.1 GEOMETRICAL OPTICS AND THE OPTICAL FIBERS

To determine the light transmission parameters of an optical fiber, there are two standard ways. The first one is based on the ray-tracing technique, which is in the frame of the geometrical optics and which will be presented in this chapter. The second one considers the wave nature of the light where the Maxwell equations in the cylindrical coordinate system will be used to treat the light propagation along an optical fiber.

The main physical principle behind the light propagation within an optical fiber in the frame of geometrical optics treatment is the well-known phenomenon of the total internal reflection. Typical optical fiber has a step-index change of the index of refraction along its cross section. There is a core with a circular cross section that is surrounded by a cladding with a ring-shaped cross section. The core is made of a glass with slightly higher refractive index than that of the cladding. Due to the total internal reflection phenomenon, the light is guided through the core. If instead of a two-layer structure, we have a single layer optical fiber, that is, simple fiber without internal structure, the guidance is also possible. However, any dirt on the fiber surface or coating that should protect the fiber can produce that some rays that propagate inside the unstructured fiber leak out leading to the very high loses. In this case, that is, when the dirt comes into contact to the unstructured fiber or any kind of a material with a similar index of refraction as of the fiber, the condition for the total internal reflection is broken down giving a rise to the leakage of the rays from the fiber and thus to the attenuation of the optical signal. By introducing the internal fiber structure, the light guidance is not any more dependent on the fiber surroundings.

As the outer fiber surface is not any more important for guiding the light in step-index fiber, we can, without losing the generality of the analysis, assume that the cladding diameter is infinitely wide, that is, we can assume that there is no outside surface. To determine the largest possible angle, with respect to the fiber axis, that provides the condition for the ray propagation along the fiber we will observe the fiber cross section as presented in Figure 5.2.

According to the Snell law, we have:

$$n_0 \sin\alpha = n_2 \sin\beta, \tag{5.1}$$

$$n_2 \sin\gamma = n_1 \sin\delta, \tag{5.2}$$

where n_0 is the air index of refraction, n_1 is the cladding index of refraction, and n_2 is the core index of refraction, where the following is fulfilled

$n_0 \leq n_1 \leq n_2$. If the fiber axis to be perpendicular to the front face, we have $\beta + \gamma = \pi/2$, and consequently:

$$\sin\beta = \sqrt{1-\sin^2\gamma}. \tag{5.3}$$

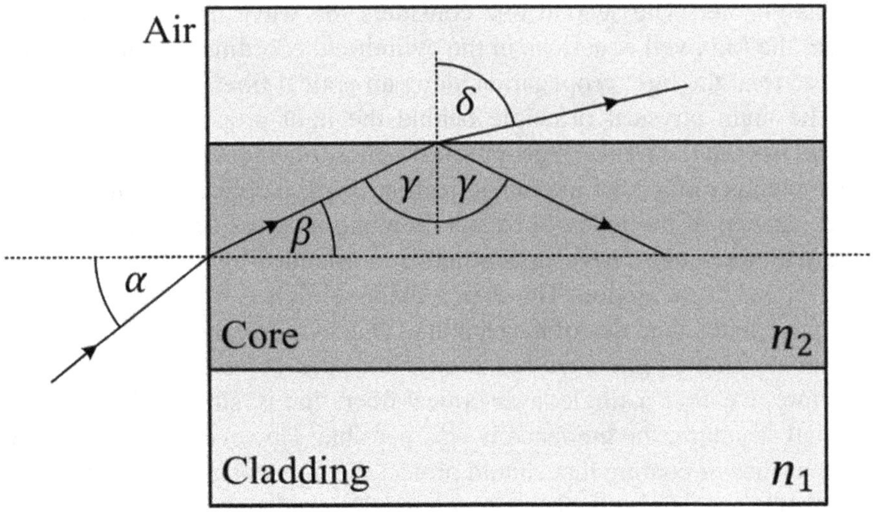

FIGURE 5.2 Cross section of the step-index fiber.

The critical angle for the total internal refraction at the core-cladding interface is given by:

$$\delta_C = 90° \Rightarrow \sin\delta_C = 1 \Rightarrow \sin\gamma_C = \frac{n_1}{n_2}. \tag{5.4}$$

The combination of eqs 5.1–5.4 gives:

$$n_0 \sin\alpha_C = n_2\sqrt{1-\left(\frac{n_1}{n_2}\right)^2} = \sqrt{n_2^2 - n_1^2}. \tag{5.5}$$

If we further take $n_0 = 1$ for air, then the critical angle α_C for rays to be guided within the fiber core is:

$$\alpha_C = \arcsin\sqrt{n_2^2 - n_1^2}. \tag{5.6}$$

The argument of the inverse sine function has a special name, that is, it is called the numerical aperture, which is often abbreviated as NA and given by:

$$NA = \sqrt{n_2^2 - n_1^2}. \tag{5.7}$$

It is obvious from eq 5.7 that the numerical aperture is a measure of the core and cladding indices of refraction difference. The largest angle of acceptance for rays impinging the fiber end facet is given by:

$$\alpha_C = \arcsin SN. \tag{5.8}$$

This angle represents at the same time the angle of the light exit cone at the fiber end when the light exits an optical fiber. The corresponding input/output cone is schematically shown in Figure 5.3.

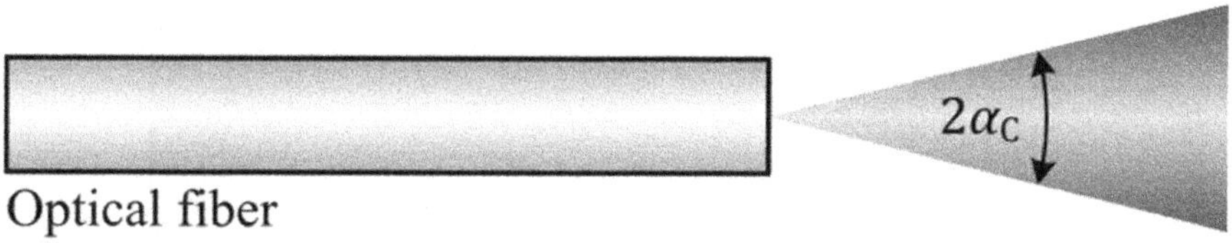

FIGURE 5.3 Acceptance and exit cone of an optical fiber.

There is another parameter that is also frequently used when speaking about an optical fiber. This parameter is given by:

$$\Delta = \frac{n_2^2 - n_1^2}{2n_2^2}. \tag{5.9}$$

The relation between NA and the Δ parameter is:

$$NA = n_2\sqrt{2\Delta}. \tag{5.10}$$

Typically, the values of core and cladding indices of refraction are very close to each other with a mutual difference of a few tenths of 1%. Bearing this in mind, eq 5.9 becomes:

$$\Delta \approx \frac{(n_2 + n_1)(n_2 - n_1)}{n_2(n_2 + n_1)} = \frac{n_2 - n_1}{n_2}. \tag{5.11}$$

According to the last relation, it is obvious why the parameter Δ is called normalized refractive index difference or normalized index step. For a typical single-mode fiber, we have $\Delta = 0.35\%$ and $n_2 = 1.45$ which leads to $NA = 0.12$, which according to eq 5.8 gives the value of $\alpha_C \approx \pm 6.9°$ for the acceptance angle.

As it was mentioned earlier, all the rays that enter the fiber end facet within the acceptance cone will propagate throughout the fiber core. The rays travel along the fiber bouncing between the core-cladding interfaces with different angles thus traveling the different path lengths until they reach the far end of the fiber. Consequently, they arrive at different times and give a rise to the modal dispersion which is illustrated in Figure 5.4.

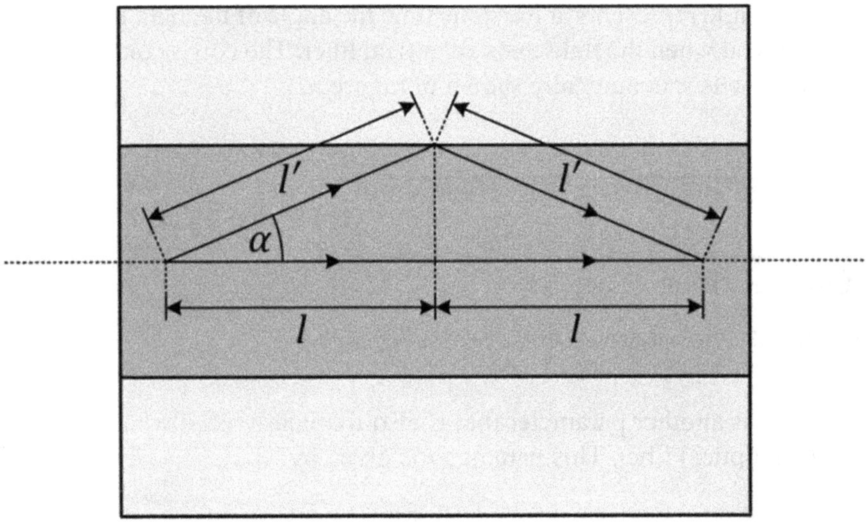

FIGURE 5.4 Modal dispersion.

According to the ray paths, shown in Figure 5.4, one can notice that the ray that impinge the core-cladding interface and travels with the angle α between the ray and the fiber axis, travels the path that is larger than the ray which travels along the fiber axis and where it is clearly valid $l' = l/\cos\alpha$. It was previously shown that $\sin\alpha \leq NA/n_2$ is valid, so if it is fulfilled $\alpha \ll 1$, we have $\sin\alpha \approx \alpha$ and $\cos\alpha \approx 1 - \alpha^2/2$, and consequently:

$$l' \approx l\left(1+\frac{NA^2}{2n_2^2}\right) \approx l(1+\Delta). \tag{5.12}$$

As the optical signal that propagates through the fiber travels along different paths, there is a scatter of propagation times which is often called delay distortion. When the optical signal is in the shape of the short pulses, the pulse propagation causes a widening of the pulse. This is not the problem until the pulse becomes wider than the pulse rate. When this happen, an intersymbol interference occurs and the transmitted message becomes undecipherable. To show that this can be a serious problem a rough calculation will be performed. If we assume that the light velocity within the fiber core is c/n_2 then the propagation time of the light in the fiber with the length L is $\tau = n_2 L/c$. According to the modal dispersion geometry shown in Figure 5.4, for the minimal time travel, we have:

$$\tau_{\min} = \frac{n_2 l}{c}, \tag{5.13}$$

where the maximal time travel is reached for the maximal ray angle and is given by:

$$\tau_{\max} = \frac{n_2 l'}{c} = \frac{n_2 l}{c}(1+\Delta). \tag{5.14}$$

The difference in time travel is equal to:

$$\delta\tau = \tau_{\max} - \tau_{\min} = \tau_{\min}\Delta. \tag{5.15}$$

The corresponding propagation time scatter is then given by:

$$\frac{\delta\tau}{\tau_{\min}} = \Delta. \tag{5.16}$$

If we take, as in the previous case, that $\Delta = 0.35\%$, then for a fiber length of 1 km, we have a pulse broadening equal to $\delta\tau \approx 17.5$ ns. This represents only a rough estimation as we assumed that only meridional rays (rays that pass through the optical fiber axis) propagate along the fiber neglecting the existence of the helical rays (rays that bounce around the optical fiber axis) that are presented in Figure 5.5. However, it is obvious that the modal dispersion can significantly broad the pulse that further limits the data rate that in our case is equal to the inverse value of the pulse broadening, that is, $1/\delta\tau \approx 57$ Mb/s. One can notice that this is a very low data rate if compared with today's standard telecom data rates especially in our case where the fiber length of only 1 km was considered. The mechanism of modal dispersion can severely hamper the use of an optical fiber for practical applications. However, there are ways to avoid the problem with modal dispersion. One of them is to use the so-called gradient-index fibers, but in the case of the highest data rates, the single-mode fibers must be used.

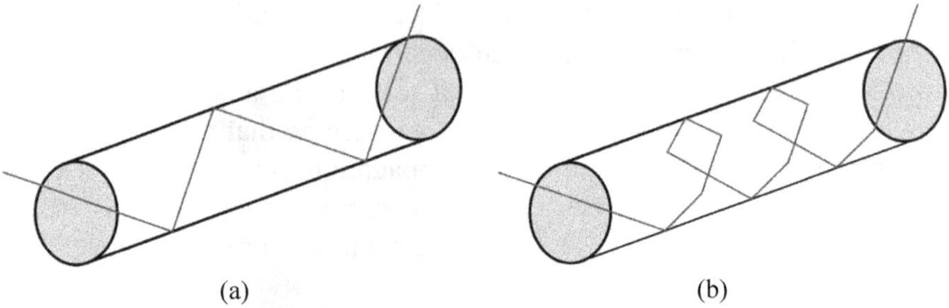

FIGURE 5.5 (a) Meridional and (b) helical rays in an optical fiber.

To suppress the modal dispersion and thus to avoid the scatter of arrival times, an optical fiber with a gradient radial profile of the refractive index must be used. Such optical fibers have the following radial profile of an index of refraction along the fiber cross section:

$$n(r) = \begin{cases} n_2\sqrt{1-2\Delta\left(\dfrac{r}{a}\right)^{\alpha}}, & |r| \le a \\ n_1, & |r| > a \end{cases}. \tag{5.17}$$

where α denotes the core radius and α is the corresponding exponential parameter. The resulting index of refraction profile for the given exponential parameter values is presented in Figure 5.6.

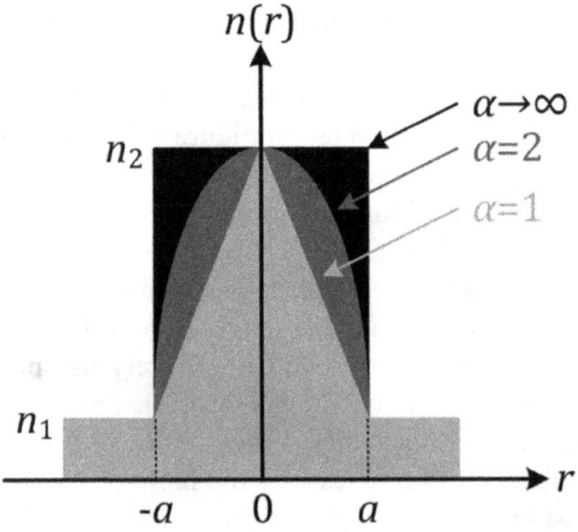

FIGURE 5.6 Some common index profiles of a gradient optical fiber.

To minimize the differences in transit time, there is an optimal profile of an index of refraction. The first approximation gives us an optimal profile for $\alpha = 2$, that is, a parabolic index profile. The rays travel curved paths in the fiber and not zigzag paths as in the case of the step-index fiber. Although the curved paths are longer then the straight paths along the fiber axis, that is, the larger the detour of the ray from the axis the longer the ray paths, the index of refraction is smaller and thus the higher is the detour ray velocity. In this way, longer paths are compensated by the higher ray velocities and consequently, the corresponding optical paths are the same. The scatter of the transit time depends on the exponential factor and for a given values of the exponential factor we have the following:

$$\delta\tau = \begin{cases} \dfrac{n_2 l}{c}\Delta, & \alpha \to \infty \\ \dfrac{n_2 l}{c}\dfrac{\Delta^2}{2}, & \alpha = 2 \end{cases} . \tag{5.18}$$

If we compare the scatter times for $\alpha \to \infty$ and $\alpha = 2$, we obtained in the case of the parabolic profile gradient fiber the reduction of the scatter time by the factor $\Delta/2 \approx 10^{-3}$, which represents a considerable improvement, where for the parabolic profiled gradient fiber, the transit time spread is reduced by about three orders of magnitude. The modal dispersion is then consequently reduced to a few tens of picosecond per kilometers. Precise calculations show that the optimal profile is not obtained for $\alpha = 2$ but for a slightly different value. This optimal value typically depends on glass type, doping material, and wavelength. Furthermore, the optimum exponent value is not the same for meridional rays and helical rays. Therefore, it is difficult to theoretically predict the optimum value and thus its significance is reduced, especially if we take into consideration the unavoidable manufacturing tolerances during the process of the fiber productions.

5.1.2 WAVE OPTICS AND THE OPTICAL FIBERS

The standard optical fiber structure implies rotational symmetry. Therefore, the Maxwell equations, which have been introduced in Chapter 1 in the Cartesian coordinate system, must be adapted in order fits better the cylindrical structure of an optical fiber. This strongly suggests the use of a cylindrical coordinate system with the corresponding cylindrical coordinates r, φ, and z. It is very well known that the Laplacian in the cylindrical coordinates is given by:

$$\nabla^2 E = \frac{1}{r}\frac{\partial}{\partial r}\left(r\frac{\partial E}{\partial r}\right)+\frac{1}{r^2}\frac{\partial^2 E}{\partial \varphi^2}+\frac{\partial^2 E}{\partial z^2}. \quad (5.19)$$

If we further introduce, a separation of the variables in the following way:

$$E(r,\varphi,z,t) = E_0\rho\zeta\tau, \quad (5.20)$$

where E_0 is the wave field amplitude and where:

$$\rho = \rho\,(r,\varphi), \quad (5.21)$$

is the field amplitude distribution in the plane normal to the z-axis, that is, across the fiber cross section,

$$\zeta = \zeta(z) = \exp(-j\beta z), \quad (5.22)$$

where β is the corresponding fiber wave number, and:

$$\tau = \tau(t) = \exp(j\omega t), \quad (5.23)$$

represents a monochromatic wave with an angular frequency ω. The variable separation can be performed due to the linearity and paraxiality of the in fiber propagating rays, which further leads that both the electric and magnetic field components are in the first approximation perpendicular to the propagation direction. Therefore, the longitudinal and transverse processes are practically decoupled. Taking the above-presented into consideration and eq 5.19, the corresponding wave equation becomes:

$$\frac{n^2}{c^2}\frac{\partial^2}{\partial t^2}(E_0\rho\zeta\tau) = \frac{1}{r}\frac{\partial}{\partial r}\left[r\frac{\partial}{\partial r}(E_0\rho\zeta\tau)\right]+\frac{1}{r^2}\frac{\partial^2}{\partial \varphi^2}(E_0\rho\zeta\tau)+\frac{\partial^2}{\partial z^2}(E_0\rho\zeta\tau). \quad (5.24)$$

The first term on the right-hand side of eq 5.24 is:

$$\frac{1}{r}\frac{\partial}{\partial r}\left[r\frac{\partial}{\partial r}(\rho\zeta\tau)\right] = \zeta\tau\frac{1}{r}\frac{\partial}{\partial r}\left(r\frac{\partial \rho}{\partial r}\right), \quad (5.25)$$

where the constant factor of the wave amplitude E_0 was omitted as it cancels out in eq 5.24 because all terms contain this factor. The second term on the right-hand side is:

$$\frac{1}{r^2}\frac{\partial^2}{\partial \varphi^2}(\rho\zeta\tau) = \zeta\tau\frac{1}{r^2}\frac{\partial^2 \rho}{\partial \varphi^2}. \quad (5.26)$$

The third term becomes:

$$\frac{\partial^2}{\partial z^2}(\rho\zeta\tau)=\rho\tau\frac{\partial^2\zeta}{\partial z^2}=-\beta^2\rho\zeta\tau. \tag{5.27}$$

The only term on the left-hand side of eq 5.24 is:

$$\frac{n^2}{c^2}\frac{\partial^2}{\partial t^2}(\rho\zeta\tau)=\rho\zeta\frac{n^2}{c^2}\frac{\partial^2\tau}{\partial t^2}=-\frac{n^2}{c^2}\omega^2\rho\zeta\tau. \tag{5.28}$$

If the vacuum wave number is denoted as k_0, then in the medium with the index of refraction n we have $k = nk_0 = n\omega/c$, so eq 5.28 becomes:

$$\frac{n^2}{c^2}\frac{\partial^2}{\partial t^2}(\rho\zeta\tau)=-k^2\rho\zeta\tau. \tag{5.29}$$

By inspecting eqs 5.25–5.29, one can notice that the factor $\zeta\tau$ is common to all terms and therefore it is cancelled out. Due to the rotational symmetry of the fiber cross section, we can further separate the variables in the following way:

$$\rho(r,\varphi)=R(r)\,\Phi(\varphi). \tag{5.30}$$

Equation 5.24 in this case becomes:

$$-k^2r^2=\frac{r}{R}\frac{\partial R}{\partial r}+\frac{r^2}{R}\frac{\partial^2 R}{\partial r^2}+\frac{1}{\Phi}\frac{\partial^2\Phi}{\partial\varphi^2}-\beta^2r^2. \tag{5.31}$$

The reorganization of the last equation gives:

$$-\frac{1}{\Phi}\frac{\partial^2\Phi}{\partial\varphi^2}=\frac{1}{R}\left[r\frac{\partial R}{\partial r}+r^2\frac{\partial^2 R}{\partial r^2}+r^2\left(k^2-\beta^2\right)R\right]. \tag{5.32}$$

The last equation contains on the left-hand side only the function $\Phi(\varphi)$, that is, it depends only on the angle φ, and where the right-hand site contains only the $R(r)$ function, that is, depends only on the radius r. As both sides are equal, they must be constant, that is, they don't depend either on φ and/or r. This constant will be denoted as p^2. In this case, eq 5.32 can be divided into two equations in the following form:

$$\frac{\partial^2\Phi}{\partial\varphi^2}+p^2\Phi=0, \tag{5.33}$$

and:

$$r^2\frac{\partial^2 R}{\partial r^2}+r\frac{\partial R}{\partial r}+r^2\left(k^2-\beta^2\right)R=p^2R. \tag{5.34}$$

If we take into consideration the following substitution $\kappa^2 = k^2 - \beta^2$, eq 5.34 becomes:

$$r^2 \frac{\partial^2 R}{\partial r^2} + r \frac{\partial R}{\partial r} + \left(\kappa^2 r^2 - p^2\right) R = 0. \tag{5.35}$$

Partial differential equation, given by eq 5.33 has the general solution in the following form:

$$\Phi(\varphi) = \Phi_0 \cos(p\varphi + \varphi_0), \tag{5.36}$$

where Φ_0 and φ_0 are the constants. Further, partial differential equation, given by eq 5.35, is in the form:

$$r^2 R'' + rR' + \left(\kappa^2 r^2 - p^2\right) R = 0. \tag{5.37}$$

The solution of this differential equation is given in the form:

$$R(r) = \rho_1 J_p(\kappa r) + \rho_2 N_p(\kappa r), \tag{5.38}$$

where $\kappa\, r$ is real ($\kappa^2 r^2 \geq 0$), or:

$$R(r) = \rho_3 I_p(\kappa r) + \rho_4 K_p(\kappa r), \tag{5.39}$$

where κr is imaginary ($\kappa^2 r^2 \geq 0$). The functions $J_p(x)$, $N_p(x)$, $I_p(x)$, and $K_p(x)$ are the corresponding Bessel functions.

To determine the constants ρ_1, ρ_2, ρ_3, and ρ_4, we must specify the geometry of the used optical fiber. Since this moment we have only assumed that fiber is of a cylindrical shape and that the core and cladding indices of refraction are just slightly different. Further, we will consider a step-index fiber with the following index of refraction distribution:

$$n = \begin{cases} n_2, & r \leq a \\ n_1, & r > a \end{cases}, \tag{5.40}$$

where it is $n_2 > n_1$ taken into account as otherwise there would be no wave propagation with small loses. In the case when the core radius is much larger than the wavelength of the propagating light, the obtained solution must follows the results obtained in the frame of the geometrical optics where the ray-tracing technique was taken into account, that is, the light is guided within the fiber core and the solution must predict that the most of the light wave is concerted in the fiber core. Also, the solution must provide that the light within the cladding will have the amplitude that decreases with the radius increase. Therefore, the parameter κr must be real in the core and imaginary in the cladding, that is, $k \geq \beta$ must be fulfilled for the core, where the opposite is valid for the cladding. Further, to have parameter κr real in the core, one must fulfill $\kappa^2 r^2 \geq 0$ or $\left(k_2^2 - \beta^2\right) r^2 \geq 0$ for $r \leq a$, where k_2 is the core wave number. The case is opposite for the cladding where the following must be fulfilled $\kappa^2 r^2 < 0$ and consequently $\left(k_1^2 - \beta^2\right) r^2 < 0$ for $r > a$, where k_1 is the cladding wave number. Following the last two inequalities, we have:

$$k_1 < \beta < k_2. \tag{5.41}$$

For the transversal components of the core and cladding wave number, the following is valid:

$$\kappa_2^2 = k_2^2 - \beta^2 \text{ and} \tag{5.42}$$

$$\kappa_1^2 = -\left(k_1^2 - \beta^2\right), \tag{5.43}$$

respectively. It is further useful to make the following substitutions:

$$u = a\kappa_2 \text{ and} \tag{5.44}$$

$$w = a\kappa_1, \tag{5.45}$$

where u and w are dimensionless, real positive quantities that are often used in the literature. The parameter u defines the progression of the phase and the parameter w defines the transverse decay of the wave amplitude. The following relations are valid for these two parameters:

$$\begin{aligned} u^2 + w^2 &= a^2\left(k_2^2 - \beta^2\right) - a^2\left(k_1^2 - \beta^2\right) \\ &= a^2\left(k_2^2 - k_1^2\right) = a^2 k_0^2\left(n_2^2 - n_1^2\right). \end{aligned} \tag{5.46}$$

so the parameter $u^2 + w^2$ represents a constant. This constant is of very high importance in the fiber optics and it is called normalized frequency or simply V number, which is given by:

$$V^2 = u^2 + w^2 = a^2 k_0^2\left(n_2^2 - n_1^2\right), \tag{5.47}$$

or equivalently:

$$V = ak_0 NA = \frac{2\pi}{\lambda_0} a\sqrt{n_2^2 - n_1^2}. \tag{5.48}$$

By using the parameters u and w, one can obtain the general solution for the step-index fiber in the following form:

$$\rho_2(r,\varphi) = C_2 J_p\left(\frac{ur}{a}\right)\cos(p\varphi + \varphi_0) \text{ and} \tag{5.49}$$

$$\rho_2(r,\varphi) = C_2 J_p\left(\frac{ur}{a}\right)\cos(p\varphi + \varphi_0), \tag{5.50}$$

where the first solution in eq 5.48 is valid for the core (subscript 2) and the second one for the cladding (subscript 1). To fulfill the boundary conditions for $r = a$ both solutions given in eqs 5.49 and 5.50 must have the same value, so the following must be fulfilled:

$$\rho_2(r = a,\varphi) = \rho_1(r = a,\varphi) \text{ and} \tag{5.51}$$

$$\frac{\partial}{\partial r}\rho_2(r = a,\varphi) = \frac{\partial}{\partial r}\rho_1(r = a,\varphi), \tag{5.52}$$

where after substituting eqs 5.51 and 5.52 into 5.49 and 5.50, we have:

$$C_2 J_p(u) = C_1 K_p(w) \text{ and} \tag{5.53}$$

$$C_2\frac{\partial}{\partial r}J_p\left(\frac{ur}{a}\right)\Bigg|_{r=a} = C_1\frac{\partial}{\partial r}K_p\left(\frac{wr}{a}\right)\Bigg|_{r=a}. \tag{5.54}$$

Based on the following identity:

$$\frac{\partial}{\partial r} = \frac{u}{a}\frac{\partial}{\partial(ur/a)} = \frac{w}{a}\frac{\partial}{\partial(wr/a)}, \tag{5.55}$$

eq 5.54 becomes:

$$C_2\frac{u}{a}J_p'(u) = C_1\frac{w}{a}K_p'(w), \tag{5.56}$$

where the prime denotes the first derivative with respect to the argument. To obtain the solutions for the constants C_1 and C_2, the following relation must be fulfilled (the determinant of the linear system of equation must be zero):

$$C_1C_2\frac{u}{a}K_p(w)J_p'(u) - C_1C_2\frac{w}{a}J_p(u)K_p'(w) = 0. \tag{5.57}$$

It is well known that for the Bessel functions is valid:

$$uJ_p'(u) = pJ_p(u) - uJ_{p+1}(u). \text{ and} \tag{5.58}$$

$$wK_p'(w) = pK_p(w) - wK_{p+1}(w). \tag{5.59}$$

The combination of eqs 5.57–5.59 finally gives:

$$\frac{J_p(u)}{uJ_{p+1}(u)} = \frac{K_p(w)}{wK_{p+1}(w)}. \tag{5.60}$$

From the last equation, one can obtain the solutions for the optical fiber fundamental modes. From the Bessel function properties, it is obvious that the right-hand side of eq 5.60 is always positive while the left-had side frequently change its sign. Therefore, there are certain values of the arguments u and w, and consequently certain values of the V number where the solution is possible.

In the following, we will analyze the case when $p = 0$ which is valid for the rotationally symmetric field across the fiber's cross section. Equation 5.60 in this case becomes:

$$\frac{J_0(u)}{uJ_1(u)} = \frac{K_0(w)}{wK_1(w)}. \tag{5.61}$$

By analyzing eq 5.61, one can notice that there are range alternations where the solutions can and cannot be found. The corresponding graphs of the left-hand and the right-hand sides of eq 5.61 are presented in Figure 5.7. To be able to find the solutions, both sides of eq 5.61 must have the same sign. Therefore, and according to Figure 5.7, the solutions exist if $0 \leq u \leq 2.405$, $3.832 \leq u \leq 5.520$, etc., that is, the ranges where the solutions are possible are marked with the gray color.

In the case when it is satisfied $p = 1$, according to eq 5.60, we have:

$$\frac{J_1(u)}{uJ_2(u)} = \frac{K_1(w)}{wK_2(w)}. \tag{5.62}$$

However, to refer to the case when $p = 0$ is satisfied, we will start from eqs 5.58 and 5.59, but write it down in a slightly different recurrence form as:

$$uJ_p'(u) = -pJ_p(u) - uJ_{p-1}(u). \text{ and} \tag{5.63}$$

$$wK_p'(w) = -pK_p(w) - wK_{p-1}(w). \tag{5.64}$$

If we combine the last equation with eq 5.57, we obtain:

$$-\frac{J_1(u)}{uJ_0(u)} = \frac{K_1(w)}{wK_0(w)}. \tag{5.65}$$

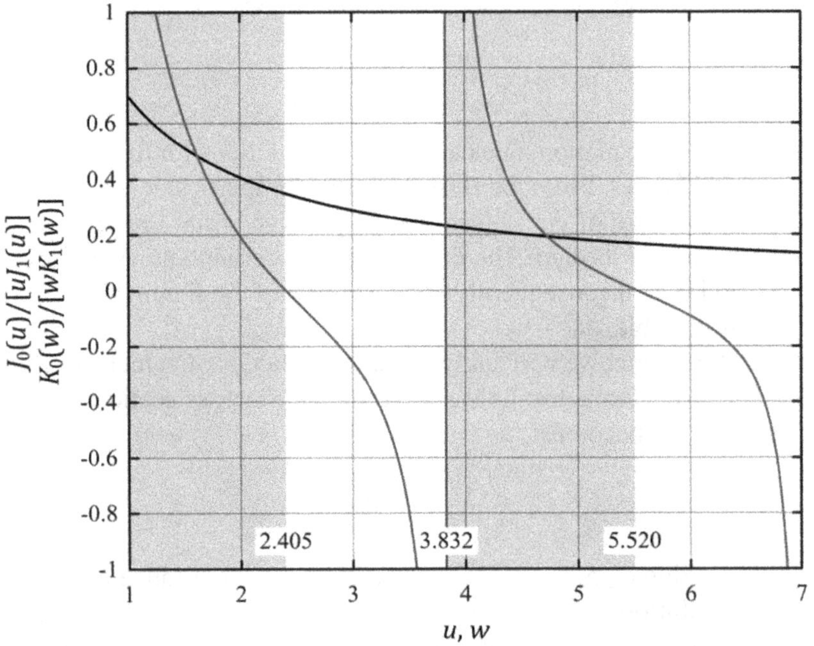

FIGURE 5.7 Graphs of the left-hand and the right-hand sides of eq 5.61.

By analyzing eq 5.65, one can also notice, like in the previous case, that there are range alternations where the solutions can and cannot be found. The corresponding graphs of the left-hand and right-hand sides of eq 5.65 are presented in Figure 5.8. To be able to find the solutions, both sides of eq 5.65 must have the same sign. Therefore, according to Figure 5.8, the solutions exist if $2.405 \leq u \leq 3.832$, $5.520 \leq u \leq 7.016$, etc., that is, the ranges where the solutions are possible are marked with the gray color.

Similar situation is for larger values of the parameter p, that is, the ranges with and without solutions are alternating. What can be concluded from the previous analysis is that for $V < 2.405$, there is only one branch of solution and for $V \geq 2.405$, there are initially two branches. The higher the value of the V number, the more branches come up.

The value of the V number of 2.405 represents the breaking point where there is more than one solution of the wave field distribution across the fiber cross section. Below this value, there is only one wave field distribution that can propagate along the fiber so it is said for the fiber to be single mode. This first mode is called the fundamental mode. Depending on the fiber geometry and index of refraction distribution along its cross section, one can calculate the fiber cutoff wavelength from the condition $V = 2.405$. If the wavelength of the propagating light is longer than the cutoff value, the fiber is single mode.

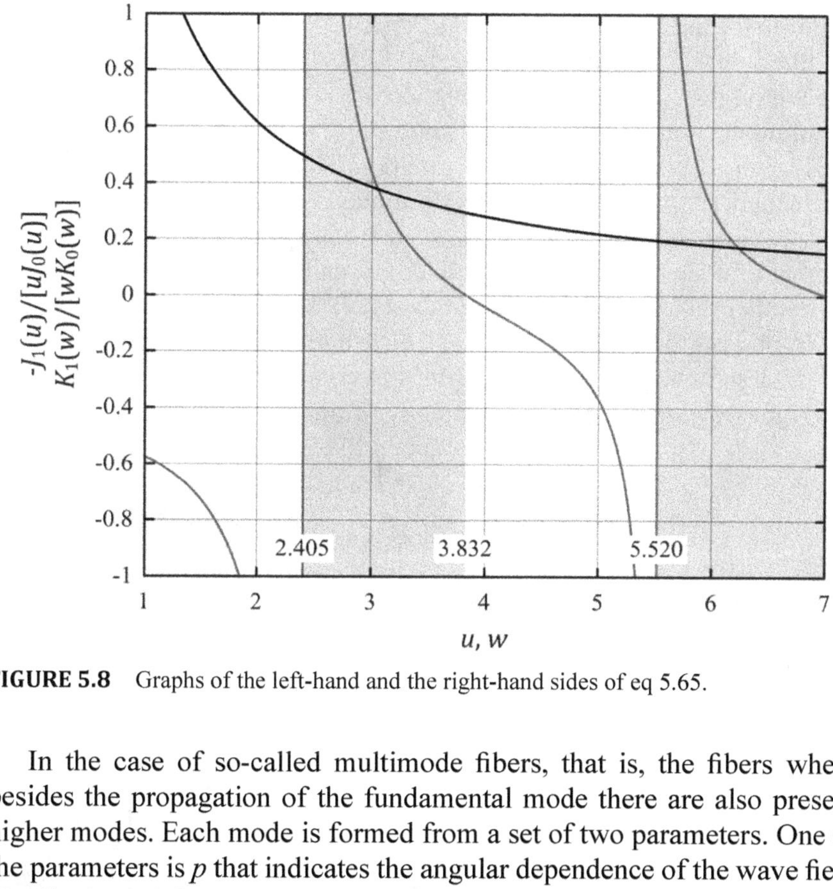

FIGURE 5.8 Graphs of the left-hand and the right-hand sides of eq 5.65.

In the case of so-called multimode fibers, that is, the fibers where besides the propagation of the fundamental mode there are also present higher modes. Each mode is formed from a set of two parameters. One of the parameters is p that indicates the angular dependence of the wave field distribution of the mode. If we have $p = 0$, the distribution is rotational invariant, that is, it has rotational symmetry. In the case where $p = 1$ is fulfilled, the wave field amplitude has a sine function dependence of the azimuth angle. Therefore, there are two zeroes at opposite positions, that is, zeroes are located along the diameter. Between the zeros, there are a positive and a negative branch or lobe. In the branches, there are located the wave field irradiance maximums. As the sign of the field phases is opposite in the lobes, the wave fields oscillate with the opposite phase.

Similarly for $p = 2$, there is four lobes as we have two full periods of the sine function around the fiber cross section. In general, we have $2p$ lobes for any value of the parameter p.

Parameter p designates that Bessel function describes the wave field distribution along the radial direction. It was found that Bessel function of the first kind Jp governs the field distribution within the core and the modified Bessel function of the second kind K_p governs the field distribution within the cladding. Bessel function of the first kind oscillates so there is unlimited number of ways to smoothly connect J_p to K_p for a given p. This set of possible solutions is marked with the parameter q, that is, the second parameter. Each possible mode is designated with "LP_{pq}" as they are essentially linearly polarized where the parameter p represents the number of node pairs in the azimuth angle, and the parameter q represents the number the possibilities along the radial coordinate.

To present the possible irradiance distribution within the fiber, Figures 5.9–5.11 show the various possibilities to connect the functions J_p and K_p in a right way and give an idea about the irradiance distribution.

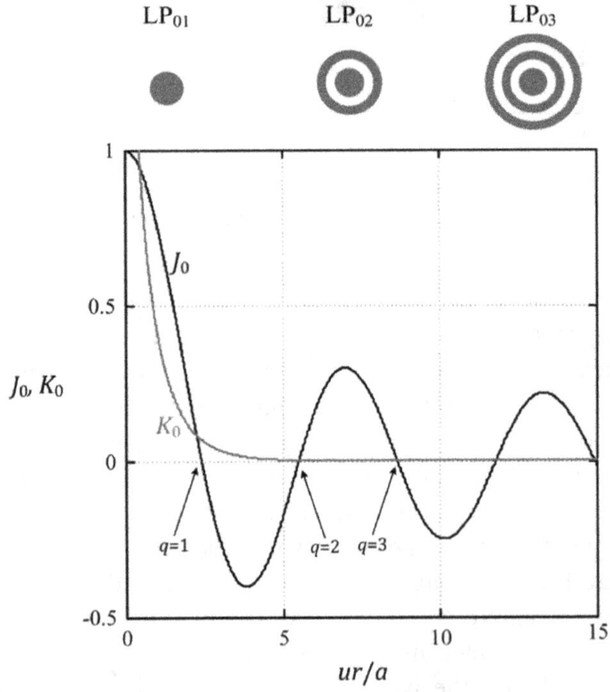

FIGURE 5.9 Radial irradiance distribution for modes with p = 0.

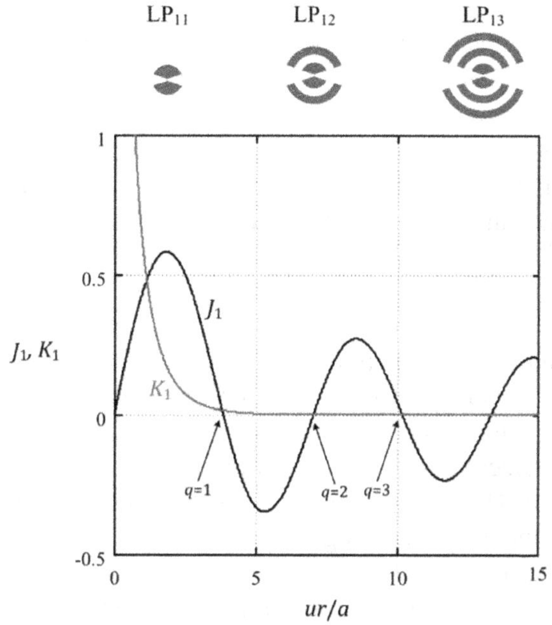

FIGURE 5.10 Radial irradiance distribution for modes with $p = 1$.

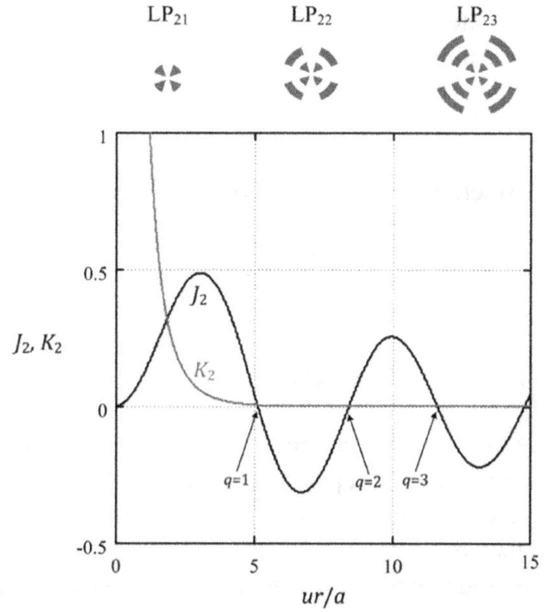

FIGURE 5.11 Radial irradiance distribution for modes with $p = 2$.

To illustrate the transition from a single-mode to a multimode fiber, we will consider a typical optical fiber with the following set of parameters: the fiber core radius a = 4 μm, the core index of refraction n_2 = 1.45, and the index difference between core and cladding $\Delta = 0.35\%$. It is well known that the fiber numerical aperture is given by $NA = n_2\sqrt{2\Delta}$, where further from the cutoff condition V = 2.405 and the definition of the V number, for the cutoff wavelength condition, we have:

$$\lambda_C = \frac{2\pi a n_2 \sqrt{2\Delta}}{2.405}. \tag{5.66}$$

By inserting the corresponding parameter values, we obtain λ_C = 1.268 μm. Therefore, this fiber is a single-mode fiber for all wavelengths longer than 1.268 μm, which then includes the 1.3 mm and the 1.55 mm range that is preferred in telecommunications. For the shorter wavelengths, the fiber is then multimode. The second mode starts to propagate for the cutoff wavelength that is obtained according to eq 5.66 where the value of V = 2.405 is changed with the value V = 3.832. One obtains the cutoff wavelength of 796 nm for the second mode. For the larger V number values, that is, the shorter wavelengths, one obtains more and more modes that can propagate.

For a given value of the V number, there is a certain number of modes that can propagate along the fiber. In the case of large V number, there is an approximate formula that estimates the number of mode as:

$$M \approx \frac{V^2}{2}, \tag{5.67}$$

for a step-index fiber. For a gradient-index fiber, it is slightly different:

$$M \approx \frac{V^2}{4}. \tag{5.68}$$

5.1.3 CHROMATIC DISPERSION

The dependence of the fiber core and cladding indices of refraction on the wavelength can affect the light propagation along the fiber. In the previous chapter, we have dealt with the phenomenon of modal dispersion where it was noticed that different modes travel with different velocities thus giving a rise to a modal dispersion phenomenon. The modal dispersion can severely affect the data transport at higher rates thus limiting the maximal allowable data rates where the signal deciphering is still achievable. Similar situation is if a multimode optical fiber is deployed in some kind of a measurement instrument that is based on a phase-sensitive detection, that is, the interferometer base measurement system. Typically, if the higher modes are present in the fiber-optic interferometer, the coherence length of the fiber propagating light is thus reduced that further can affect the measurement if the interferometer branches lengths are misbalanced.

Single-mode fibers successfully overcome the problem of the modal dispersion. However, there is still a chromatic dispersion that is also present in the single-mode fibers that can significantly influence the characteristic of the fiber-optic data links at very high rates. Therefore, the phenomenon of the chromatic dispersion will be analyzed in the following. The cause of the chromatic dispersion is certainly the very well-known phenomenon of the index of refraction of the glass material from which the fiber is made of dependence on the wavelength. As there is no ideally monochromatic light, that is, each signal's and thus the optical signal's spectrum is consisted of at least a very narrow range of frequencies, the chromatic dispersion occurs. Due to the limited spectral width, different spectral components propagate with different velocity giving a rise to differential transit time at the fiber end and thus to signal distortion. There are three contributing factors that are hidden behind the wavelength dependence of the refractive index collectively called chromatic dispersion:

- Material dispersion, whose contribution comes directly from the wavelength dependence of the glass material index of refraction. This type of dispersion is not specific to fibers only but can be generally found in any bulk glass and it is independent of the geometry.
- Waveguide dispersion, whose contribution comes from the particular geometry of the optical fibers where the ratio of optical powers that propagate through the fiber core and cladding depends on the wavelength. As the light spectrum has different wavelengths, we have a crossover from mostly core index to mostly cladding index, which results in a contribution to the wavelength dependence of the effective index.
- Profile dispersion, whose contribution comes from the normalized refractive index difference dependence on the wavelength. The main cause of this type of dispersion is that the core and cladding indexes of refraction do not vary in the same way. However, this type of dispersion is often much smaller than the material and waveguide dispersion.

To characterize the chromatic dispersion, it is useful to specify the amount of the effect per unit distance. Therefore, in general case, for the chromatic dispersion, as well as for the modal and polarization mode dispersion, we can write:

$$D = \frac{1}{L}\delta\tau. \tag{5.69}$$

where δτ represents the difference of propagation time after fiber length L. Unit that is commonly used is ps/km. In the case of the chromatic dispersion, which includes material, waveguide, and profile dispersion, we have:

$$D = \frac{1}{L}\frac{\mathrm{d}\tau}{\mathrm{d}\lambda}, \tag{5.70}$$

where the parameter D contains of three components:

$$D = D_{\mathrm{M}} + D_{\mathrm{W}} + D_{\mathrm{P}}, \tag{5.71}$$

representing the corresponding material D_{M}, waveguide D_{W}, and profile D_{P} dispersions. The chromatic dispersion represents the propagation time difference per distance and per wavelength difference. Therefore, the unit that is commonly used is ps/(nm · km).

If we consider a plane monochromatic wave with the angular frequency ω and the wave number β, it propagates along the fiber with phase velocity given by:

$$v_{\mathrm{p}} = \frac{\omega}{\beta}, \tag{5.72}$$

whereas the signal propagation velocity is governed by the group velocity given by:

$$v_{\mathrm{g}} = \frac{\mathrm{d}\omega}{\mathrm{d}\beta}. \tag{5.73}$$

Therefore, the signal time of travel will be:

$$\tau = \frac{L}{v_{\mathrm{g}}} = L\frac{\mathrm{d}\beta}{\mathrm{d}\omega} = L\frac{\mathrm{d}\beta}{\mathrm{d}\lambda}\frac{\mathrm{d}\lambda}{\mathrm{d}\omega}. \tag{5.74}$$

Since we have:

$$\beta = nk_0 = \frac{2\pi n}{\lambda}, \tag{5.75}$$

it is valid:

$$\frac{\mathrm{d}\beta}{\mathrm{d}\lambda}=2\pi\frac{\mathrm{d}}{\mathrm{d}\lambda}\left(\frac{n}{\lambda}\right)=\frac{2\pi}{\lambda^2}\left(\lambda\frac{\mathrm{d}n}{\mathrm{d}\lambda}-n\right), \tag{5.76}$$

and also:

$$\frac{d\lambda}{d\omega}=-2\pi\frac{c}{\omega^2}. \tag{5.77}$$

The combination of eqs 5.74–5.77 gives:

$$\tau=\frac{L}{c}\left(n-\lambda\frac{\mathrm{d}n}{\mathrm{d}\lambda}\right). \tag{5.78}$$

Equation 5.78 represents the wavelength dependence of the group (signal) propagation time along the fiber as a function of the glass index of refraction and wavelength. The term in the brackets represents the group index of refraction that is given as:

$$n_{\mathrm{g}}=n-\lambda\frac{\mathrm{d}n}{\mathrm{d}\lambda}. \tag{5.79}$$

Bearing in mind that according to the Sellmeier equation, the index of refraction decreases with the wavelength, we have $\mathrm{d}n/\mathrm{d}\lambda<0$ and thus $n_{\mathrm{g}}>n$. According to eq 5.78 for the scatter of arrival times at the fiber end, the following is obtained:

$$\delta\tau=\frac{\mathrm{d}\tau}{\mathrm{d}\lambda}\delta\lambda=\frac{L}{c}\delta\lambda\frac{\mathrm{d}}{\mathrm{d}\lambda}\left(n-\lambda\frac{\mathrm{d}n}{\mathrm{d}\lambda}\right)=-\frac{L}{c}\lambda\frac{\mathrm{d}^2n}{\mathrm{d}\lambda^2}\delta\lambda. \tag{5.80}$$

The corresponding contribution of the material dispersion D_{M} to the dispersion coefficient is given by:

$$D_{\mathrm{M}}=\frac{1}{L}\frac{\mathrm{d}\tau}{\mathrm{d}\lambda}=-\frac{\lambda}{c}\frac{\mathrm{d}^2n}{\mathrm{d}\lambda^2}. \tag{5.81}$$

Sometimes, especially in the cases when the signal occupies a broad spectral range and when the dispersion variation on the wavelength must be taken into account, it is of interest to introduce the dispersion slope as:

$$S_{\mathrm{M}}=\frac{\mathrm{d}D_{\mathrm{M}}}{\mathrm{d}\lambda}. \tag{5.82}$$

Besides the material dispersion, there is also waveguide dispersion in the fibers. For the step-index fiber, the waveguide dispersion, the following is valid:

$$D_W = -\frac{Vn_2\Delta}{c\lambda}\frac{d^2}{dV^2}(bV), \tag{5.83}$$

where we have:

$$b = \frac{\beta^2 - k_1^2}{k_2^2 - k_1^2}, \tag{5.84}$$

which is given without explicitly deriving. In the case where V number takes large values, the dimensionless quantity b tends to unity where at the cutoff of each mode there is $b = 0$ is valid.

The profile dispersion occurs when one takes into account that the normalized refractive index difference is not constant but depends slightly on wavelength. This contribution is typically small and can be neglected. In Figure 5.12, it is presented the typical overall chromatic dispersion ($D = D_M + D_W + D_P$) dependence on the wavelength.

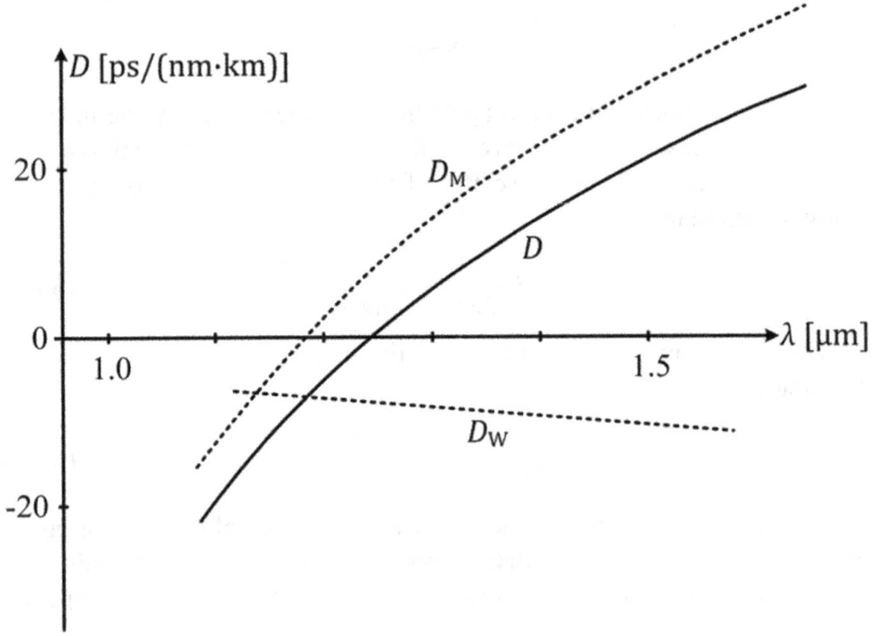

FIGURE 5.12 Total dispersion as a result from the material contribution of the core and from the waveguide contribution.

Further, we will analyze the impact of the chromatic dispersion on the propagation of a light pulse, which will be considered as a monochromatic wave multiplied, that is, modulated by a corresponding envelope function. Typically, a Gaussian function has been chosen for the envelope function. Therefore, the temporal profile of the pulse irradiance is:

$$I(t) = I_0 \exp\left[-\left(\frac{t}{T}\right)^2\right], \tag{5.85}$$

where I_0 is the peak irradiance and T the pulse duration. Sometimes as a pulse duration, the full width at half maximum (FWHM) has been taken into account. In the case of a Gaussian-shaped pulse, the relation between the FWHM τ and the pulse duration T is given by:

$$\tau = 2\sqrt{\ln 2}T. \tag{5.86}$$

When a Gaussian-shaped pulse propagates along the fiber length L, both pulse parameters duration and peak irradiance are modified. It is not difficult to obtain the following relation for the pulse duration:

$$T_{\mathrm{L}} = T\sqrt{1+\left(\frac{L}{l_{\mathrm{D}}}\right)^2}, \tag{5.87}$$

where for the dispersion length l_{D} is valid:

$$l_{\mathrm{D}} = \frac{T^2}{|\beta_2|}, \tag{5.88}$$

and where the β_2 parameter is obtained as:

$$\beta_2 = \left.\frac{\mathrm{d}^2\beta}{\mathrm{d}\omega^2}\right|_{\omega=\omega_0}, \tag{5.89}$$

that is, as the second derivative of the propagation constant $\beta(\omega) = \omega n(\omega)/c$ for the given central angular frequency ω_0. According to eq 5.87 after passing the fiber length, which is equal to the dispersion length, the pulse duration is increased by a factor $\sqrt{2}$. In the case of a very long fiber length, that is, when L ≫ lD is satisfied, one obtain the following pulse duration at the fiber end:

$$T_{\mathrm{L}} = \frac{|\beta_2|L}{T}, \tag{5.90}$$

where an interesting situation is obtained where the shorter the pulse duration at one end of the fiber, the longer the pulse duration at the other fiber end.

5.1.4 POLARIZATION MODE DISPERSION

Polarization mode dispersion is a result of the pulse broadening due to polarization of a fiber propagating optical pulse and the birefringence of the fiber. An optical pulse that travels along the fiber may consist of different polarization states where each state corresponds to a slightly different index of refraction. There are many reasons for different indices of refraction for different polarization states such as mechanical stress in the fiber, fiber bends, variations in the fiber structure, and the deviation from the circularity of the fiber. Therefore, the different polarization states experience different indices of refraction thus propagating with different velocities, causing pulse broadening and thus polarization mode dispersion (PMD). In Figure 5.13, the illustration of the polarization mode dispersion is presented.

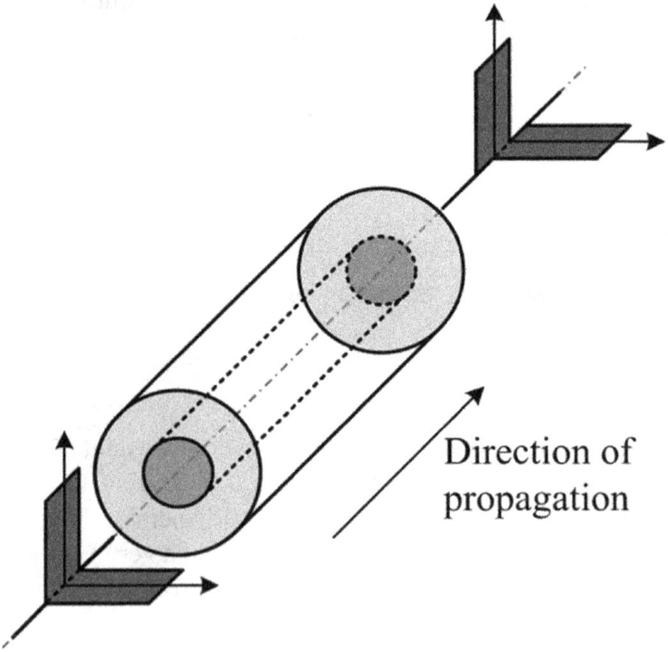

FIGURE 5.13 Polarization mode dispersion as a result of fiber birefringence.

We can consider that an optical pulse consists of two orthogonal polarizations. Due to the fiber birefringence, two modes travel at different group velocities. Therefore, they start to separate with the pulse propagation along the fiber. It should be pointed out that unlike other dispersion phenomena, the polarization mode dispersion has a random nature. The reason for this lies in the fact that PMD causes many different physical phenomena, which generally have random nature such as stress and temperature. However, PMD effects are small and become important at very high data rates (10 Gbps and above). As an effect of the PMD random nature, the delay between two orthogonal modes accumulates along the distance, but in opposite with other deterministic dispersions where the time effects are linearly added along the fiber, in the case of PMD, the addition of the time delay of each elementary fiber length is random and thus the total time delay is proportional to the square root of the fiber length:

$$\tau_{\text{PMD}} = D_{\text{PMD}}\sqrt{L} \tag{5.91}$$

where D_{PMD} is the polarization dispersion coefficient, given in ps/km$^{1/2}$, and L is the fiber length. The polarization dispersion coefficient is typically a function of the fiber construction and has the value somewhere in the range of 0.1 – 1 ps/km$^{1/2}$.

The polarization state of the propagating wave cannot be maintained along the standard fiber. Therefore, to maintain the polarization state, one can try to reduce the birefringence. However, this can be hard to achieve, even if the built-in tension can be suppressed during the manufacturing process, there is still no control over bending the fiber by the users. As it is not possible to suppress the birefringence to the negligible levels, one can try to do quite opposite by making the fibers intentionally with large birefringence. One can make such fibers by introducing an asymmetric core structure such as elliptic core or insert additional structural elements that break the fiber circular symmetry. By inserting elements with a slightly different thermal expansion coefficient, where during the cooling of the glass of the fiber, a mechanical stress has been built into the fiber, the symmetry of the fiber is broken. Typical structures of such polarization-maintaining fibers are presented in Figure 5.14.

5.1.5 FIBER LOSSES

As mentioned at the beginning of this chapter, there was a long road in reducing the fiber loses from the former value of 20 dB/km to the today's typical value of 0.2 dB/km, which represents the fundamental limit. This limit is determined by three factors:

- the material resonances in the ultraviolet,
- the material resonances in the infrared (IR), and
- the Rayleigh scattering as a result of the glass statistical structure.

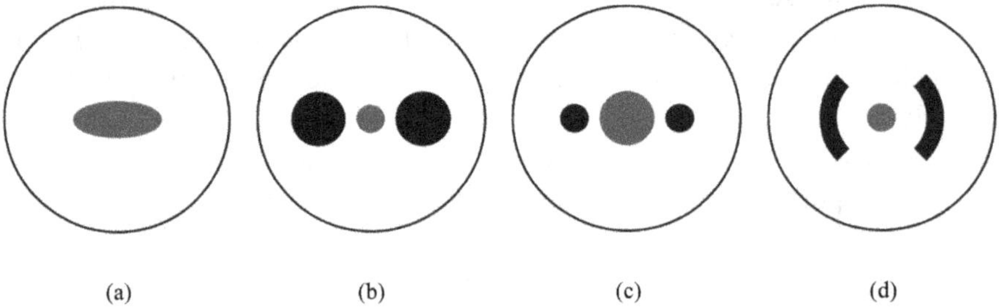

FIGURE 5.14 Polarization-maintaining fiber structures: (a) elliptical core, (b) panda, (c) pits, and (d) bowtie fiber.

It is rather difficult task to theoretically determine the lowest possible loss of an optical fiber. By considering the bulk glass data, one can derive that the minimal loss is 0.114 dB/km that corresponds to the energy loss of 2.6 % km. However, this theoretical level hasn't achieved yet, but the loss of 0.2 dB/km has been routinely obtained since the late 1980s.

In Figure 5.15 is presented the wavelength dependence of a standard optical fiber loss. One can notice that due to the impurity molecules in the glass there is a local loss peaks between the second and third windows at about at 0.95, 1.25, and 1.38 μm. This loss peaks are caused by the water molecules vibrations that can be found embedded in the glass. Besides the water, there are other impurities which can contribute the overall loss. Typically, among these impurities are metals Fe, Cu, Co, Cr, Ni, and Mn. For example, the copper (Cu) concentration of only 10^{-9} can cause loss of several tenths of dB/km.

As can be seen from Figure 5.15 IR absorption tail causes attenuation for the wavelengths longer than 1.6 μm. The lowest loss of about 0.18 dB/km is obtained for the wavelength region of 1.55 μm for the standard optical fiber. This loss is very close to the fundamental scattering limit.

5.2 FIBER-OPTIC COMMUNICATION SYSTEMS

Any fiber-optic communication system consists at least of three basic elements: optical fiber, optical transmitter (source), and optical detector (receiver). Therefore, their limiting parameters will determine the characteristics of any fiber-optic communication system, such as the highest possible data rate, longest possible optical links, and highest possible coverage. Depending on the service, which can be offered based on the given the fiber-optic communication system, we can distinguish different network topologies. Therefore, from the network topology point of view there are three basic categories: point-to-point links, distribution networks, and local-area networks, which will be in the focus of this chapter.

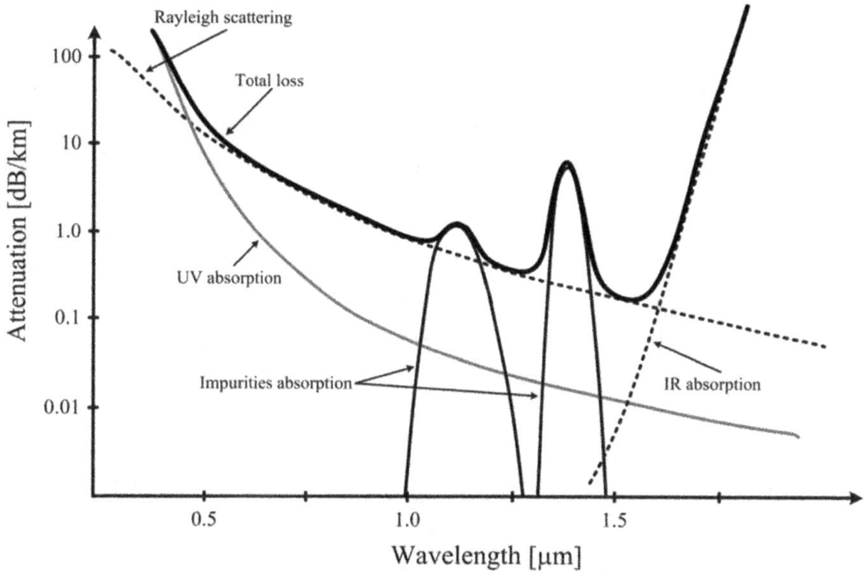

FIGURE 5.15 Wavelength dependence of a standard optical fiber loss.

5.2.1 POINT-TO-POINT LINKS

The simplest possible lightwave communication systems are certainly point-to-point links. Their main role is to transport information from one point to another, where the information, which is in the form of a digital bitstream, must be transported as accurately as possible. The length of such link varies from less than a kilometer, that is, the short-haul links to thousands of kilometers, that is, the long-haul links, which depends on the links actual application. As example for a short-haul link, we can use an optical data links used to connect the computers and terminals within the same building or between two relatively adjacent buildings. Such links do not require low loss and/or wide bandwidth of optical fibers. For a long-haul link, for example, undersea lightwave systems, the use of low losses and a large bandwidth of optical fibers are of paramount importance where the reducing the overall operating cost is one of the main issues.

If the data link lengths exceed a certain value, typically in the range of 20 – 100 km, which depends on the transmitter wavelength, it is necessary to compensate for the fiber losses. Otherwise, the signal becomes too weak to be detected with a certain given probability. To compensate the fiber losses there are two schemes that are commonly used that are presented in Figure 5.16. At the early stages of the optical links development, the regenerators are exclusively used for the loss compensations. The regenerator consists of the receiver–transmitter pair, which detects the incoming optical signal, recovers the electrical bitstream, and finally converts it back into optical form. Since the invention of the optical amplifiers, which has happened around 1990, fiber losses are compensated by using optical amplifiers. Optical amplifier amplifies the optical bitstream directly without converting it in the electric domain. Optical amplifiers are especially useful for the wavelength-division multiplexed (WDM) lightwave systems where they can manage the amplification of many channels simultaneously.

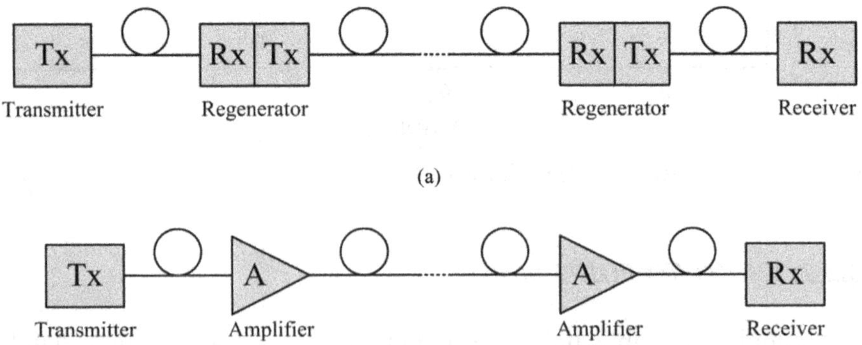

(a)

(b)

FIGURE 5.16 The compensation of point-to-point fiber links losses with the help of (a) regenerators and (b) optical amplifiers.

5.2.2 DISTRIBUTION NETWORKS

When we have a large number of customers to whom the information must not only be transmitted but also distributed the point-to-point network topology cannot provide such service. Typical examples are local-loop distribution of the telephone services and the broadcast of the video channels over cable television (CATV—**C**ommon-**A**ntenna **T**ele**V**ision). There are two topologies for the distribution networks that are presented in Figure 5.17. For the hub topology, the corresponding channel distribution occurs at the central locations, which are commonly named hubs. This type of network is commonly called metropolitan-area networks (MANs) as the hubs are typically located in the major cities. In this case, fiber role is similar as in the case of point-to-point networks. Telephone networks use hub topology for distribution of voice channels within a city.

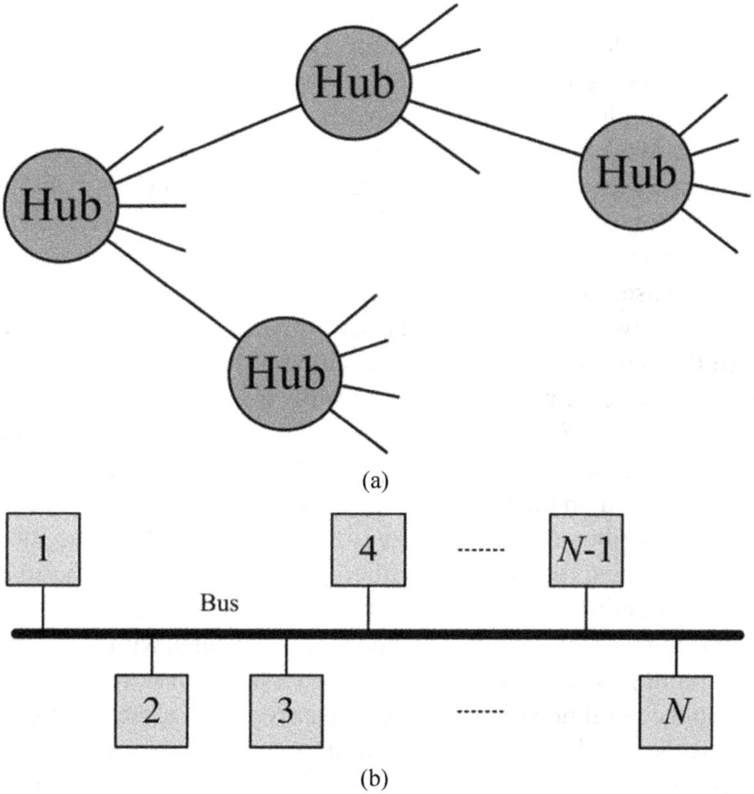

FIGURE 5.17 Distribution networks with (a) hub topology and (b) bus topology.

Bus topology is based on a single fiber cable that carries the multi-channel optical signal. The distribution is performed by using an optical tap, which redirects a small portion of the power of the optical signal to each subscriber. Typically, CATV systems use bus topology. The use of an optical fiber allows distribution of a large number of channels due to its large bandwidth compared with coaxial cables. Moreover, the distribution of the high-definition television (HDTV) channels requires the lightwave transmission due to a large bandwidth of each video channel. The main problem of the bus topology is that the signal loss increases exponentially with the number of used taps and thus limits the number of subscribers connected to the same optical bus.

5.2.3 LOCAL AREA NETWORKS

In the cases when there is a large number of users within a relatively small area, for example, a university campus or a large company facility, etc. where the users are interconnected in a way that each user can access the network randomly to transmit data to any other user, we have a so-called local area network (LAN). The transmission distances in the optical access networks of a local subscriber loop are relatively short (less than 10 km), so the fiber losses are not an issue for LAN applications. The advantage of using optical access over other techniques is the large bandwidth of the fiber optic–based communication systems.

There are two main topologies used for the LAN applications, the ring and the star topology that are presented in Figure 5.18. The adjacent nodes are connected by point-to-point links to form a closed ring in the ring topology case. Each node transmits and receives the data where the corresponding transmitter–receiver pair acts also as a repeater. The fiber-optic LANs based on the ring topology have standardized interface known as the fiber distributed data interface (FDDI), which operates at 100 Mb/s (data rate) by using the multimode fibers and the transmitters based on the 1.3 mm light-emitting diodes (LEDs).

Regarding the star topology, all nodes are connected with the help of the point-to-point links to a central node called a hub or a star. Depending on whether the central node is an active or passive device, the star topology–based LANs can be further divided into two groups: active-star and passive-star networks. The active-star configuration converts all incoming optical signals into the electrical domain through optical receivers. The corresponding electrical signal is then distributed to the node transmitters. In the case of the passive-star configuration, the distribution has been performed place in the optical domain through the directional couplers.

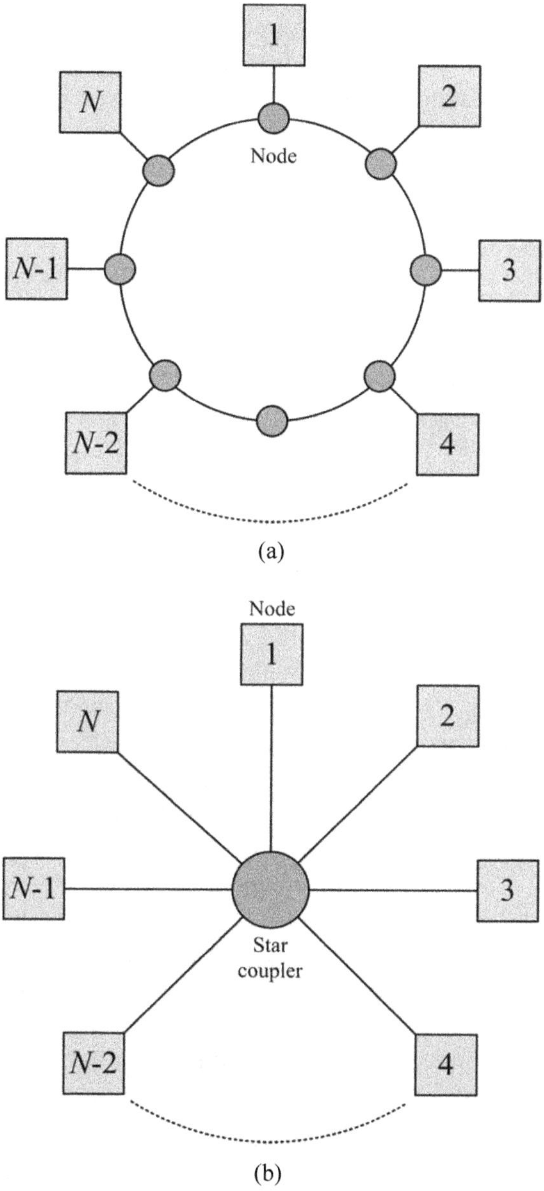

FIGURE 5.18 Local-area networks with (a) ring topology and (b) star topology.

5.2.4 FIBER-OPTIC NETWORK DESIGN CONSIDERATION

To design a fiber-optic communication system one requires a clear insight into the limitations that are imposed by the fiber loss, dispersion, and nonlinear effects. As the wavelength affects the fiber properties, the right choice of the operating wavelength is a major design issue. The two most important characteristics of any fiber-optic communication system are the bit rate and the transmission distance. Therefore, here it will be discussed how the fiber loss and dispersion affect the data rate and transmission distance.

Fiber losses play an important role in the system design except in the case of some short-haul fiber links. Let P_T be the optical power launched by the transmitter in the optical fiber. To decipher the data stream with the bit rate B, the receiver at the other end of the fiber-optic link requires a minimum average power P_R. The corresponding maximal fiber length between the transmitter and the receiver is given by:

$$L = \frac{1}{\alpha}\log_{10}\left(\frac{P_T}{P_R}\right), \tag{5.92}$$

where α is the fiber cable net loss (in dB/km), including splice and connector losses. Bearing in mind that $P_R = Nh\nu B$ where $h\nu$ is the corresponding photon energy and N is the average number of photons per bit required by the receiver to achieve the minimum bite error rate (BER), the length L decreases logarithmically as B increases at a given operating wavelength. If the transmitted power is taken to be $P_T = 1$ mW at $\lambda = 1.55$ μm, the minimum required number of photons per bit is $N = 10$ for a bit error rate of BER $= 10^{-9}$, so the maximal distance than can be reached with such a system is $L_{max} = 245$ km where the bit rate of B = 10 Gbps and fiber cable loss of $\alpha = 0.2$ dB/km was assumed.

To ensure that enough power will reach the receiver and provide reliable signal detection, a proper power budget of an optical link must be calculated during the design process. The receiver sensitivity represents the minimum average power required by the receiver, which must be provided by the transmitter with the given emitting power. The power budget is much simpler to present in decibel units with optical powers expressed in dBm units and thus we have:

$$P_T = P_R + C_L + M, \tag{5.93}$$

where C_L is the total channel loss and M is the system margin. With the system aging, the component parameters may degrade during their lifetime or other unforeseen events thus reaching the low levels of the receiver captured power, which can be even below its sensitivity. Therefore, the system margin introduces a certain reserve in the design where the parameter degradation cannot affect the receiver sensitivity. Typical value of the system margin that has been taken during the system design is 4 – 6 dB.

The channel loss C_L includes all possible losses along the link such as connector and splice losses, which can be written as following:

$$C_L = \alpha L + \alpha_C N_C + \alpha_S N_S, \tag{5.94}$$

where α_C and α_S are the connector and splice losses, respectively, and N_C and N_S are the connector and splice number along the link, respectively. Sometimes the splice loss is included with the specified fiber cable loss. The connector loss must include the connectors at the transmitter and receiver ends as well as other connectors if used within the fiber link.

Typically connectors for a multimode fiber have losses in the range of 0.2 – 0.5 dB per connector, whereas the connectors for the single-mode fiber, which are factory made and fusion spliced, will have losses of 0.1 – 0.2 dB. Field terminated single-mode connectors may have higher losses somewhere in the range of 0.5 – 1.0 dB.

Splicing the multimode fibers are typically performed with the mechanical splices, although there is a possibility of fusion splicing as well. Due to the large core fusion splicing has about the same loss as mechanical splicing. However, the fusion is more reliable in adverse environments. Typical value of multimode splice losses is in the range 0.1 – 0.5 dB. Fusion splicing of a single-mode fiber typically has less than 0.05 dB.

In the case of a step-index multimode fiber, eq 5.15 provides an approximation for the upper limit for the product of the maximal bit rate B and the fiber length L. According to eq 5.15, where the relation for the difference in time travel is given, we have:

$$B \cdot L = \frac{c}{n_2 \Delta}, \tag{5.95}$$

where for the values $n_2 = 1.46$ and $\Delta = 0.01$ even at a low bit rate of only 1 Mbps, we have that such multimode fiber-optic systems are dispersion-limited where their transmission distance is limited to below than 5km. Therefore, the multimode step-index fibers are very rarely used for designing the fiber-optic communication systems. Much better situation regarding the dispersion-limited fiber-optic communication system is when gradient optical fibers are used. In the case of a gradient multimode fiber, eq 5.18 provides an approximation for the upper limit for the product of the maximal bit rate B and the fiber length L. According to eq 5.18, where the relation for the difference in time travel is given, we have:

$$B \cdot L = \frac{2c}{n_2 \Delta^2}. \tag{5.96}$$

By comparing eqs 5.95 and 5.96, one has considerable improvement in the product of the maximal bit rate and the fiber length of two orders of magnitude if $\Delta = 0.01$ in the case of gradient fibers. The first generation of fiber-optic telecommunication systems took advantage of such an improvement and used graded-index fibers where the first commercial system operated at a bit rate of 45 Mbps with a repeater spacing of less than 10 km.

The second generation of the fiber-optic communication systems uses mainly single-mode fibers near the minimal dispersion wavelength of about 1.31 μm. The main limiting factor in these systems is the pulse broadening induced by dispersion due to the large source spectral width. For the product of the maximal bit rate and the fiber length, we have:

$$B \cdot L = \frac{1}{4|D|\Delta\lambda}, \tag{5.97}$$

where $\Delta\lambda$ is the source spectral width and D is the total dispersion of the fiber. The value of $|D|$ mainly depends on how close we are to the zero-dispersion wavelength of the fiber and is typically $\sim 1\,\text{ps/(km} \cdot \text{nm)}$. For the spectral width of the source of $\Delta\lambda = 2$ nm, we have $|D| \cdot \Delta\lambda \sim 2\,\text{ps/km}$, so finally is obtained $B \cdot L = 125$ Gbps · km. Therefore, the second-generation systems are mainly loss-limited up to the bit rates of about 1 Gbps.

The third and fourth generations of lightwave systems operate near 1.55 μm where the minimal fiber loss occurs. In this spectrum range, fiber dispersion becomes a major limitation as $D \sim 2$ ps/(km · nm) for the standard silica fibers. The relation that gives the limit for the corresponding bit rate and the fiber length of such a communication system is given by:

$$B^2 \cdot L = \frac{1}{16|\beta_2|}, \tag{5.98}$$

where β_2 is the second derivative of the propagation constant. Typically, the 1.55 mm operating ligthwave systems become dispersion-limited only for $B > 5$ Gbps.

5.2.5 COHERENT FIBER-OPTIC COMMUNICATION SYSTEMS

Coherent detection is based on the combining of the received optical signal coherently with a continuous-wave (CW) optical field, which is generated by the local oscillator LO, before it impinges the photodetector D as presented in Figure 5.19. At the receiver R output, a corresponding electrical signal is obtained. Local oscillator generates a CW field by using a narrow linewidth laser.

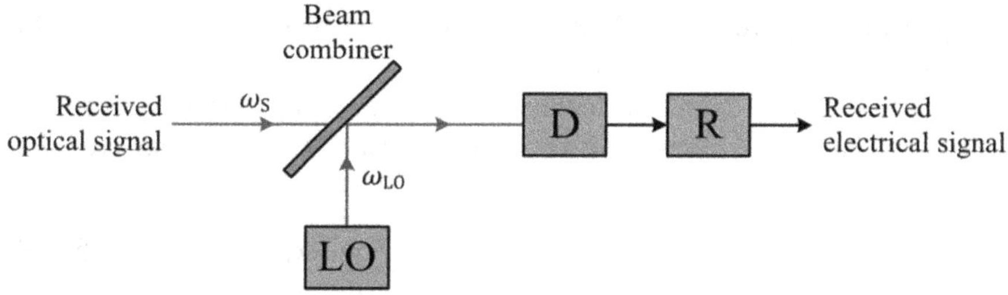

FIGURE 5.19 Coherent detection of an optical signal.

To gain an insight how the mixing of the received optical signal with the local oscillator output can improve the receiver performance, we will start from the following optical signals in complex domain:

$$E_S = E_{S0}\exp\left[-j\left(\omega_S t + \varphi_S\right)\right], \tag{5.99}$$

where ω_S is the carrier frequency, E_{S0} is the amplitude, and φ_S is the phase. The optical field associated with the local oscillator is given similarly as:

$$E_{LO} = E_{LO0}\exp\left[-j\left(\omega_{LO} t + \varphi_{LO}\right)\right], \tag{5.100}$$

where E_{LO0}, ω_{LO}, and φ_{LO} are the corresponding amplitude, frequency, and phase of the local oscillator, respectively. We will further assume that both fields E_S and E_{LO} have identical polarization. As the photodetectors are sensitive to the captured optical power, the optical power incident at the photodetector is given by:

$$P(t) = K\left|E_S + E_{LO}\right|^2, \quad (5.101)$$

where K is proportionality constant. Equations 5.99–5.101 further gives:

$$P(t) = P_S + P_{LO} + 2\sqrt{P_S P_{LO}}\cos\left(\omega_{IF}t + \varphi_S - \varphi_{LO}\right), \quad (5.102)$$

where we have:

$$P_S = \left|E_S\right|^2, P_{LO} = \left|E_{LO}\right|^2, \omega_{IF} = \omega_S - \omega_{LO}, \quad (5.103)$$

where ω_{IF} is the intermediate angular frequency. In the case when $\omega_S \neq \omega_{LO}$ is fulfilled, the optical signal is demodulated in two stages. In the first stage, the carrier frequency is converted to an intermediate frequency $\nu_{IF} = \omega_{IF}/(2\pi)$, typically in the range of 0.1 – 5 GHz, before the signal is demodulated to the baseband. Depending on the intermediate frequency value, there are two different coherent detection techniques. If the intermediate frequency is equal to zero, we have a homodyne detection technique and if not the heterodyne detection technique.

The homodyne detection technique has the local oscillator frequency identical with the corresponding signal carrier frequency, so we have $\omega_{IF} = 0$. In this case, eq 5.102 becomes:

$$P(t) = P_S + P_{LO} + 2\sqrt{P_S P_{LO}}\cos\left(\varphi_S - \varphi_{LO}\right), \quad (5.104)$$

where for the electrical current of the photodetector, we have:

$$i(t) = \Re P(t) = \Re\left[P_S + P_{LO} + 2\sqrt{P_S P_{LO}}\cos\left(\varphi_S - \varphi_{LO}\right)\right], \quad (5.105)$$

where $\Re$ is the detector responsivity. Typically, it is fulfilled $P_{LO} \gg P_S$, and $P_S + P_{LO} \approx P_{LO}$, where the last term in eq 5.105 carries the transmitted information. If the local oscillator phase is locked to the signal phase $\varphi_S = \varphi_{LO}$, then the homodyne signal is given by:

$$i(t) = 2\Re\sqrt{P_S P_{LO}}. \quad (5.106)$$

If compared with the direct detection of an optical signal, where the photodetector current is given by $i_{DD}(t) = \Re P_S(t)$, the homodyne detection provides an electrical current which power is increased by a factor $4P_{LO} / \langle P_S(t)\rangle$, where $\langle P_S(t)\rangle$ denotes the received optical signal average power. As the local oscillator power is much larger than the received optical signal power ($P_{LO} \gg P_S$), the power enhancement can exceed 20dB. Although the corresponding detector shot noise is also increased, the signal-to-noise ratio (SNR) is enhanced by a large factor.

One of the advantages of the coherent detection is hidden in the last term of eq 5.105 where the phase dependence of the photodiode current can be used to transmit the phase modulated signal. A disadvantage of the homodyne technique also originates from the phase sensitivity. As the phase of the local oscillator fluctuates, it can influence the overall system noise. However, the signal and local oscillator phase difference can be kept constant through an optical phase-locked loop (PLL) thus eliminating the corresponding local oscillator phase noise. It is worth to mention that the implementation of an optical PLL is not a simple task.

Heterodyne detection technique uses the local oscillator with the frequency that differs from the signal carrier frequency such that the intermediate frequency is in the microwave region ($\nu_{IF} \sim$ 1GHz). The photocurrent in the case of a heterodyne detection is given by:

$$i(t) = \Re\left(P_S + P_{LO}\right) + 2\Re\sqrt{P_S P_{LO}}\cos\left(\omega_{IF}t + \varphi_S - \varphi_{LO}\right). \tag{5.107}$$

As in the case of the homodyne detection, we also have the following conditions fulfilled: $P_{LO} \gg P_S$, and $P_S + P_{LO} \approx P_{LO}$. The DC component, which is contained in the first term of eq 5.107, can be easily removed with the help of a simple band-pass filter. In this case, the AC component of the photocurrent is given with:

$$i_{AC}(t) = 2\Re\sqrt{P_S P_{LO}}\cos\left(\omega_{IF}t + \varphi_S - \varphi_{LO}\right). \tag{5.108}$$

Information can be transmitted as in the case of the homodyne detection through the amplitude, phase, or frequency modulation of the optical carrier. The local oscillator amplifies the received signal by a large factor thus increasing the SNR. However, the SNR improvement is lower by a factor of approximately 3 dB if compared with the homodyne detection. The origin of the reduction in SNR can be found in eqs 5.106 and 5.108 where the average power of the heterodyne photocurrent is reduced by a factor 2 when $i_{AC}^2(\mathrm{t})$ is averaged over a full cycle at the intermediate frequency.

Although there is a 3 – dB penalty, the receiver design is much simpler as there is no need for the PLL. The fluctuations of the carrier and local oscillator phases should be controlled by using narrow linewidth semiconductor lasers for both sources. Therefore, this feature makes the heterodyne detection technique quite suitable for practical implementation in the coherent fiber-optic communication systems.

Considerable signal-to-noise enhancement is one of the most prominent advantages of the coherent detection techniques. To determine the signal-to-noise improvement, an analysis of a noise of such a system will be performed. There are two main noise sources of a receiver the shot noise and the thermal noise. The corresponding variance of the photocurrent fluctuation is given by:

$$\left\langle i_{\mathrm{n}}^{2}\right\rangle = \left\langle i_{\mathrm{nS}}^{2}\right\rangle + \left\langle i_{\mathrm{nT}}^{2}\right\rangle, \tag{5.109}$$

where $\left\langle i_{\mathrm{nS}}^{2}\right\rangle$ is the variance of the photocurrent fluctuation caused by the shot noise and $\left\langle i_{\mathrm{nT}}^{2}\right\rangle$ is the variance of the photocurrent fluctuation caused by the thermal noise for which is valid:

$$\left\langle i_{\mathrm{nS}}^{2}\right\rangle = 2q\left(i + I_{\mathrm{D}}\right)B, \tag{5.110}$$

$$\left\langle i_{\mathrm{nT}}^{2}\right\rangle = \frac{4k_{\mathrm{B}}TF}{R_{\mathrm{L}}}B, \tag{5.111}$$

where q is the proton charge, i is the photocurrent, I_{D} is the photodetector dark current, B is the receiver bandwidth, k_{B} is the Boltzmann constant, T is the receiver absolute temperature, F is the receiver noise factor, and R_{L} is the receiver load resistance. The corresponding SNR is obtained by dividing the average signal power by the average noise power. Therefore, in the heterodyne detection case, we have the following:

$$SNR = \frac{\left\langle i_{\mathrm{AC}}^{2}(t)\right\rangle}{\left\langle i_{\mathrm{n}}^{2}\right\rangle} = \frac{\Re^{2} P_{\mathrm{LO}} \left\langle P_{\mathrm{S}}(t)\right\rangle}{q\left(\Re P_{\mathrm{LO}} + I_{\mathrm{D}}\right)B + \left(2k_{\mathrm{B}}TF / R_{\mathrm{L}}\right)B}. \tag{5.112}$$

In the case of a homodyne detection, the SNR is larger by a factor of 2 if $\varphi_{\mathrm{S}} = \varphi_{\mathrm{LO}}$ is assumed. The power of the local oscillator power can be easily controlled and it can be made large enough that the receiver noise is dominated by shot noise, or $\left\langle i_{\mathrm{nS}}^{2}\right\rangle >> \left\langle i_{\mathrm{nT}}^{2}\right\rangle$, so we have:

$$P_{\mathrm{LO}} >> \frac{2k_{\mathrm{B}}TF}{q\Re R_{\mathrm{L}}}. \tag{5.113}$$

If the dark current contribution is negligible ($I_{\mathrm{D}} \ll \Re P_{\mathrm{LO}}$), the SNR is then given by:

$$SNR \approx \frac{\Re\left\langle P_{\mathrm{S}}(t)\right\rangle}{qB}. \tag{5.114}$$

5.3 BASIC CONCEPTS OF FIBER-OPTIC SENSING SYSTEMS

In parallel with the fiber-optic communication technology, the fiber-optic sensor technology becomes the major user of technology associated with the optoelectronic and fiber-optic communications industries. Moreover, many of these components were often developed for fiber-optic sensor applications. Along with the price reduction and the quality improvements of the optoelectronic and fiber-optic components, the possibility of substituting the traditional sensors for various physical quantities such as temperature, pressure, vibration, linear and angular position, strain, humidity, and concentrations for sensor applications has been enhanced. The advantage of the fiber-optic sensors over classical sensors one can see in their ability to be lightweight, of very small size, passive, resistant to electromagnetic interference, high sensitive, etc. In the following section, the basic types of fiber-optic sensors will be reviewed and followed by the description of how these sensors can be applied.

5.3.1 FIBER-OPTIC SENSOR BASIC TOPOLOGIES

Typically, fiber-optic sensors are roughly grouped into two basic classes usually called the extrinsic, or hybrid, fiber-optic sensors and intrinsic, or all-fiber, sensors. The simplified model of an extrinsic fiber-optic sensor is presented in Figure 5.20. The input fiber leads the light into the light modulator which imprints the information onto the propagating light beam as a response of environmental effects. The information can be stored in the intensity, phase, frequency, polarization, spectral content, etc. of a propagating wave. Finally, the output optical fiber further carries the impressed information toward an optoelectronic processing unit where the information can be extracted.

If the input optical fiber also acts as the output fiber and where the environmental effect impresses information onto the light beam while it is in the fiber, we have an intrinsic or an all-fiber sensor that is presented in Figure 5.21.

In its simplest possible form, the fiber-optic sensor is usually based on the intensity modulation. For example, a simple intensity modulated fiber-optic sensor aimed for the vibration measurement is presented in Figure 5.22 where a simple closure of the vibration sensor consists of two optical fibers held in close proximity to each other. The light, which input fiber emits in the free space, expands into a cone of light, the angle of which depends on the numerical aperture of the fiber. The amount of light, which is captured by the output fiber, depends on its acceptance angle (numerical aperture) and the distance *d* between the optical fibers. When the distance between them has been changed (due to the vibration for example), the power transmitted through the output fiber has been changed in the same manner thus giving the rise of the intensity modulation.

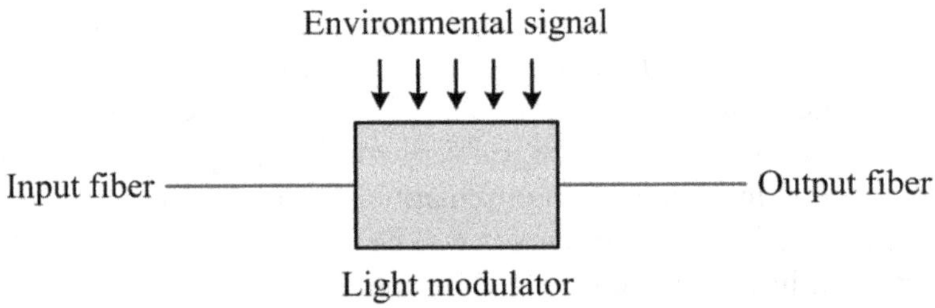

FIGURE 5.20 Extrinsic-fiber optic sensor.

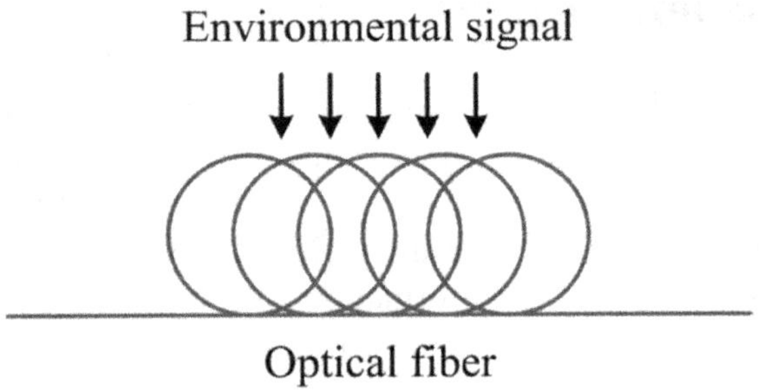

FIGURE 5.21 Intrinsic fiber-optic sensor.

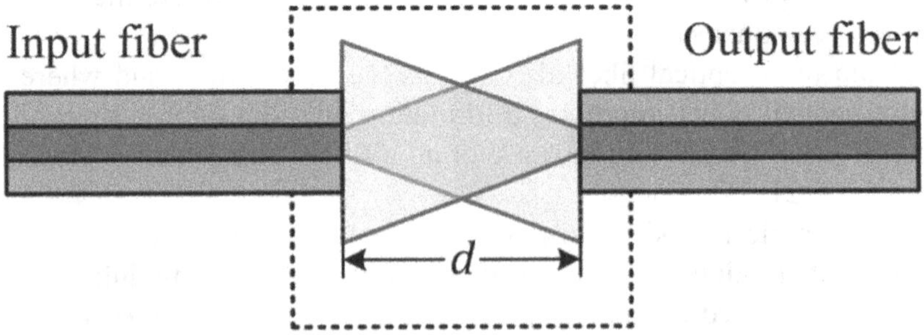

FIGURE 5.22 Simple fiber-optic sensor for vibration measurement.

Similar solution to the previous one is presented in Figure 5.23. Instead of an in-line fiber, here a mirror has been flexibly mounted to the object in which vibration is to be measured. As the mirror vibrates along with the object, the effective separation between the optical fibers shifts modulates the captured intensity.

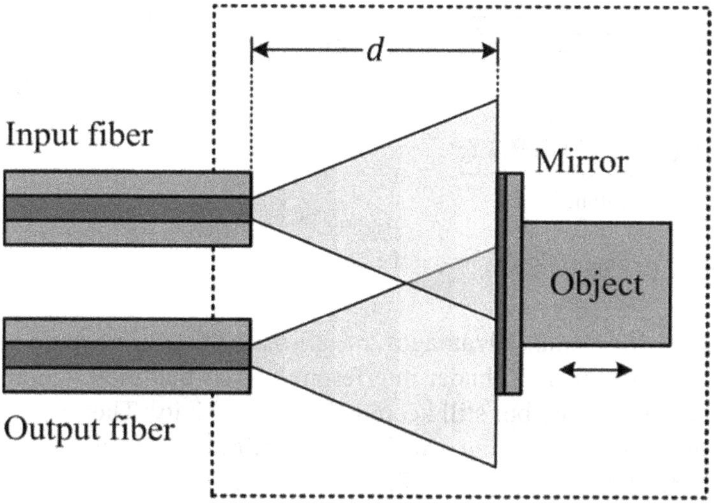

FIGURE 5.23 Simple fiber-optic sensor for vibration measurement by using a mirror.

To measure lateral displacement, a simple translation sensor can be configured as in presented in Figure 5.24. The input fiber, which has been illuminated with the help of a light source LS, illuminates further two output fibers. The light form the output fibers have been captured with the photodetectors PD1 and PD2. The photocurrents difference is proportional to the small translation displacements and thus we have:

$$i_{\mathrm{PD1}} - i_{\mathrm{PD2}} \propto d, \tag{5.115}$$

where i_{PD1} and i_{PD2} are the photocurrents and d is the lateral displacement.

5.3.2 BASIC CONCEPTS OF INTERFEROMETRIC FIBER-OPTIC SENSORS

Based on different interferometer topologies such as Michelson, Mach–Zehnder, ring resonator, and Sagnac interferometers, one can construct highly sensitive fiber-optic sensors aimed for measurement versatile physical quantities. Therefore, this section will be devoted to a brief review of different fiber-optic interferometer basic topologies for sensor applications and their main characteristics.

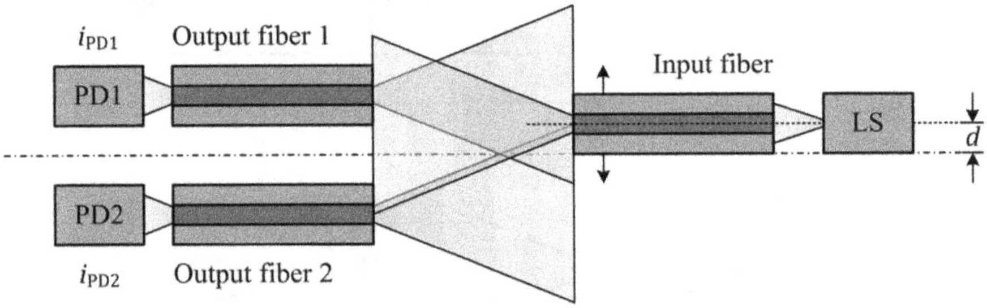

FIGURE 5.24 Simple fiber-optic translation sensor.

One of the main advantages of all-fiber interferometers, such as Michelson and Mach–Zehnder interferometers, is that their geometry can be extremely flexible but still keeping high sensitivity. Therefore, there is a possibility of constructing a wide variety of high-performance elements and fiber arrays, as presented in Figure 5.25 where are presented the linear array, the planar array, the gradient element, and the omnidirectional element.

To interrogate the phase change along the sensing fiber where the specifically designed sensing elements are employed, there is a variety of interferometric topologies. One of the most common is certainly the fiber-optic Mach–Zehnder interferometer that is presented in Figure 5.26. As presented, the fiber-optic Mach–Zehnder interferometer consists of a light source, a coupler module, a transducer, and a homodyne demodulator. Usually, the light source is an isolated laser diode with a long coherence length. The light from the source is divided by the beam splitter, that is, the coupler module to illuminate both arms of the interferometer. The transducer senses an environmental effect by exposing one of the interferometer arms to the measured parameter. In this way, the optical path difference between the arms corresponds to the physical quantity to be measured. Finally, a homodyne demodulator is used to measure the optical path length difference and thus the corresponding physical quantity.

The sensitivity of the Mach–Zehnder interferometers varies as a function of the relative phase of the light beams in two interferometer arms,

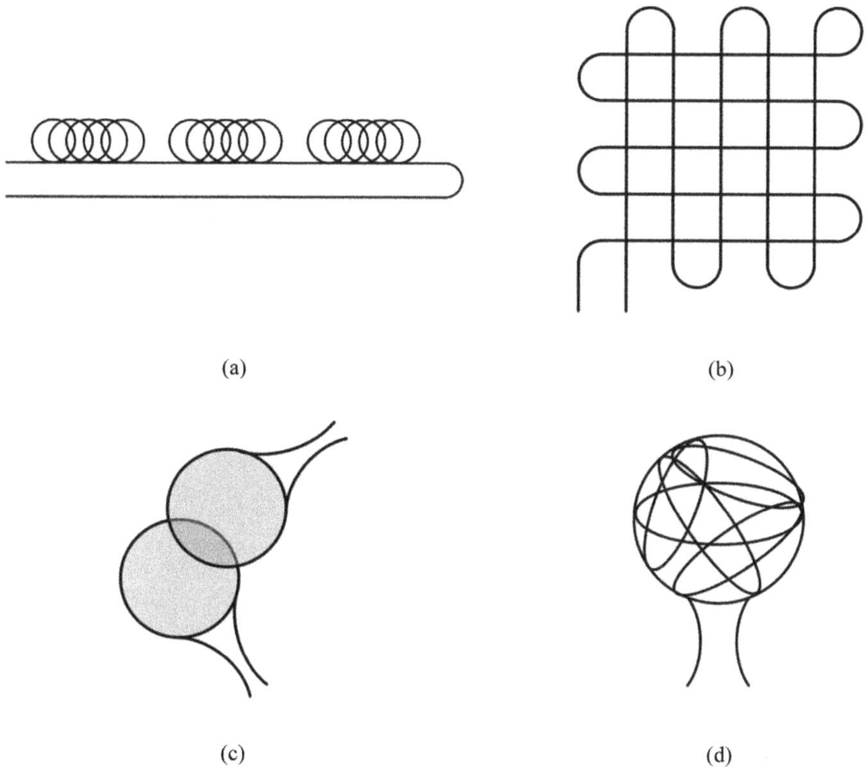

FIGURE 5.25 Flexible topology of interferometric fiber-optic sensors: (a) linear array, (b) planar array, (c) gradient element, and (d) omnidirectional element.

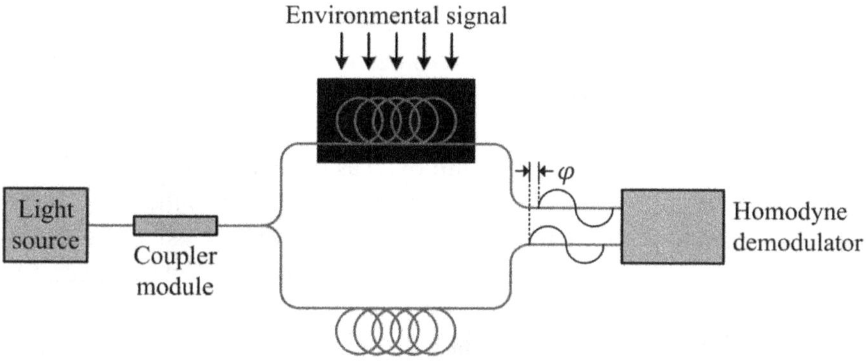

FIGURE 5.26 Fiber-optic Mach–Zehnder sensing interferometer basic elements.

as it is presented in Figure 5.27. Due to other interfering environmental parameters such as temperature, pressure, and humidity, which can cause the additional phase shift and thus the signal fading, one can introduce, for example, a piezoelectric fiber stretcher into one of the arms and adjust the relative path length of the two arms in a way to gain the highest possible sensitivity. If the interferometer is set to the highest possible sensitivity, it is usually said that the interferometer is in the quadrature.

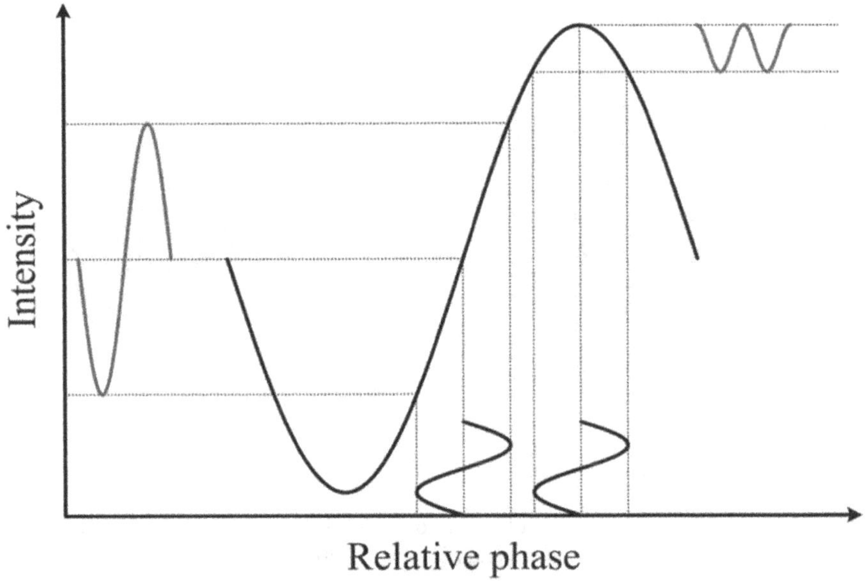

FIGURE 5.27 The sensitivity of the Mach–Zehnder interferometer.

The fiber-optic Michelson interferometer is in many respects similar to the Mach–Zehnder interferometer. In the case of the Michelson interferometer, instead of a homodyne demodulator, at the end of the interferometer arms, the mirrors have been employed as presented in Figure 5.28. As a result, a very high level of the light is back-reflected to the light source that can degrade the system performance. However, by using the diode-pumped YAG (***Y****ttrium* ***A****luminum* ***G****arnet*) ring lasers as light sources, this problem can be overcame. Moreover, if the phase conjugate mirrors are used instead of a standard mirror at the arms ends, the polarization fading can be suppressed.

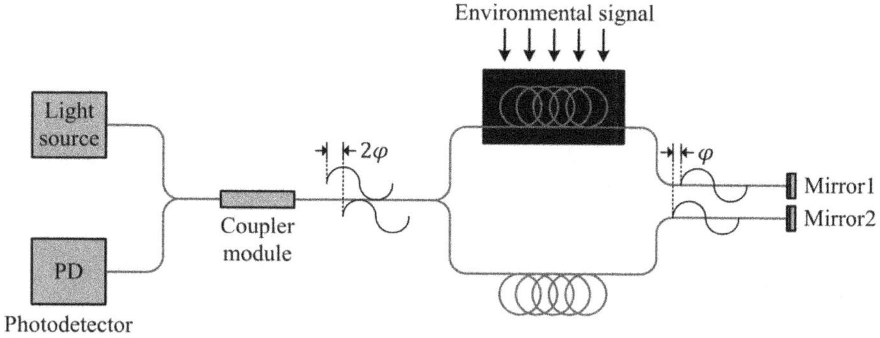

FIGURE 5.28 Fiber-optic Michelson sensing interferometer basic elements.

One of the most prominent characteristic of the fiber-optic sensors is their possibility to be multiplexed. Therefore, a large number of whether intrinsic or extrinsic sensor can be located along a single fiber-optic line. In other to address each sensor at the same fiber, several techniques such as time, frequency, wavelength, coherence, polarization, and spatial multiplexing can be employed. An example of multiplexed fiber-optic sensors has been presented in Figure 5.29 where the low-coherence light source (LCS) has been used to illuminate the sensors (S_k). This system uses the coherence multiplexing, where the corresponding back-reflected light spectrum from each sensor has been interrogated with the help of an optical spectrometer (OS).

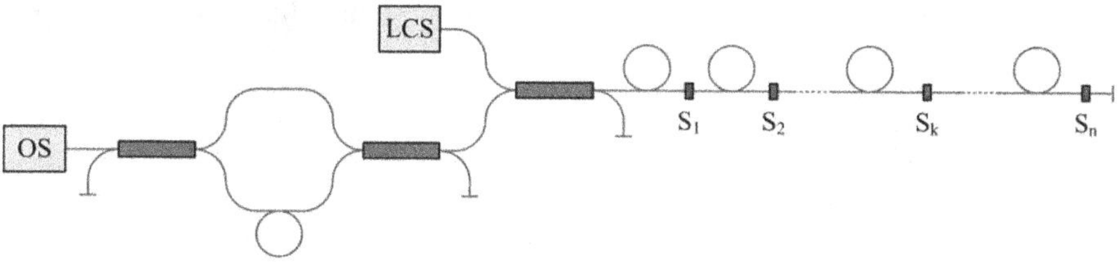

FIGURE 5.29 Multiplexing of fiber-optic sensors.

Typical physical quantities that can be measured with the multiplexed fiber-optic sensors are, for example, temperature, pressure, and strain. The measured parameter typically modulates the cavity length of the intrinsic or extrinsic fiber-optic sensor thus producing different modulation patterns of the channeled spectrum. When the light from all the sensors has been captured by the OS, each sensor can be extracted as it produces different modulation "frequency" of its channeled spectrum. Simply by applying the Fourier transform of the captured optical spectrum and finding the positions of the local peaks, one can measure the corresponding cavity length and thus measure the physical quantity.

KEYWORDS

- **extrinsic fiber-optic sensor**
- **intrinsic fiber-optic sensor**
- **fiber-optic communication system**
- **fiber optics**
- **fiber-optic sensing system**

REFERENCES

Agrawal, G. P. *Fiber-Optic Communications Systems*, 3rd ed.; John Wiley & Sons, Inc.: New York, NY, 2002.

Anderson, D. R.; Johnson, L.; Bell, F. G. *Troubleshooting Optical-Fiber Networks*, 2nd ed.; Elsevier, Inc.: Amsterdam, 2004.

Azadeh, M. *Fiber Optics Engineering*; Springer Science+Business Media, LLC: New York, NY, 2009.

Bass, M.; Van Stryland, E. W., Eds. *Fiber Optics Handbook: Fiber, Devices, and Systems for Optical Communications*; The McGraw-Hill Companies, Inc.: New York, NY, 2002.

Becker, P. C.; Olsson, N. A.; Simpson, J. R. *Erbium-Doped Fiber Amplifiers: Fundamentals and Technology*; Academic Press: San Diego, CA, 1999.

Brillant, A. *Digital and Analog Fiber Optic Communications for CATV and FTTx Applications*; Society of Photo-Optical Instrumentation Engineers: Bellingham, 2008.

DeCusatis, C. *Handbook of Fiber Optic Data Communication*, 2nd ed.; Academic Press: San Diego, CA, 2002.

DeCusatis, C., Ed. *Fiber Optic Data Communication: Technological Trends and Advances*; Academic Press: San Diego, CA, 2002.

DeCusatis, C. M.; DeCusatis, C. J. S. *Fiber Optic Essentials*; Elsevier Inc.: Amsterdam, 2006.

Dorf, R. C., Ed. *Broadcasting and Optical Communication Technology*; Taylor & Francis Group, LLC: Boca Raton, FL, 2006.

Elliott, B.; Gilmore, M. *Fiber Optic Cabling*, 2nd ed.; Newnes: Oxford, 2002.

Forestieri, E., Ed. *Optical Communication Theory and Techniques*; Springer Science + Business Media, Inc.: Boston, MA, 2005.
Glišić, B.; Inaudi, D. *Fibre Optic Methods for Structural Health Monitoring*; John Wiley & Sons Ltd: Chichester, 2007.
Goure, J-P.; Verrier, I. *Optical Fibre Devices*; IOP Publishing Ltd: Bristol, 2002.
Hecht, J. *City of Light: The Story of Fiber Optics: The Story of Fiber Optics*; Oxford University Press, Inc.: Oxford, 2004.
Iizuka, K. *Elements of Photonics Volume II For Fiber and Integrated Optics*; John Wiley & Sons, Inc.: New York, NY, 2002.
Kaminow, I. P.; Li, T. *Optical Fiber Telecommunications IV A – Components*; Elsevier Science, USA: San Diego, CA, 2002.
Kaminow, I. P.; Li, T.; Willner, A. E. *Optical Fiber Telecommunications V A Components and Subsystems*; Elsevier Inc.: Amsterdam, 2008.
Kaminow, I. P.; Li, T.; Willner, A. E. *Optical Fiber Telecommunications V B Systems and Networks*; Elsevier Inc.: Amsterdam, 2008.
Lam, C. F. *Passive Optical Networks: Principles and Practice*; Elsevier Inc.: Amsterdam, 2007.
Manojlović, L. M. *Optics and Applications*; Arcler Press: Oakville, Canada, 2018.
Measures, R. M. *Structural Monitoring with Fiber Optic Technology*; Academic Press: San Diego, CA, 2001.
Mitschke, F. *Fiber Optics: Physics and Technology*; Springer-Verlag: Berlin Heidelberg, 2009.
Noé, R. *Essentials of Modern Optical Fiber Communication*; Springer-Verlag: Berlin Heidelberg, 2010.
Powers, J. *An Introduction to Fiber Optic Systems*, 2nd ed.; McGraw-Hill Professional: New York, NY, 1993.
Senior, J. *Optical Fiber Communications*; Prentice-Hall International, Inc.: London, 1985.
Thyagarajan, K.; Ghatak, A. *Fiber Optic Essentials*; John Wiley & Sons, Inc.: Hoboken, NJ, 2007.
Tricker, R. *Optoelectronic and Fiber Optic Technology*; Newnes: Oxford, 2002.
Yin, S.; Ruffin, P. B.; Yu, F. T. S., Eds. *Fiber Optic Sensors,* 2nd ed.; Taylor & Francis Group, LLC: Boca Raton, FL, 2008.
Yu, F. T. S.; Yin, S., Eds. *Fiber Optic Sensors*; Marcel Dekker, Inc.: New York, NY, 2002.

CHAPTER 6

Low-Coherence Fiber-Optic Sensor Principle of Operation

ABSTRACT

The basic configuration of a fiber-optic sensor based on low-coherence interferometry consists essentially of two interferometers, one located at the measurement point and the other at the reading point. These two interferometers can be of different or the same type and are connected by one or two optical fibers. The principle of operation can be most simply introduced as in the case of two regularly coupled Michelson interferometers one being the sensing interferometer and one being the receiving interferometer. The optical path difference of the sensing interferometer, which changes under the influence of the some physical parameter, is measured by a receiving interferometer, in which a linear change of the difference of the optical paths is caused by a suitable scanning mechanism. At the moment when the optical path difference of the receiving interferometer becomes equal to the optical path difference of the sensing interferometer is recognized as the center of the left or right lateral coherence region in the interferogram. In this way, the indirect, remote reading of the optical path difference in the sensor interferometer gives a value that is uniquely related to the value of the physical parameter.

The basic configuration of a fiber-optic sensor based on low-coherence interferometry consists essentially of two interferometers, one located at the measurement point and the other at the reading point. These two interferometers can be of different or the same type and are connected by one or two optical fibers. The principle of operation can be most simply shown on the configuration of two regularly coupled Michelson interferometers, as shown in Figure 6.1.

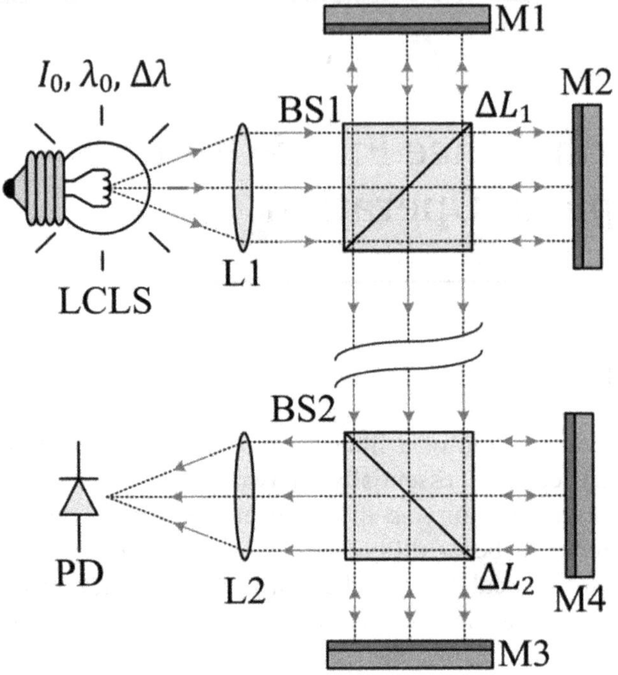

FIGURE 6.1 Two coupled Michelson interferometers with LCLS being the low-coherence light source; PD being the photodetector; L1 and L2 being the collimating lenses; BS1 and BS2 being the beam splitters; and M1, M2, M3, and M4 being the mirrors.

The first interferometer, which consists of the beam splitter BS1 and the mirrors M1 and M2, launches a low-coherence parallel light beam with the irradiance I_0 into the first interferometer. We will assume a Gaussian spectral distribution of the low-coherence light source (LCLS) with the central wavelength λ_0 and spectral width $\Delta\lambda$. The two beams are reflected from the mirrors M1 and M2, recombined at beam splitter BS1, and launched again into another interferometer, which consists of the beam splitter BS2 and mirrors M3 and M4. The optical path difference of the first interferometer ΔL_1 and of the second interferometer ΔL_2 in both cases is greater than the light source coherence length. To simplify the analysis without losing the generality, we will assume that the mirrors are perfect having the 100% reflectivity of all mirrors, the splitting ratios of both beam splitters are 50%: 50% and that all components are spectrally and polarization insensitive and do not have losses. The intensity of the radiation passing through both interferometers and falling on the photodetector (PD) is then given by:

$$I = \frac{I_0}{8}\left\{\begin{array}{l} 1+\left|\gamma_{11}\left(\Delta L_1\right)\right|\cos\left(k_0\Delta L_1\right)+\left|\gamma_{11}\left(\Delta L_2\right)\right|\cos\left(k_0\Delta L_2\right) \\ +\frac{1}{2}\left|\gamma_{11}\left(\Delta L_1+\Delta L_2\right)\right|\cos\left[k_0\left(\Delta L_1+\Delta L_2\right)\right] \\ +\frac{1}{2}\left|\gamma_{11}\left(\Delta L_1-\Delta L_2\right)\right|\cos\left[k_0\left(\Delta L_1-\Delta L_2\right)\right] \end{array}\right\}, \tag{6.1}$$

where $k_0 = 2\pi/\lambda_0$ is the wavenumber for the central wavelength λ_0, $|\gamma_{11}(\Delta s)|$ is the degree of coherence of two waves with the optical path difference of Δs. For the Gaussian light source, the degree of coherence can be written as:

$$\left|\gamma_{11}\left(\Delta s\right)\right| = \exp\left[-\left(\frac{\Delta s}{L_C}\right)^2\right], \tag{6.2}$$

where L_C is the LCLS coherence length.

It can be noticed that if the optical path difference in the interferometer is greater than the coherence length of the LCLS, the degree of coherence is very small. Then, the influence of members containing $|\gamma_{11}(\Delta L_1)|$, $|\gamma_{11}(\Delta L_2)|$, and $|\gamma_{11}(\Delta L_1 + \Delta L_2)|$ in eq 6.1 is negligible. The relatively pronounced interference effects occur, however, if the difference in the optical path difference of the two interferometers ΔL_1 and ΔL_2 is within the coherence length: $|\Delta s| = |\Delta L_1 - \Delta L_2| \leq L_C$.

In the position of the mirror of the second interferometer when $\Delta L_2 - \Delta L_1$ is fulfilled, the degree of coherence and contrast of the interferogram are the greatest. Following the photodiode signal as we move the mirror of the first interferometer, we will notice the maximum amplitude at the moment when the difference of optical paths in that interferometer becomes equal to the optical path difference in the second interferometer (and vice versa).

By measuring the optical path difference in one interferometer at the position when the contrast of the interferogram is the greatest, one obtains the information about the optical path difference in the second interferometer, which can be remote and connected to the first by the optical fiber.

In order for the two interferometer configuration to represent a sensor for different physical parameters, it is necessary to convert the physical parameter into the optical path difference in the interferometer located at the measuring site (sensor interferometer). In the second interferometer located at the reading point (receiving interferometer), a mechanism for scanning and measuring the position of the scanning mirror shall be provided. Measuring the position of the scanning mirror at the time of maximum signal indirectly determines the value of the physical parameter.

A typical interferometric signal obtained by moving one of the mirrors in the receiving interferometer, while keeping the optical path difference in the sensor interferometer of about 100 μm, is shown in Figure 6.2.

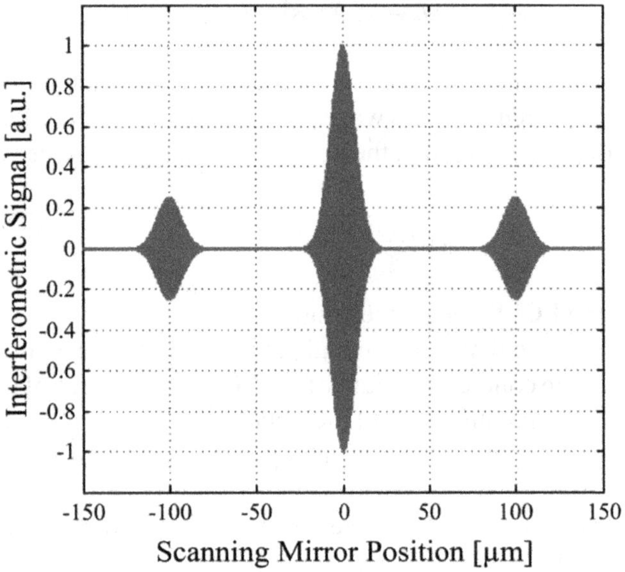

FIGURE 6.2 A typical interferogram obtained by moving one of the mirrors in the receiving interferometer. The optical path difference of sensor interferometer is 100 μm.

In Figure 6.2, three peaks are observed, that is, three areas of coherence:

- the central region with the highest peak, which corresponds to the zero optical path difference in the receiving interferometer, when $\Delta L_2 = 0$;
- left lateral peak, at −100 μm, corresponding to condition $\Delta L_2 - \Delta L_1 = 0$; and
- right lateral peak, at +100 μm, corresponding to the condition $\Delta L_1 - \Delta L_2 = 0$.

Due to the parity of the degree of coherence function, the interferogram is always symmetrical, except in the case of strong dispersion in fibers and/or interferometers.

A fiber-optic sensor having LCLS generally consists of a sensor and receiving interferometer that are interconnected by one or more optical fibers, as shown in Figure 6.3.

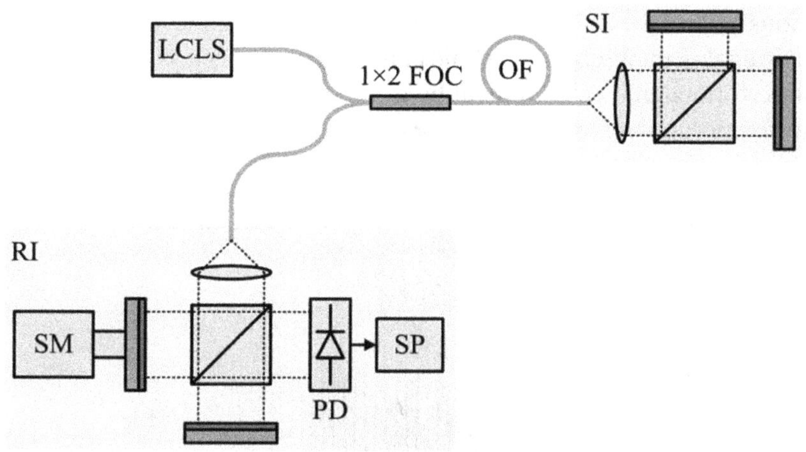

FIGURE 6.3 Typical fiber-optic sensing system with low-coherence light source: LCLS, Low-coherence light source; FOC, fiber-optic coupler; OF, optical fiber; SI, sensor interferometer; RI, receiving interferometer; SM, scanning mechanism; PD, photodetector; and SP, signal processing.

The optical path difference of the sensor interferometer, which changes under the influence of the some physical parameter, is measured by a receiving interferometer, in which a linear change of the difference of the optical paths is caused by a suitable scanning mechanism. At the moment when the optical path difference of the receiving interferometer becomes equal to the optical path difference of the sensor interferometer is recognized as the center of the left or right lateral coherence region in the interferogram. In this way, the indirect, remote reading of the optical path difference in the sensor interferometer gives a value that is uniquely related to the value of the physical parameter.

6.1 ALGORITHMS FOR SIGNAL PROCESSING OF LOW-COHERENCE INTERFEROGRAMS

The processing of a low-coherence interferogram signal means procedures applied to a basic interferometer signal, which determine the exact moment, or position of the mirror of a receiving interferometer, in which the optical path differences of the sensor and receiving interferometer are identical. This moment-position is exactly defined by the maximum intensity of the centerline in the low-coherence interferogram, as shown in Figure 6.4. Various techniques have been developed to determine it, ranging from simple analog methods of threshold comparison to very complex algorithms that perform signal development by different orthogonal functions, which require microprocessor signal processing.

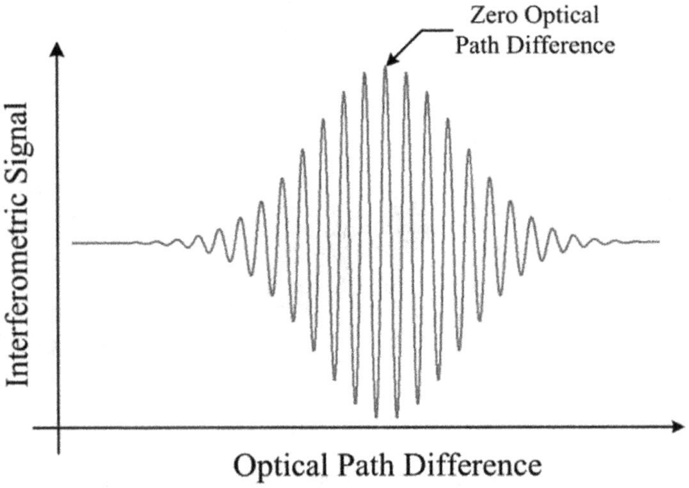

FIGURE 6.4 Typical signal form of a low-coherence interferogram.

6.1.1 *THRESHOLD COMPARISON METHOD*

Threshold comparison is a method based on the constant comparison of the signal with the preset threshold, which is set directly below the maximum value of the central interferometric line. The threshold can also be partially dynamically set, with constant measurement of the maximum value and adjustment of the threshold to a value, for example, a few percent/ppm (or millivolts) below that value.

This is the simplest method, which, however, can only be applied in cases with relatively large signal-to-noise ratios. If multiple light source combinations are used that give sharp envelope coherence zones, when the intensity difference between the central and adjacent interference fringes is large, a smaller signal-to-noise ratio may also be employed. The method enables very fast signal processing, which is performed in "real time", that is, immediately after the scan is complete, the position of the central maximum is known in the receiving interferometer. Therefore, it is possible to average more measurements over a shorter time interval, especially if the interferometer image is detected by a linear CCD array, which allows rapid scanning.

To calculate the required signal-to-noise ratio, we observe the difference between the amplitude value of the normalized interferogram signal in the central maximum V_0 and the amplitude in the left or right lateral maximum V_1 within the observed coherence region. For the Gaussian form of the coherence function, this difference is:

$$\Delta V_{01} = V_0 - V_1 = 1 - \exp\left[-\left(\frac{2\lambda}{L_C}\right)^2\right], \tag{6.3}$$

where L_C is the coherence length of the LCLS. We find that if the noise is less than the difference of the heights of the two maxima, the central maximum can be accurately determined by this method of comparison. The minimum signal-to-noise ratio is therefore given by:

$$\text{SNR}_{\min} = -20 \log_{10} \Delta V_{01}. \tag{6.4}$$

Based on this assumption, the theoretical curves as shown in Figure 6.5 show the minimum required signal-to-noise ratio as a function of the central wavelength and the coherence length of a Gaussian light source. The curves are shown for three wavelengths, which correspond to the usual so-called windows in fiber-optic communications, on which practically all commercially available light sources operate.

The accuracy with which the central maximum is determined depends directly on how far the level at which the comparison is made is below the level of the central maximum. The closer the comparison level is to the top, the higher the accuracy. However, the required signal-to-noise ratio is then higher, so a trade-off must be made. The hysteresis introduced into the comparator, which is necessary if the processing at this stage is done at an analogue level, also directly affects the accuracy of the measurement.

The precision of this technique is high if the signal-to-noise ratio is sufficient to allow the comparator to accurately identify the required central maximum. The conditions of linear scanning and position reading in the receiving interferometer are, as with other methods and algorithms, implied. Comparator hysteresis in analog processing can affect precision, but in a predictable way. This influence can therefore be eliminated by correction factors, the introduction of which results in the impairment of the basic advantages of this technique: complication and slowing down of signal processing.

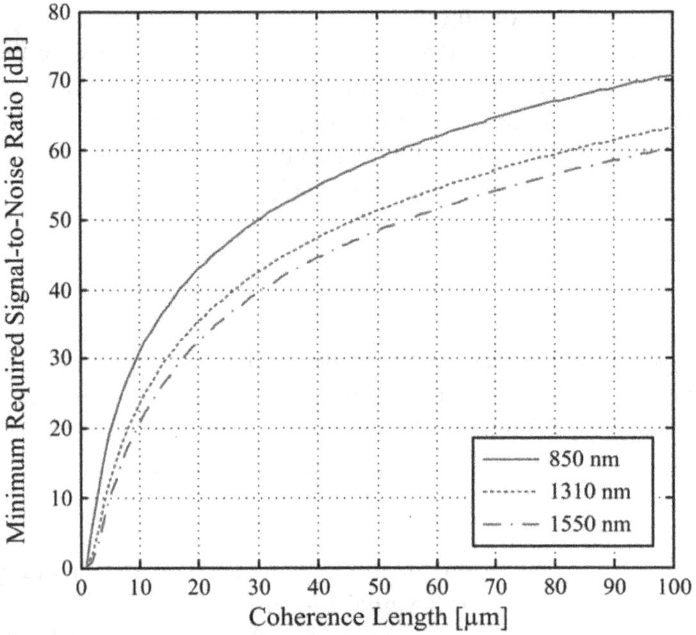

FIGURE 6.5 The minimum signal-to-noise ratio required to apply the threshold comparison method as a function of light source coherence length. The parameter is the wavelength of radiation.

Very similar to this method is a technique called fringe visibility measurement, which can have a very high resolution but is still based on the detection of the intensity difference of two adjacent fringes and therefore also requires a large signal-to-noise ratio.

6.1.2 *ENVELOPE COHERENCE FUNCTION METHOD*

One of the most natural ways to solve the problem of finding the center of the coherence zone is to approximate the maximum values in a low-coherence interferogram with an envelope corresponding to the coherence function. Since the form of the coherence function is known in advance, fitting comes down to finding two parameters: one that defines the required center of the coherence zone and the other that is related to the intensity of the interferogram. In some measurement configurations, in which there is no change in the intensity of the measuring beam with the measured size, even the latter parameter can be considered constant.

To obtain the required envelopes, local maxima must first be isolated in a low-coherence interferogram. At local maximum points, the first derivative of the interferogram curve is zero and the second derivative has a negative value. Using this criterion, the pairs of coordinates in the basic interferogram signal are extracted, which now constitute a new data set. The points thus obtained are then anchored in the form of the coherence function. The sum of squared deviations is used as a criterion for finding the envelope parameters that best approximate the measurement points.

If the pairs of real obtained measurement points are (x_d, y_d), where x_d is the position on the interferogram and y_d is the intensity of the local maximum in the interferogram, the required parameters (vector K) are obtained by finding the minimum of the sum of the square deviations:

$$\min_K \left\| \sum_i \left[F(K, x_{di}) - y_{di} \right]^2 \right\|. \tag{6.5}$$

6.1.3 *CENTROID ALGORITHM*

The centroid algorithm involves finding the center of gravity of the interferogram. The algorithm can be combined with other procedures, primarily by identifying the closest local maximum in the interference pattern. Also, the distribution of light intensities in the vicinity of the central maximum can be approximated by the best sine function (multiplied by the envelope of the coherence zone) and thus more accurately determine the position of the central maximum.

The centroid algorithm and the least-squares algorithm require a signal-to-noise ratio of less than 20dB than that required for the threshold comparison methods, which are based on the intensity differences of the fringes. Newer modifications to the least-squares algorithm allow very fast signal processing with very low signal-to-noise ratio.

Denote the distribution of light in the interferogram by $I(x)$, where x is the spatial or temporal coordinate in the interferogram obtained either by mechanical scanning or by reading the signals of a linear CCD sensor. The center of gravity of this image is obtained from:

$$\overline{x} = \frac{\int_{-\infty}^{+\infty} xI(x)\,\mathrm{d}x}{\int_{-\infty}^{+\infty} I(x)\,\mathrm{d}x}. \tag{6.6}$$

If the interferogram is obtained by a linear CCD array consisting of N pixels, or by sampling with A/D converters at N points during mechanical scanning, the integrals in eq 6.6 go into summation at those points:

$$\overline{x} = \frac{\sum_{i=1}^{N} x_i I_i}{\sum_{i=1}^{N} I_i}. \tag{6.7}$$

The number of points by which summation is performed usually ranges from 512 to 2048 for CCD linear sensors, while for mechanical scanning it can be much higher, of the order of over 10,000.

The main advantage of this method is that the result is obtained in a very short time. The sums in eq 6.7 are found immediately during the scan, that is, CCD array readings. At the end of the scan, only two more numbers are needed. Local maximum identification is also performed relatively quickly, as local maximum positions are determined during scanning, in the order in which they appear.

For multicoherence zones interferogram, the center of gravity computation area must first be subdivided into several zones, where some coherence zones are expected to emerge. While this may seem like a huge problem at first glance, the position of the central maximum (which is usually immobile) and the extent to which the lateral maximum is expected to occur are most commonly known in practice. It is not uncommon for scans to be performed only in the zone of the lateral maximum, or that only the zone in which it moves in a given dynamic range is formed on a CCD linear array.

6.1.4 ALGORITHM WITH PHASE-SHIFTED INTERFEROGRAMS

Classical phase-shifted interferometric algorithms, such as algorithms with three, four, or five shifts, can also be applied to a set of low-coherence interferograms. Interferograms can be obtained by temporal, spatial or so-called parallel phase shifts. An efficient nonlinear algorithm has been demonstrated, which is very suitable for the rapid calculation of time phase–shifted two-dimensional interferograms, such as appear with, for example, surface profiling. Given that a set of phase-shifted algorithms is obtained by special, rather complex techniques, and as this issue has been widely addressed in the literature, this type of algorithm is not specifically analyzed here.

6.1.5 WAVELET TRANSFORM ALGORITHM

Developing signals by different orthogonal functions is a powerful mathematical tool that is widely used in signal processing. The most famous and, in signal processing, certainly the most applied technique is Fourier analysis. Developing a signal in the Fourier series gives the frequency content of the signal, which is very useful in many applications. Fourier's development, however, loses information about the timing of the signal. If an event occurs within the total time interval to which Fourier development applies (e.g., the appearance of an impulse), it will be very difficult to determine exactly when that event occurred. Reducing the interval at which the analysis is performed increases the accuracy and precision of the location of an "irregular" occurrence in the signal, but this has its disadvantages if longer lasting phenomena are also observed.

Although at first glance it seems that Fourier transform is absolutely unsuitable for processing interferogram signals, there are a number of authors who have successfully applied this technique. For example, one uses the fast Fourier transform (FFT) algorithm to analyze classical interferogram signals. Applying the Fourier transform to the signal of a low-coherence interferogram, a resolution of 20 nm in the dynamic range of 140 μm was achieved.

Algorithms based on the generation of quadrature signals by a phase-generated carrier (e.g., variant with homodyne detection or synthetic heterodyne detection), or by diffraction grating, have been successfully adapted for low-coherence interferometric signals. The frequency spectrum of interferogram signals with a sinusoidal phase-generated carrier was also analyzed. By processing the signal according to an algorithm based on the relationships among the frequency components, a system resolution of up to 4 pm is predicted. However, the scan is based on the movement of the mechanical part, which must be made with very high precision and very stable motion.

Many of the disadvantages of Fourier analysis, in particular the inability to locate events, which is the main task in processing a low-coherence interferogram, are overcome by wavelet transform application. This transformation has previously been successfully applied generally in optics, interferometry, and low-coherence interferometry.

The theory and techniques of wavelet transformations have been given in detail in numerous literatures where only the basic concepts will be outlined here. Wavelet is a term we associate with a time-bound waveform of very different forms, which has to satisfy certain conditions to be called "wavelet". Such are, for example, conditions that its shape is continuous, that it has passes through zero, and that it tends to zero when the time coordinate is increasing to one side or the other, etc. There are a number of different wavelet forms, the most famous of which are Daubechies, biorthogonal, Meyer, Coiflet, etc.

Wavelet analysis consists of developing a test signal in these forms that are scaled in duration, that is, more or less stretched along the time axis relative to the basic shape. The differently stretched wavelet forms shift over time with respect to the signal being tested, and the mutual product of the two forms integrates over the entire time axis. This is how the so-called wavelet coefficients of the test signal $f(t)$ are obtained:

$$C_f(a,b) = \int_{-\infty}^{+\infty} f(t)\Psi_{a,b}(t)\,\mathrm{d}t, \tag{6.8}$$

where parameter a is related to wavelet form scaling by time axis, while parameter b represents wavelet form position on time axis, that is, the time position at which the signal $f(t)$ is observed. The basic wavelet form is:

$$\Psi_{a,b}(t) = \frac{1}{\sqrt{a}}\Psi\left(\frac{t-b}{a}\right). \tag{6.9}$$

The wavelet coefficient $Cf(a,b)$ quantifies the similarity of the test signal $f(t)$ at each of its time instants with the wavelet form of the corresponding scale. Wavelet coefficients are in the general case of complex form, with modulus $M_{a,b}$ and phase $\Phi_{a,b}$. The modulus corresponding to the particular wavelet form scale, defined by parameter a, shows the degree of overlap of this waveform with the observed signal at time instant b. The wavelet coefficient phase represents the relative phase difference between the wavelet form of a given scale with the observed signal, at time instance b.

6.2 A MODIFIED CENTROID ALGORITHM

As stated in Section 6.1.3 for the centroid algorithm, the main advantage of this algorithm is the high-speed signal processing as well as the relatively low required signal-to-noise ratio to provide high accuracy in identifying the central maximum of the interferogram. These two reasons usually determine the choice of the centroid algorithm as the basic algorithm for measuring the optical path difference in low-coherence interferometry based on monitoring the center of the coherence zone in real time. The center of the coherence zone is monitored on the basis of the error signal, which, on the other hand, is obtained by analogous processing of a low-coherence interference signal, as shown in Figure 6.6.

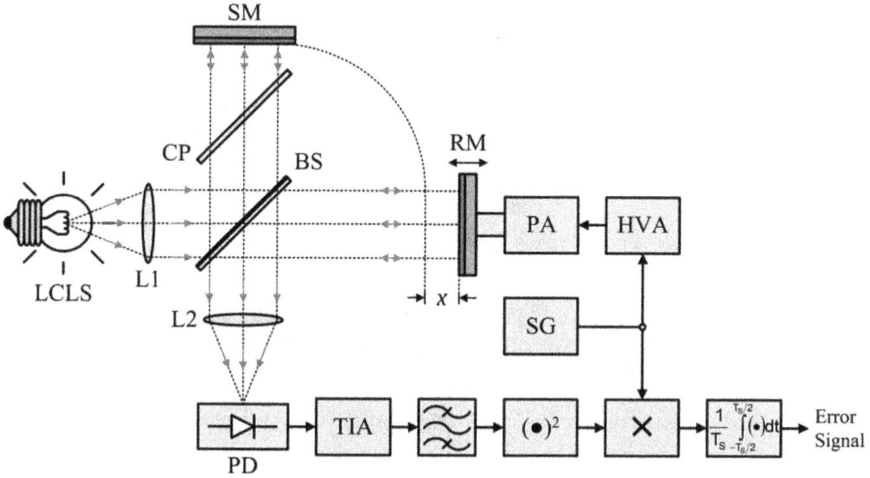

FIGURE 6.6 Finding error signals using analog processing of a low-coherence interference signal. LCLS, low-coherence light source, L1 and L2 lenses; CP, compensation plate; BS, beam splitter; SM, sensor mirror; RM, reference mirror; PA, piezo actuator; HVA, high-voltage amplifier; SG, signal generator; PD, photodetector ; TIA, transimpedance amplifier.

The principle of operation of analog signal processing for the purpose of obtaining an error signal is based on the constant scanning of a part of the coherence zone of a LCLS using a piezo actuator (PA) in the reference arm of the Michelson interferometer. PA, alternately scanning the reference mirror (RM) using a high-voltage amplifier (HVA) the inputs of which are fed a periodic signal (sawtooth, triangular, sinusoidal, or other suitably selected periodic voltage signal) of frequency fs with zero mean from the signal generator (SG). On the other hand, the low-coherence interference current signal obtained on the PD is amplified by a low-noise transimpedance amplifier (TIA). The amplified voltage signal is then filtered using a bandpass filter to eliminate the DC component and suppress noise. The filtered voltage signal is further processed by a squaring circuit (or some other rectifying circuit such as an absolute value circuit) and the signal so processed is multiplied by the reference signal from the SG. From the output of the multiplier circuit, a voltage signal is fed to the input of the integration circuit, at the output of which an error signal is obtained that is proportional to the optical path difference of the two arms x (sensor and reference) of the Michelson interferometer. The optical path difference shown in Figure 6.6 is valid $x = l_S - l_R < 0$, where l_S and l_R are the optical path lengths of the sensor and reference arms, respectively. To gain a better understanding of the principle of obtaining an error signal proportional to the optical path difference of the two arms of the Michelson interferometer in Figure 6.7, the time shapes of relevant voltage signals at characteristic points are given.

As the simplest form of voltage signal from the SG, a sawtooth voltage signal shape was taken into account. The corresponding reference signal linearly changes over the time interval $t \in [-T_S/2, +T_S/2]$, where $T_S = 1/fs$ is the sawtooth signal period, with one period of this signal shown in Figure 6.7(a1), (b1), and (c1) (these signals are identical in all three cases). Mechanical scanning in the reference arm of the Michelson interferometer generates PD current signals that, after amplification and filtering, take the form as shown in Figure 6.7(a2), (b2), and (c2), where the first column shows the situation for $x < 0$, the second column for $x = 0$, and the third column for $x > 0$. The third row shows the squared signals in the same situations, while the fourth row shows the squared signals multiplied by the reference signal from the SG, that is, a sawtooth signal from the SG. Since all these signals are periodic with the same period, after integration over a period of time, we get an error signal that is shown as a quasi-DC signal in the fifth row in Figure 6.7. As can be seen from this figure, the error signal $e = e(x)$ is proportional to the optical path difference. The error signal can be used as a control signal in the feedback loop to position the PA in the center of the coherence zone. This can be done by combining the error signal with the signal from the SG. This closes the feedback loop, which always has the role of positioning the PA in the center of the coherence zone. This achieves that each relatively slow time variable displacement of the sensor mirror

(SM) in the sensor arm is accompanied by an identical displacement of the RM in the reference arm. Since the position of the PA is proportional to the supply voltage, we can measure the displacement in the sensor branch by measuring the quasistationary voltage at the input of the PA.

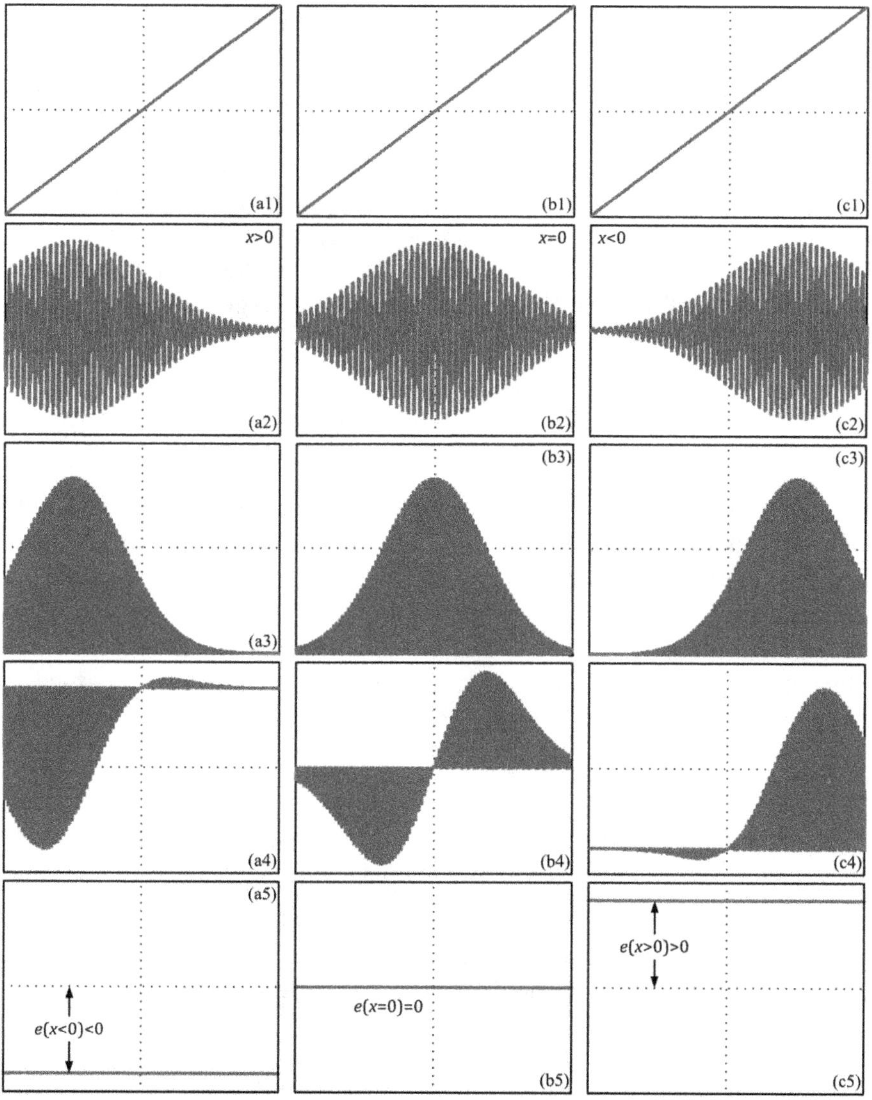

FIGURE 6.7 Voltage signals obtained by scanning the Michelson interferometer at characteristic points for (a) $x < 0$, (b) $x = 0$, and (c) $x > 0$.

6.2.1 SENSITIVITY OF THE MODIFIED CENTROID ALGORITHM WITH LINEAR SCANNING

Based on the algorithm presented earlier, we will find the functional dependence of the error signal e on the optical path difference x. The following is a mathematical analysis that will give the functional dependence of these two quantities. We will assume that we have a LCLS with a central wavelength λ and a coherence length L_C and that the coherence zone has a Gaussian shape, which is the case for superluminescent diodes, where L_C represents one half of the $1/e$ width of the coherence zone. In addition, we will assume that the PA performs an ideal linear scan of the RM in the range $z \in [-D, +D]$ at a constant velocity v_{PA}, where $2D$ is the total path that transits the RM. If the optical path difference is equal to x at the output of the bandpass filter, we have the normalized voltage signal $u_F(z)$ given by the following expression:

$$u_F(z) = \exp\left\{-\left[\frac{2(z-x)}{L_C}\right]^2\right\}\cos\left[\frac{4\pi}{\lambda}(z-x)\right]. \tag{6.10}$$

All the voltage signals that we have at the outputs of corresponding elements in the sensor system are functions of time, however, as a constant speed scan is performed, $z = v_{PA}t$, so that the temporal and spatial coordinates are equivalent (i.e., linearly proportional), all signals given depending on the spatial coordinate. The voltage signal from the filter output is led to a squaring circuit, the output of which produces a voltage signal whose normalized value $u_S(z)$ is given by the following expression:

$$u_S(z) = u_F^2(z) = \frac{1}{2}\exp\left[-8\left(\frac{z-x}{L_C}\right)^2\right]\left\{1+\cos\left[\frac{8\pi}{\lambda}(z-x)\right]\right\}. \tag{6.11}$$

As we said earlier, the effective normalized voltage signal from the generator signal $u_{SG}(z)$ is a periodic signal with a $2D$ spatial period and has a sawtooth shape given by the following expression:

$$u_{SG}(z) = \frac{z}{D}, z \in [-D, +D]. \tag{6.12}$$

At the output of the multiplication circuit, we have a normalized voltage signal $u_M(z)$ given by the following expression:

$$u_{\mathrm{M}}(z)=u_{\mathrm{S}}(z)\cdot u_{\mathrm{SG}}(z)=\frac{z}{2D}\exp\left[-8\left(\frac{z-x}{L_{\mathrm{C}}}\right)^{2}\right]$$
$$\left\{1+\cos\left[\frac{8\pi}{\lambda}(z-x)\right]\right\},\ z\in[-D,+D]. \tag{6.13}$$

The signal from the output of the multiplication circuitry is further fed to the input of the integration circuit, which performs integration, as shown in Figure 6.7, during the time interval $t\in[-T_{\mathrm{S}}/2,+T_{\mathrm{S}}/2]$, which is further equivalent to the integration of the signal $u_{\mathrm{M}}(z)$ in the spatial domain in the range $z\in[-D,+D]$. At the output of the integration circuit, an error signal is generated that depends on the optical path difference, as shown in the following expression:

$$e(x)=\frac{1}{2D}\int_{-D}^{+D}u_{\mathrm{M}}(z)\,\mathrm{d}z=\frac{1}{4D^{2}}\int_{-D}^{+D}z\cdot\exp\left[-8\left(\frac{z-x}{L_{\mathrm{C}}}\right)^{2}\right]\left\{1+\cos\left[\frac{8\pi}{\lambda}(z-x)\right]\right\}\mathrm{d}z. \tag{6.14}$$

Solving the integral given in eq 6.14 yields:

$$e(x)\approx\frac{L_{\mathrm{C}}^{2}}{4D^{2}}\left\{\begin{array}{l}\frac{1}{16}\left\{\begin{array}{l}\exp\left[-8\left(\frac{D+x}{L_{\mathrm{C}}}\right)^{2}\right]\\ \left[1+\cos\left(2\sqrt{2}a\frac{D+x}{L_{\mathrm{C}}}\right)\right]-\exp\left[-8\left(\frac{D-x}{L_{\mathrm{C}}}\right)^{2}\right]\left[1+\cos\left(2\sqrt{2}a\frac{D-x}{L_{\mathrm{C}}}\right)\right]\end{array}\right\}\\ +\sqrt{\frac{\pi}{2}}\frac{x}{4L_{\mathrm{C}}}\left[\operatorname{erf}\left(2\sqrt{2}\frac{D+x}{L_{\mathrm{C}}}\right)+\operatorname{erf}\left(2\sqrt{2}\frac{D-x}{L_{\mathrm{C}}}\right)\right]-\frac{a}{16}\int_{-2\sqrt{2}\frac{D+x}{L_{\mathrm{C}}}}^{2\sqrt{2}\frac{D-x}{L_{\mathrm{C}}}}\exp\left(-u^{2}\right)\sin(au)\,\mathrm{d}u\end{array}\right\} \tag{6.15}$$

where erf(x) is the error function. As can be seen from eq. 6.15, this dependence is very complex. Since the error signal is used as a control signal in a closed-loop circuit, all with the aim of eliminating the optical path difference, for the optical path difference, it is usually worth: $x\propto L_{\mathrm{C}},D$. For this reason, the following can be written for the error signal:

$$e(x)\approx S\frac{x}{L_{\mathrm{C}}}, \tag{6.16}$$

where S is the proportionality coefficient of the relative displacement x/L_{C} (displacement relative to the length of the coherence zone) for which the following applies:

$$S = L_C \left.\frac{\partial e(x)}{\partial x}\right|_{x=0}. \tag{6.17}$$

The value of this proportionality coefficient will be found by finding the first derivative of the error signal given by eq 6.15. In this case, we have:

$$S = \frac{2L_C}{4D^2}\int_{-D}^{+D}\exp\left[-8\left(\frac{z}{L_C}\right)^2\right]\left\{2\left(\frac{z}{L_C}\right)^2\left[1+\cos\left(\frac{8\pi}{\lambda}z\right)\right]+\frac{\pi}{\lambda}z\Delta\sin\left(\frac{8\pi}{\lambda}z\right)\right\}. \tag{6.18}$$

If we now introduce the following substitution $u = z/L_C$, then we get the following:

$$S = 2\left(\frac{L_C}{D}\right)^2\left[\begin{array}{l}2\int_{-\frac{D}{L_C}}^{\frac{D}{L_C}}u^2\exp\left(-8u^2\right)du + 2\int_{-\frac{D}{L_C}}^{\frac{D}{L_C}}u^2\exp\left(-8u^2\right)\\ \cos\left(8\pi\frac{L_C}{\lambda}u\right)du + \pi\frac{L_C}{\lambda}\int_{-\frac{D}{L_C}}^{\frac{D}{L_C}}u\exp\left(-8u^2\right)\sin\left(8\pi\frac{L_C}{\lambda}u\right)du\end{array}\right]. \tag{6.19}$$

As can be seen from eq 6.19, the dependence of the proportionality coefficient S on the central wavelength λ, the coherence length L_C, and the length of the $2D$ scan zone is very complex. Since the complete subintegral function is an even function, eq 6.19 can be simplified and written as:

$$S = 4\left(\frac{L_C}{D}\right)^2\left[\begin{array}{l}2\int_{0}^{\frac{D}{L_C}}u^2\exp\left(-8u^2\right)du + 2\int_{0}^{\frac{D}{L_C}}u^2\exp\left(-8u^2\right)\cos\left(8\pi\frac{L_C}{\lambda}u\right)\\ du + \pi\frac{L_C}{\lambda}\int_{0}^{\frac{D}{L_C}}u\exp\left(-8u^2\right)\sin\left(8\pi\frac{L_C}{\lambda}u\right)du\end{array}\right]. \tag{6.20}$$

By solving all integrals in eq 6.20, we have:

$$S = \frac{\sqrt{2}}{8p^2}\left[\frac{\sqrt{\pi}}{2}\operatorname{erf}\left(2\sqrt{2}p\right) + s(p,q) - \frac{1}{2p}\exp\left(-8p^2\right)\left[1+\cos\left(8\pi pq\right)\right]\right], \tag{6.21}$$

where the parameters p, q, and $s(p,q)$ are given by:

$$p = \frac{D}{L_C},\ q = \frac{L_C}{\lambda},\ s(p,q) = \int_0^{2\sqrt{2}p}\exp\left(-t^2\right)\cos\left(2\pi\sqrt{2}qt\right)dt \tag{6.22}$$

As for LCLSs, for example, the superluminescent diode is almost always satisfied $L_C \gg \lambda$ or $q \gg 1$, then due to the averaging process of the integration, where $q \gg p$ is also satisfied, the following is obtained:

$$\frac{\sqrt{\pi}}{2}\operatorname{erf}\left(2\sqrt{2}p\right)=\int_0^{2\sqrt{2}p}\exp\left(-t^2\right)\mathrm{d}t >> \int_0^{2\sqrt{2}p}\exp\left(-t^2\right)$$
$$\cos\left(2\pi\sqrt{2}qt\right)\mathrm{d}t = s\left(p,q\right). \tag{6.23}$$

In this case, the expression for the sensitivity of the modified centroid algorithm can be simplified, so that eq 6.21 becomes:

$$S \approx \frac{\sqrt{2\pi}}{16}\frac{\operatorname{erf}\left(2\sqrt{2}p\right)}{p^2}-\frac{1}{2p}\exp\left(-8p^2\right)\left[1+\cos\left(8\pi pq\right)\right]. \tag{6.24}$$

It is now interesting to show the functional dependence of the sensitivity S on the size of the scanned zone D, based on eq 6.21. In Figure 6.8, this dependence is shown as well as the dependence of the individual terms of eq 6.21 for the typical value of the parameter q from $q = L_C/\lambda = 10$, with the following being considered:

$$A=\frac{\sqrt{2\pi}}{16}\frac{\operatorname{erf}\left(2\sqrt{2}p\right)}{p^2},\ B=\frac{\sqrt{2}}{8p^2}s\left(p,q\right),\ C=-\frac{1}{2p}\exp\left(-8p^2\right)$$
$$\left[1+\cos\left(8\pi pq\right)\right],\ S=A+B+C \tag{6.25}$$

As can be seen in Figure 6.8, the red line shows the dependence of parameter B on the parameter p, which is much smaller than the other two parameterss, and the approximation given by eq 6.24 is fully justified. What is very important to note with Figure 6.8 is that for small values of parameter p ($p < 0.35$), the sensitivity changes sign very quickly depending on the parameter value. If we use a LCLS in our sensor system, such as superluminescent diode, then due to change, for example, the supply current or the junction temperature may change the central wavelength or change the size of the coherence zone. For this reason, the parameter $q = L_C/\lambda$ can vary over time, which can cause significant variations as well as changes in the sensitivity sign S. On the other hand, since the error signal is proportional to the sensitivity as well as the optical path differences, the change of the sign of the error signal for the same value of the error signal in the closed-loop sensor system may lead to instability of the entire sensor system.

Based on eq 6.24, the sensitivity can vary within the limits given in the following expression regardless of the value of parameter q:

$$\frac{\sqrt{2\pi}}{16}\frac{\operatorname{erf}\left(2\sqrt{2}p\right)}{p^2}-\frac{1}{2p}\exp\left(-8p^2\right)\le S\le\frac{\sqrt{2\pi}}{16}\frac{\operatorname{erf}\left(2\sqrt{2}p\right)}{p^2} \tag{6.26}$$

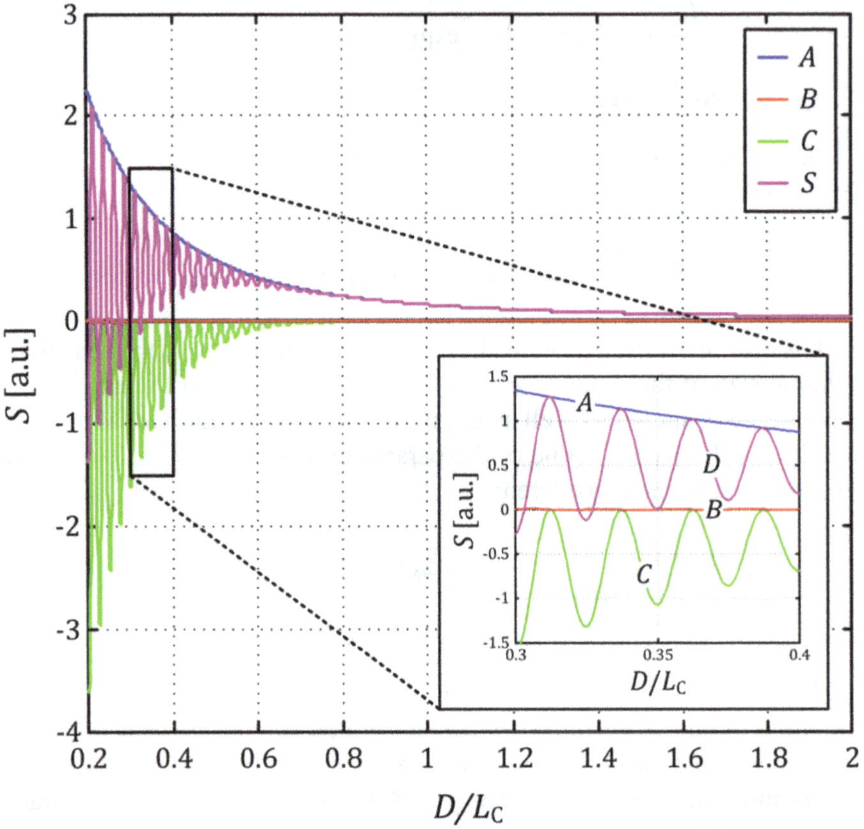

FIGURE 6.8 Functional dependence of sensitivity S on the size of the scanned zone D for $q = L_C/\lambda = 10$.

In Figure 6.9, the sensitivity range depending on the parameter p is shown for all possible values of the parameter q. The range is defined by two curves that represent the curves of the function given by eq 6.24. The gray zone indicates the area where the sensitivity value S changes very quickly with the change of parameter p. For this reason, it is necessary to avoid the operation of our sensor system in the parameter range $p < 0.35$, since changing the parameter q can lead to a change in the sensitivity of our sensor system, which can very easily lead to instability in the operation of the entire system. For the values of the parameter $p > 0.35$, as can be seen from Figure 6.9, the sensitivity value can only take positive values for any value of parameter q, and as a minimum value of parameter D/L_C can be adopted $(D/L_C)_{min} = 0.35$. It is also interesting to note with Figure 6.9 that the maximum sensitivity in the worst case when the parameter q takes the value from the bottom of the anvelope (minimum sensitivity) is reached when the parameter p takes a value of $p_{opt} = (D/L_C)_{opt} = 0.554$, which also represents the optimal choice of the value of this parameter when we have the case of the linear scan of the coherence zone. In this case, we have the maximum sensitivity value, in the worst case of $S_{max} = 0.3419$.

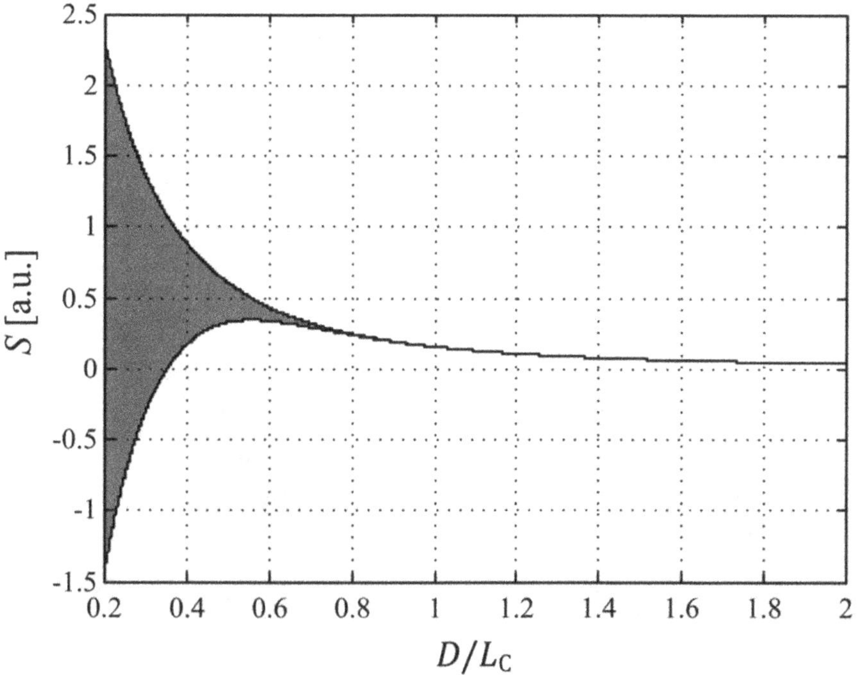

FIGURE 6.9 Sensitivity range depending on parameter p for all possible values of parameter q.

As noted earlier, for values of $p > 0.35$, the sensitivity can only take positive values. However, in some cases when the abovementioned condition is fulfilled, changes in the q parameter (i.e., the conditions of operation of the LCLS) can result in significant changes in the sensitivity value. On the other hand, significant changes in the sensitivity value can lead to instability of the entire sensor system realized in the feedback if the sensitivity can change in a relatively wide range. For this reason, in

Figure 6.10, the relative change of sensitivity $\delta S(p)$ depending on the parameter p is shown as follows:

$$\ddot{a}S(p) = \frac{S_{max}(p) - S_{min}(p)}{1/2\left[S_{max}(p) - S_{min}(p)\right]}, \quad (6.27)$$

where $S_{max}(p)$ and $S_{min}(p)$ represent the values from the upper and lower anvelope of the sensitivity function, respectively. As can be seen from Figure 6.10, for $p > 0.4$, the relative change in sensitivity is less than approximately 100%, so this range can be used without fear of a significant disturbance in the stability of the operation of the entire sensor system.

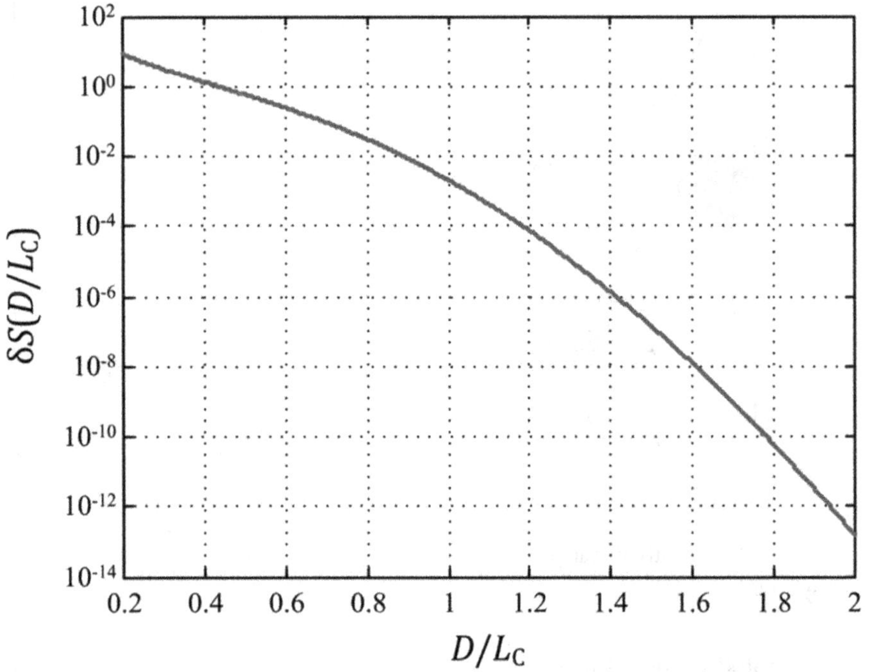

FIGURE 6.10 Relative change of sensitivity $\delta S(D/L_C)$ depending on parameter D/L_C.

Finally, it would be interesting to return to eq 6.15 again and show the dependence of the error function $e(x)$ on the optical path difference x. The following expression is given in the extension in a slightly more compact form:

$$e(r) \approx \frac{1}{64p^2}\left\{\begin{array}{l} \exp\left[-8(p+r)^2\right]\left\{1+\cos\left[8\pi q(p+r)\right]\right\} - \exp\left[-8(p-r)^2\right] \\ \left\{1+\cos\left[8\pi q(p-r)\right]\right\} - +2\sqrt{2\pi}r \\ \left\{\operatorname{erf}\left[2\sqrt{2}(p+r)\right]+\operatorname{erf}\left[2\sqrt{2}(p-r)\right]\right\} \\ -2\pi\sqrt{2}q\int_{-2\sqrt{2}(p+r)}^{2\sqrt{2}(p-r)} \exp\left(-u^2\right)\sin\left(2\pi\sqrt{2}qu\right)du \end{array}\right\}, \quad (6.28)$$

where $p = D/L_C$, $q = L_C/\lambda$ and $r = x/L_C$. In Figure 6.11, the dependence of the error function on the relative difference of the optical paths x/L_C for different values of the size of the scanning zone is shown, for $p = D/L_C =$ 0.25, 0.554, 1, and 2 and for a typical parameter value $q = L_C/\lambda = 10$. As can be seen from Figure 6.11, and especially in the section shown in the same diagram, for a parameter $p = D/L_C$ that is less than the critical value of $p = 0.35$, the error value fluctuates significantly around the point $x = 0$. This oscillation can cause instability in the operation of the entire sensor system, and thus the inability to measure the difference of optical paths in this way. For parameter values $p = 0.35$ in a relatively wider range (e.g., $-0.2 < x/L_C < 0.2$ for $p = D/L_C = 0.554$), the error function does not oscillate, that is, the error function is monotonically increasing, so this signal can be used as an error signal in the feedback system, to eliminate that error. In Figure 6.11, it can also be observed that as the value of the parameter p increases, the sensitivity of the error function decreases, but the linearity of such measurement of the difference of optical paths is larger as well as the measurement range is larger. Based on the above, if we want a higher sensitivity of the measuring system, it is necessary to choose as small value of parameter p as possible but not less than say $p = 0.5$, since further reduction of this parameter does not result in a significant increase in sensitivity, and there is a risk of irregular measurement due to oscillations of the error function.

The error function in addition to the parameter p also depends on the parameter q. The dependence of the error function for the different values of the parameter $q = 5$, 10, 20, and 40 and the parameter $p = 0.25$, 0.554, 1, and 2 (green—1, blue—2, red—3, and purple—4 curve, respectively) and of the relative differences of the x/L_C optical paths is shown in the four diagrams in Figure 6.12. As can be seen from this figure, with an increase in the parameter q, there is a significant increase in the oscillation of the sensitivity function, especially in cases where we have $p < 0.35$. For the values of the parameter $p < 0.35$ around the point $x = 0$, the magnitude of the oscillations decreases significantly, as noted earlier, that is, for very small values of the optical path difference, the sensitivity is always the same sign if the previous condition is satisfied.

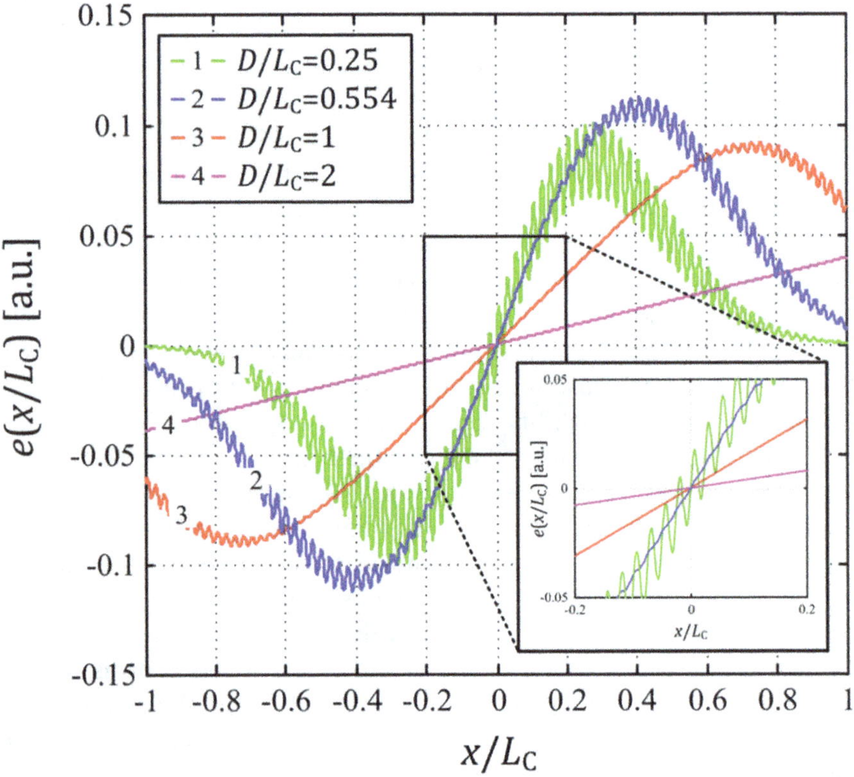

FIGURE 6.11 Dependence of the error function e on the relative difference of the optical paths x/L_C for different values of the size of the scan zone, for $p = D/L_C = 0.25$, 0.554, 1 and 2 (green—1, blue—2, red—3, and purple—4 curve, respectively) and for the typical parameter value $q = L_C/\lambda = 10$.

6.2.2 OPTICAL PATH DIFFERENCE MEASUREMENT ERROR OF LINEAR SCANNING

Measurement errors of any physical quantity can generally be divided into two large groups, systematic (deterministic) errors, which can usually be easily eliminated by calibration of the entire measurement system, and random (stochastic) errors that cannot be easily eliminated and are always

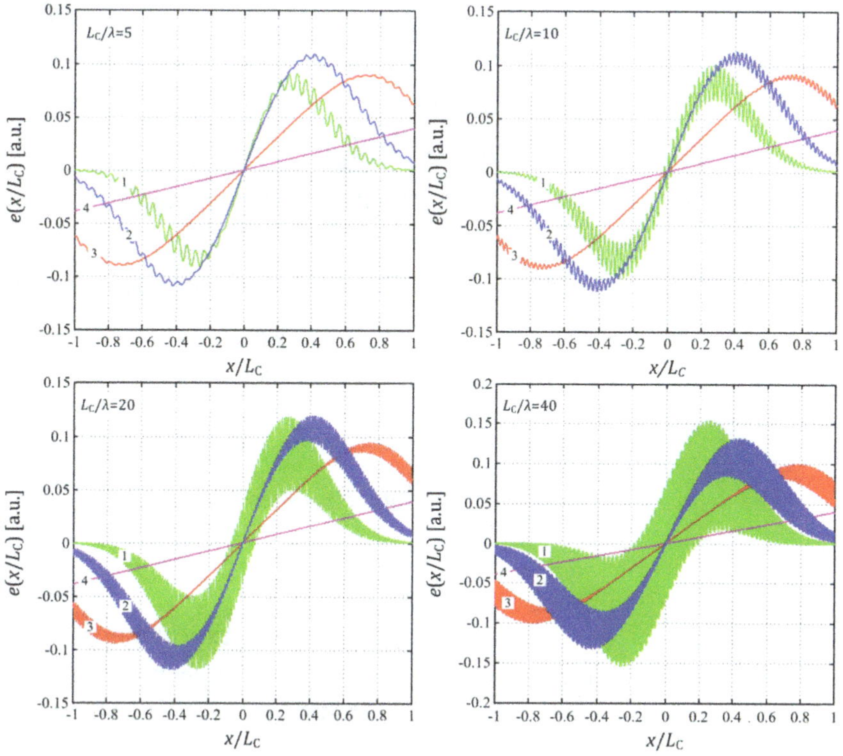

FIGURE 6.12 The dependence of the error function on the different values of parameter $q = 5$, 10, 20, and 40 and parameter $p = 0.25$, 0.554, 1, and 2 (green—1, blue—2, red—3, and purple—4 curve, respectively) and of relative optical path differences x/L_C.

present at each measurement. For this reason, the subject matter of the analysis in this chapter is the incidental errors of the presented sensor system, which are usually caused by thermal noise and noise of the PD and TIA in the analog signal processing circuit. In case the optical path difference is equal to x and in the presence of noise on the PD and TIA, at the output of the bandpass filter we have the normalized voltage signal given by the following expression:

$$\hat{u}_F(z) = u_F(z) + n(z) = \exp\left\{-\left[\frac{2(z-x)}{L_C}\right]^2\right\}\cos\left[\frac{4\pi}{\lambda}(z-x)\right] + n(z). \quad (6.29)$$

where $n(z)$ is the band-limited additive white Gaussian noise with zero mean and standard deviation σ_n and where for the signal-to-noise ratio is valid: $\mathrm{SNR} = 1/\sigma_n^2$. Such a voltage signal from the filter output is led to a squaring circuit, the output of which produces a voltage signal whose normalized value is given by the following expression:

$$\hat{u}_{\mathrm{S}}(z) = \hat{u}_{\mathrm{F}}^2(z) = u_{\mathrm{F}}^2(z) + 2u_{\mathrm{F}}(z)n(z) + n^2(z). \tag{6.30}$$

At the output of the multiplying circuit, we have a normalized voltage signal given by the following expression:

$$\hat{u}_{\mathrm{M}}(z) = \hat{u}_{\mathrm{F}}^2(z)u_{\mathrm{SG}}(z) = u_{\mathrm{F}}^2(z)u_{\mathrm{SG}}(z) + 2u_{\mathrm{F}}(z)n(z)u_{\mathrm{SG}}(z) + n^2(z)u_{\mathrm{SG}}(z). \tag{6.31}$$

The output of the integration circuit generates an error signal depending on the optical path difference, as shown in the following expression:

$$\hat{e}(x) = \frac{1}{2D}\int_{-\mathrm{D}}^{+\mathrm{D}} \hat{u}_{\mathrm{M}}(z)\,\mathrm{d}z = e(x) + \Delta e(x) \tag{6.32}$$

where the displacement measurement error $\Delta e(x)$ is given by:

$$\Delta e(x) = \frac{1}{D}\int_{-D}^{+D} n(z)u_{\mathrm{F}}(z)u_{\mathrm{SG}}(z)\,\mathrm{d}z + \frac{1}{2D}\int_{-D}^{+D} n^2(z)u_{\mathrm{SG}}(z)\,\mathrm{d}z. \tag{6.33}$$

The ensemble average value (mathematical expectation) of the displacement measurement error $\overline{\Delta e(x)}$, where $\bar{\bullet}$ represents the ensemble average operator (the mathematical expectation operator) is easily shown to be worth:

$$\overline{\Delta e(x)} = \frac{\overline{n(z)}}{D}\int_{-D}^{+D} u_{\mathrm{F}}(z)u_{\mathrm{SG}}(z)\,\mathrm{d}z + \frac{\overline{n^2(z)}}{2D}\int_{-D}^{+D} u_{\mathrm{SG}}(z)\,\mathrm{d}z \equiv 0. \tag{6.34}$$

For the measurement error variance σ_e^2, based on eq 6.34, the following applies:

$$\sigma_e^2 = \overline{\left[\Delta e(x) - \overline{\Delta e(x)}\right]^2} = \overline{\left[\Delta e(x)\right]^2} = \frac{1}{D}\int_0^{2D}\left[R_N(\zeta) - N(z)^2\right]\left(1 - \frac{\zeta}{2D}\right)\mathrm{d}\zeta, \tag{6.35}$$

where $R_N(\zeta)$ is an autocorrelation function of the random process $N(z)$ and $\langle\bullet\rangle$ is the averaging operator along spatial coordinate (or equivalently to the time averaging operator), for which the following applies:

$$N(z) = n(z)u_{\mathrm{SG}}(z)\left[u_{\mathrm{F}}(z) + n(z)\right]. \tag{6.36}$$

and

$$R_N(\zeta) = \langle N(z)\cdot N(z+\zeta)\rangle. \tag{6.37}$$

For the mean value of the random variable $N(z)$, the following applies:

$$\langle N(z)\rangle = \langle n(z)u_F(z)u_{SG}(z)\rangle + \langle n^2(z)u_{SG}(z)\rangle = \langle n(z)\rangle$$
$$\langle u_F(z)\rangle\langle u_{SG}(z)\rangle + \langle n^2(z)\rangle\,\langle u_{SG}(z)\rangle = 0, \tag{6.38}$$

since $\langle n(z)\rangle = 0$ and $\langle u_{SG}(z)\rangle = 0$, that is, the signal from the SG does not have a DC component. Based on eqs 6.38 and 6.35, we have:

$$\sigma_e^2 = \frac{1}{D}\int_0^{2D} R_N(\zeta)\left(1-\frac{\zeta}{2D}\right)\mathrm{d}\zeta. \tag{6.39}$$

Next, it is necessary to find the autocorrelation function $R_N(\zeta)$ of the random variable $N(z)$. The following applies to the autocorrelation function:

$$R_N(\zeta) = \lim_{L\to+\infty}\frac{1}{2L}\int_{-L}^{+L} N(z)\leq\cdot N(z+\zeta)\,\mathrm{d}\zeta, \tag{6.40}$$

which further gives:

$$R_N(\zeta) = \lim_{L\to+\infty}\frac{1}{2L}\int_{-L}^{+L}$$
$$\left[n(z)n(z+\zeta)\right]\left[u_F(z)u_F(z+\zeta)\right]\left[u_{SG}(z)u_{SG}(z+\zeta)\right]$$
$$\mathrm{d}\zeta + \lim_{L\to+\infty}\frac{1}{2L}\int_{-L}^{+L}\left[n^2(z)n^2(z+\zeta)\right]\left[u_{SG}(z)u_{SG}(z+\zeta)\right]\mathrm{d}\zeta. \tag{6.41}$$

As both deterministic functions $u_F(z)$ and $u_{SG}(z)$ are periodic with the same period $2D$, so are the functions $u_1(z) = \left[u_F(z)u_F(z+\zeta)\right]\left[u_{SG}(z)u_{SG}(z+\zeta)\right]$ and $u_2(z) = u_{SG}(z)u_{SG}(z+\zeta)$ also periodic with the same period, so these functions can be represented as:

$$u_1(z) = \sum_{k=-\infty}^{+\infty} u_1'(z+2kD) \text{ and} \tag{6.42}$$

$$u_2(z) = \sum_{k=-\infty}^{+\infty} u_2'(z+2kD), \tag{6.43}$$

where for the functions $u_1'(z)$ and $u_2'(z)$ the following is valid:

$$u_1'(z) = \begin{cases} u_1(z), & z \leq |D| \\ 0, & z > D \end{cases} \text{ and} \tag{6.44}$$

$$u_2'(z) = \begin{cases} u_2(z), & z \leq |D| \\ 0, & z > D \end{cases}. \tag{6.45}$$

Based on eqs 6.41–6.43 and if we now adopt without losing the generality that $L = (2N + 1)D$, we obtain for the autocorrelation function:

$$R_N(\zeta)=\frac{1}{D}\lim_{N\to+\infty}\frac{1}{2(2N+1)}\int_{-(2N+1)D}^{+(2N+1)D}$$

$$\left[n(z)n(z+\zeta)\right]u_1(z)\mathrm{d}\zeta+\frac{1}{D}\lim_{N\to+\infty}\frac{1}{2(2N+1)}\int_{-(2N+1)D}^{+(2N+1)D}$$

$$\left[n^2(z)n^2(z+\zeta)\right]u_2(z)\mathrm{d}\zeta. \tag{6.46}$$

or equivalently:

$$R_N(\zeta)=\frac{1}{D}\lim N\to+\infty\frac{1}{2(2N+1)}\int_{-(2N+1)D}^{+(2N+1)D}\left[n(z)n(z+\zeta)\right]$$

$$\left[\sum_{k=-\infty}^{+\infty}u_1'(z+2kD)\right]\mathrm{d}\zeta+\frac{1}{D}\lim_{N\to+\infty}\frac{1}{2(2N+1)}$$

$$\int_{-(2N+1)D}^{+(2N+1)D}\left[n^2(z)n^2(z+\zeta)\right]\left[\sum_{k=-\infty}^{+\infty}u_2'(z+2kD)\right]\mathrm{d}\zeta. \tag{6.47}$$

If we now include the first subintegral terms in the sums in eq 6.47, we obtain:

$$R_N(\zeta)=\frac{1}{D}\lim_{N\to+\infty}\frac{1}{2(2N+1)}\int_{-(2N+1)D}^{+(2N+1)D}$$

$$\left\{\sum_{k=-\infty}^{+\infty}\left[n(z)n(z+\zeta)\right]u_1'(z+2kD)\right\}\mathrm{d}\zeta+\frac{1}{D}\lim_{N\to+\infty}\frac{1}{2(2N+1)}$$

$$\int_{-(2N+1)D}^{+(2N+1)D}\left\{\sum_{k=-\infty}^{+\infty}\left[n^2(z)n^2(z+\zeta)\right]u_2'(z+2kD)\right\}\mathrm{d}\zeta. \tag{6.48}$$

For the variable z in subintegral expression, the following applies: $-(2N+1)D\le z\le+(2N+1)D$, since the integration takes place in the same range. Accordingly, the values of all subintegral functions outside this range are irrelevant, that is, they can be identically equal to zero, and therefore, as can be seen from Figure 6.13, for the spatial displacement z_0 of the basic functions (indicated by the red curve in Figure 6.13) which is greater than the absolute value of $(2N+1)D$ (or $|z_0| > 2N+2)D$) the product of subintegral functions which originate from a random process (indicated by the blue curve in Figure 6.13) and spatially displaced basic functions is identically equal to zero and does not contribute to the integration process. Based on the earlier, for the absolute value of the parameter k (which must be an integer) from the eq 6.48, the following values are true: $2|k|D\le(2N+2)D$, or equivalently: $|k|\le N+1$, and so, therefore, for the range of values of the parameter k in which the summation is made, it obtains: $k \in \left[-(N+1), N+1\right]$.

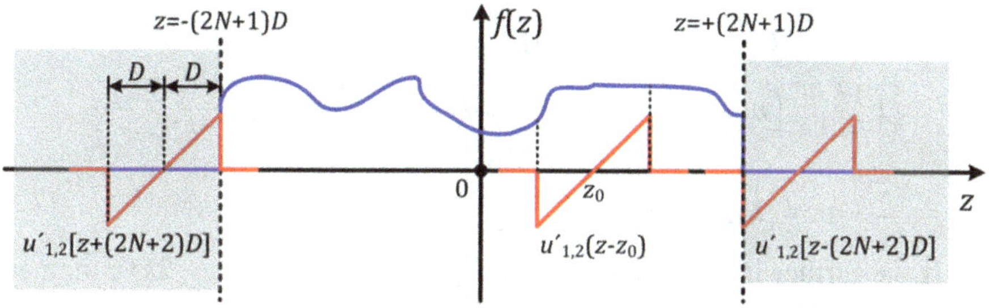

FIGURE 6.13 The calculation of the value of the subintegral function given in eq 6.48.

Since for the values $|k| > N+1$ the subintegral function does not contribute to the integration process, eq 6.48 becomes:

$$R_N(\zeta) = \frac{1}{D}\lim_{N\to+\infty}\frac{1}{2(2N+1)}\int_{-(2N+1)D}^{+(2N+1)D} \left\{\sum_{k=-(N+1)}^{N+1}\left[n(z)n(z+\zeta)\right]u_1'(z+2kD)\right\}\mathrm{d}\zeta + \frac{1}{D}\lim_{N\to+\infty}\frac{1}{2(2N+1)} \int_{-(2N+1)D}^{+(2N+1)D}\left\{\sum_{k=-(N+1)}^{N+1}\left[n^2(z)n^2(z+\zeta)\right]u_2'(z+2kD)\right\}\mathrm{d}\zeta. \tag{6.49}$$

If we now change the order of integration and summation, eq 6.49 becomes:

$$R_N(\zeta) = \frac{1}{D}\lim_{N\to+\infty}\frac{1}{2(2N+1)}\sum_{k=-(N+1)}^{N+1} \left\{\int_{-(2N+1)D}^{+(2N+1)D}\left[n(z)n(z+\zeta)\right]u_1'(z+2kD)\mathrm{d}\zeta\right\} + \frac{1}{D}\lim_{N\to+\infty}\frac{1}{2(2N+1)} \sum_{k=-(N+1)}^{N+1}\left\{\int_{-(2N+1)D}^{+(2N+1)D}\left[n^2(z)n^2(z+\zeta)\right]u_2'(z+2kD)\mathrm{d}\zeta\right\}. \tag{6.50}$$

Taking into consideration the maximum values of the parameter k and the size of the integration area, it can be observed that for the values of the parameter $|k| = N+1$, we have only a partial integration of the stochastic signal due to the smaller integration area than the length of the functions $u_1'(z)$ and $u_2'(z)$. However, this does not affect the final result, because the process of seeking the limit value makes these two elements disappear. Furthermore, since the functions are in general different from zero only in the range $-D \le z + 2kD \le D$ and $-(2k+1)D \le z \le -(2k-1)D$, respectively, and outside this range identically equal to zero, then eq 6.50 becomes:

$$R_N(\zeta) = \frac{1}{D}\lim_{N\to+\infty}\frac{1}{2(2N+1)}\sum_{k=-(N+1)}^{N+1}$$
$$\left\{\int_{-(2k+1)D}^{-(2k-1)D}\left[n(z)n(z+\zeta)\right]u_1'(z+2kD)\,\mathrm{d}\zeta\right\} + \frac{1}{D}\lim_{N\to+\infty}\frac{1}{2(2N+1)}$$
$$\sum_{k=-(N+1)}^{N+1}\left\{\int_{-(2k+1)D}^{-(2k-1)D}\left[n^2(z)n^2(z+\zeta)\right]u_2'(z+2kD)\,\mathrm{d}\zeta\right\}. \tag{6.51}$$

If we further introduce the following substitution $t = z + 2kD$, then eq 6.51 becomes:

$$R_N(\zeta) = \frac{1}{D}\lim_{N\to+\infty}\frac{1}{2(2N+1)}$$
$$\sum_{k=-(N+1)}^{N+1}\left\{\int_{-D}^{D}\left[n(t-2kD)n(t-2kD+\zeta)\right]u_1'(t)\,\mathrm{d}t\right\} + \frac{1}{D}\lim_{N\to+\infty}$$
$$\frac{1}{2(2N+1)}\sum_{k=-(N+1)}^{N+1}\left\{\int_{-D}^{D}\left[n^2(t-2kD)n^2(t-2kD+\zeta)\right]u_2'(t)\,\mathrm{d}t\right\}. \tag{6.52}$$

As the noise signal $n(z)$ is assumed to be ergodic and stationary, the values of this stochastic signal at different instants $n(t - 2kD)$ (spatial or equivalent in time) can be replaced by a corresponding ensemble of independent stochastic variable $n_k(t)$ with identical statistics. This is possible if the noise signal samples are independent of each other, which is fulfilled since the size of the scan area (or equivalent to the sampling time) is much larger than the correlation length (or equivalent to the correlation time) of this signal, which is proportional to the bandwidth of the band-pass filter used. In this case, eq 6.52 becomes:

$$R_N(\zeta) = \frac{1}{2D}\lim_{N\to+\infty}\frac{1}{2N+1}\sum_{k=1}^{2N+3}\left\{\int_{-D}^{D}\left[n_k(t)n_k(t+\zeta)\right]u_1'(t)\,\mathrm{d}t\right\}$$
$$+\frac{1}{2D}\lim_{N\to+\infty}\frac{1}{2N+1}\sum_{k=1}^{2N+3}\left\{\int_{-D}^{D}\left[n_k^2(t)n_k^2(t+\zeta)\right]u_2'(t)\,\mathrm{d}t\right\}. \tag{6.53}$$

As can be seen from eq 6.53, the boundary value operator as a result gives the mean value per ensemble, so that eq 6.53 becomes:

$$R_N(\zeta) = \frac{1}{2D}\overline{\int_{-D}^{D}\left[n_k(t)n_k(t+\zeta)\right]u_1'(t)\,\mathrm{d}t}$$
$$+\frac{1}{2D}\overline{\int_{-D}^{D}\left[n_k^2(t)n_k^2(t+\zeta)\right]u_2'(t)\,\mathrm{d}t}, \tag{6.54}$$

that is, by passing the mean operator per ensemble through the integral and knowing that the signal $n_k(t)$ is stochastic, and the signals $u_1'(t)$ and $u_2'(t)$ deterministic, eq 6.54 becomes:

$$R_N(\zeta) = \overline{n_k(t)n_k(t+\zeta)}\left[\frac{1}{2D}\int_{-D}^{D} u_1'(t)\mathrm{d}t\right]$$
$$+\overline{n_k^2(t)n_k^2(t+\zeta)}\left[\frac{1}{2D}\int_{-D}^{D} u_2'(t)\mathrm{d}t\right]. \tag{6.55}$$

As random processes $n_k(t)$ are assumed to be ergodic and stationary, their mean values by ensemble and spatial coordinate are identical and replacing $z = t$ eq 6.55 becomes:

$$R_N(\zeta) = n(z)n(z+\zeta)\left[\frac{1}{2D}\int_{-D}^{D} u_1'(z)\mathrm{d}z\right]$$
$$+\langle n^2(z)n^2(z+\zeta)\rangle\left[\frac{1}{2D}\int_{-D}^{D} u_2'(z)\mathrm{d}z\right], \tag{6.56}$$

or equivalently:

$$R_N(\zeta) = R_n(\zeta)\left[\frac{1}{2D}\int_{-D}^{D} u_1'(z)\mathrm{d}z\right] + R_{n^2}(\zeta)\left[\frac{1}{2D}\int_{-D}^{D} u_2'(z)\mathrm{d}z\right], \tag{6.57}$$

where $R_n(\zeta)$ and $R_{n^2}(\zeta)$ are also the autocorrelation functions of the random processes $n(z)$ and $n^2(z)$, respectively. For the autocorrelation function of a random variable at the output of a square detector (as we have in our case) if we bring a random variable at the input of a square detector with a Gaussian distribution and a zero mean value, it is worth the following:

$$R_{n^2}(\zeta) = R_n^2(0) + 2R_n^2(\zeta). \tag{6.58}$$

Now substituting eqs 6.58 into 6.57 yields:

$$R_N(\zeta) = R_n(\zeta)\left[\frac{1}{2D}\int_{-D}^{D} u_1'(z)\mathrm{d}z\right] + \left[R_n^2(0) + 2R_n^2(\zeta)\right]\left[\frac{1}{2D}\int_{-D}^{D} u_2'(z)\mathrm{d}z\right] \tag{6.59}$$

Based on the Wiener-Khinchin theorem the following is valid for the autocorrelation function in the time domain $R_n(\tau)$ of the random process $n(z)$:

$$R_n(\tau) = \frac{p_n}{4\pi}\int_{-\infty}^{+\infty} \left|H_F(j\omega)\right|^2 \exp(j\omega\tau)\mathrm{d}\omega, \tag{6.60}$$

where $\tau = \zeta / v_{PA}$, p_n is the effective unilateral (defined only for $f \geq 0$) power spectral density of the noise at the output of the TIA and $H_F(j\omega)$ is the transfer function of the band-pass filter in the frequency domain. To find the autocorrelation function of the random variable $N(z)$, it is necessary to solve the integrals in eq 6.59. Let's start with the following integral:

$$I_2(\zeta) = \frac{1}{2D}\int_{-D}^{D} u_2'(z)\,\mathrm{d}z = \frac{1}{2D}\int_{-D}^{D} u_{SG}(z)\,u_{SG}(z+\zeta)\,\mathrm{d}z. \qquad (6.61)$$

Since the signal from the SG is a periodic function with period D, then it makes sense to solve the above integral only for $|\zeta| \leq D$ since the integral $I_2(\zeta)$ is also a periodic function with the same period. Based on Figure 6.14 (the positive sign denotes the part where the product $u_{SG}(z)$ $u_{SG}(z + \zeta)$ is greater than zero while the negative sign indicates the part where that product is less than zero) and eq 6.12, the integral becomes:

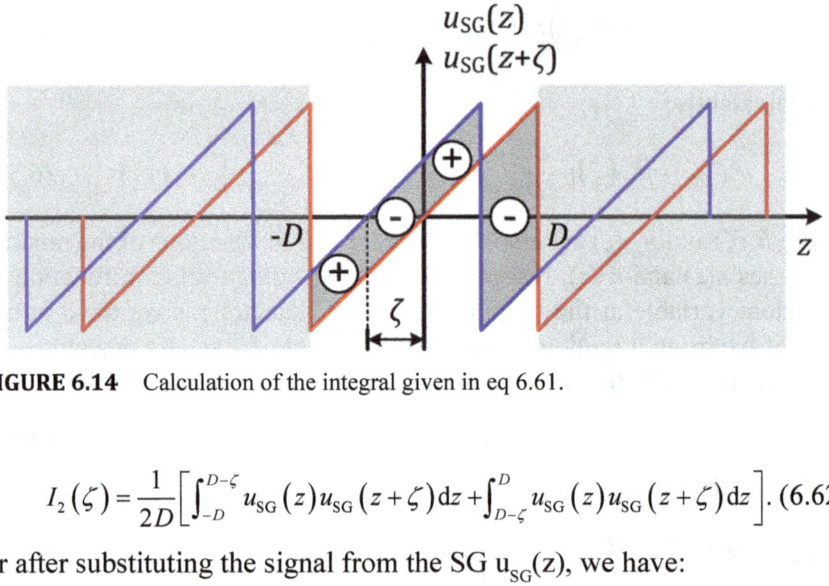

FIGURE 6.14 Calculation of the integral given in eq 6.61.

$$I_2(\zeta) = \frac{1}{2D}\left[\int_{-D}^{D-\zeta} u_{SG}(z)\,u_{SG}(z+\zeta)\,\mathrm{d}z + \int_{D-\zeta}^{D} u_{SG}(z)\,u_{SG}(z+\zeta)\,\mathrm{d}z\right]. \qquad (6.62)$$

or after substituting the signal from the SG $u_{SG}(z)$, we have:

$$I_2(\zeta) = \frac{1}{2D^3}\left[\int_{-D}^{D-\zeta} z(z+\zeta)\,\mathrm{d}z + \int_{D-\zeta}^{D} z(z+\zeta-2D)\,\mathrm{d}z\right]. \qquad (6.63)$$

The procedure for solving this integral is very simple, so in the end, it is obtained:

$$I_2(\zeta) = \frac{\zeta^2}{2D^2} - \frac{\zeta}{D} + \frac{1}{3}. \qquad (6.64)$$

It is now necessary to solve the much more complicated integral given by the following expression:

$$I_1(\zeta)=\frac{1}{2D}\int_{-D}^{D}u_1'(z)\mathrm{d}z=\frac{1}{2D}\int_{-D}^{D}\left[u_{\mathrm{F}}(z)u_{\mathrm{F}}(z+\zeta)\right]\cdot\left[u_{\mathrm{SG}}(z)u_{\mathrm{SG}}(z+\zeta)\right]\mathrm{d}z. \tag{6.65}$$

The integral given by eq 6.65 can be solved on a similar principle to the previous integral, except that solving this integral is much more complex. To avoid solving this integral on a full scale, we first consider eqs 6.59 and 6.39. Substituting eqs 6.59 into 6.39 yields:

$$\sigma_e^2=\frac{1}{D}\left[\begin{array}{l}\int_0^{2D}R_n(\zeta)I_1(\zeta)\left(1-\frac{\zeta}{2D}\right)\mathrm{d}\zeta+2\int_0^{2D}R_n^2(\zeta)I_2(\zeta)\left(1-\frac{\zeta}{2D}\right)\mathrm{d}\zeta\\+\int_0^{2D}R_n^2(0)I_2(\zeta)\left(1-\frac{\zeta}{2D}\right)\mathrm{d}\zeta\end{array}\right]\cdot \tag{6.66}$$

As mentioned in this chapter, the correlation length l_n of the random process $n(z)$ is much smaller than the size of the scanning zone D, that is, $l_n \ll D$, because we assumed that the bandwidth of the filter is relatively large to pass the high-frequency interferometric signal ($L_C \gg \lambda$, that is, the frequency of the interferometric signal is much higher than the scanning frequency). Based on this assumption, we can write for the autocorrelation function $R_n(\zeta)$ of the random process $n(z$:

$$R_n(\zeta)\approx\begin{cases}\sigma_n^2, & \zeta\le l_n\\ 0, & \zeta>l_n\end{cases}, \tag{6.67}$$

where $R_n(0)\equiv\sigma_n^2$ is the variance of the random process is $n(z)$. Substituting eqs 6.67 into 6.66 and considering that is $l_n \ll D$ valid, eq 6.66 becomes:

$$\sigma_e^2\approx\frac{1}{D}\left[\sigma_n^2 l_n I_1(0)+2\sigma_n^4 l_n I_2(0)+\sigma_n^4\int_0^{2D}I_2(\zeta)\left(1-\frac{\zeta}{2D}\right)\mathrm{d}\zeta\right], \tag{6.68}$$

If we now substitute the values for the integral $I_2(\zeta)$ and calculate the integral in the third term of the sum, we get:

$$\sigma_e^2\approx\frac{l_n}{D}\sigma_n^2\left[I_1(0)+\frac{2}{3}\sigma_n^2\right]\approx I_1(0)\frac{l_n}{D}\sigma_n^2, \tag{6.69}$$

whereby calculation the integral in the third term of the sum in eq 6.68 is identically equal to zero and where is adopted: SNR $\gg$ 1 (SNR = $1/\sigma_n^2$).

Now we need to find the integral $I_1(0)$ for which, on the basis of eq 6.65, the following is true:

$$I_1(0) = \frac{1}{2D}\int_{-D}^{D} u_F^2(z)\,u_{SG}^2(z)\,\mathrm{d}z = \frac{1}{4D^3}\int_{-D}^{D} z^2 \exp\left[-8\left(\frac{z}{L_C}\right)^2\right]\left[1+\cos\left(\frac{8\pi}{\lambda}z\right)\right]\mathrm{d}z, \tag{6.70}$$

whereby we have adopted that $x = 0$, since we are only interested in the situation when the difference of optical paths is kept at zero because only in this case we do not have a systematic error in our measuring system. Since the subintegral function is even and adopting the shift $u = z/L_C$, the integral in eq 6.70 becomes:

$$I_1(0) = \frac{L_C^3}{2D^3}\int_0^{\frac{D}{L_C}} u^2 \exp(-8u^2)\left[1+\cos\left(8\pi\frac{L_C}{\lambda}u\right)\right]\mathrm{d}u \approx \frac{L_C^3}{2D^3}\int_0^{\frac{D}{L_C}} u^2\exp(-8u^2)\mathrm{d}u, \tag{6.71}$$

whereby we adopted a similar approximation as in eq 6.23. Based on the integrals solved in the previous section, the integral eq 6.71 becomes:

$$I_1(0) \approx \frac{1}{32p^3}\left[\frac{\sqrt{2\pi}}{8}\operatorname{erf}\left(2\sqrt{2}p\right) - p\exp\left(-8p^2\right)\right], \tag{6.72}$$

whereby the parameter p is: $p = D/L_C$. If we now combine eqs 6.16 and 6.69 for the standard deviation in measuring the difference of optical paths σ_x we get:

$$\sigma_x = \frac{L_C}{S(p,q)}\sigma_e \approx \frac{L_C}{S(p,q)}\sqrt{\frac{N(p)}{D}l_n\sigma_n^2}, \tag{6.73}$$

where $N(p) = I_1(0)$, $S(p, q)$ is the sensitivity given by eq 6.24. As we defined in τ (eq 6.60), $\tau = \zeta / v_{PA}$, this also holds true for the correlation of τ_n random process $n(z)$: $\tau_n = l_n / v_{PA}$. Between the correlation time τ_n of the random process $n(z)$ (that is, in the time domain $n(t)$) and the bandwidth of the band-pass filter B_F, there is a simple connection: $\tau_n B_F = 1$, where we have the correlation length l_n (or equivalently the correlation time τ_n) and the bandwidth of the filter B_F defined as the effective values, that is:

$$l_n = \frac{1}{R_n(0)}\int_0^{+\infty}\left|R_n(\zeta)\right|\mathrm{d}\zeta, \tag{6.74}$$

$$\tau_n = \frac{1}{R_n(0)}\int_0^{+\infty}\left|R_n(\tau)\right|\mathrm{d}\tau \quad \text{and} \tag{6.75}$$

$$B_{\mathrm{F}} = \frac{1}{\left|H_{\mathrm{F}}(0)\right|^2} \int_0^{+\infty} \left|H_{\mathrm{F}}(j\omega)\right|^2 \mathrm{d}f, \ \omega = 2\pi f. \tag{6.76}$$

For the variance σ_n^2 of the random signal $n(z)$, the following applies:

$$\sigma_n^2 = p_n \int_0^{+\infty} \left|H_{\mathrm{F}}(j\omega)\right|^2 \mathrm{d}f. \tag{6.77}$$

Combining eqs 6.77 and 6.76 and assuming, without loss of generality, that $|H_{\mathrm{F}}(j\omega)|^2 = 1$ (both the useful signal and the noise pass through the same filter, so that if the filter has some amplification or attenuation, it is equally for both useful signal and noise), eq 6.77 becomes: $\sigma_n^2 = p_n B_{\mathrm{F}}$. Further, taking into account the relationship between noise correlation time and filter bandwidth, as well as correlation length and correlation time, eq 6.77 becomes:

$$\sigma_n^2 = p_n \frac{v_{\mathrm{PA}}}{l_n}. \tag{6.78}$$

Substituting eqs 6.78 into 6.73 for a measurement error, we get:

$$\sigma_x \approx L_{\mathrm{C}} \frac{\sqrt{N(p)}}{S(p,q)} \sqrt{\frac{v_{\mathrm{PA}}}{D} p_n}. \tag{6.79}$$

As the scan is performed by a periodic sawtooth signal with a period T_{S} and a constant speed v_{PA}, and in the $2D$ range, this is true for the size of the scan area: $2D = v_{\mathrm{PA}} T_{\mathrm{S}}$. Substituting the value for the size of the scanning zone into eq 6.79 finally for the measurement error, we get:

$$\sigma_x \approx L_{\mathrm{C}} \sqrt{\frac{2p_n}{T_{\mathrm{S}}}} \frac{\sqrt{N(p)}}{S(p,q)} = L_{\mathrm{C}} \sqrt{2 f_{\mathrm{S}} p_n} \frac{\sqrt{N(p)}}{S(p,q)}, \tag{6.80}$$

where $f_{\mathrm{S}} = 1/T_{\mathrm{S}}$ is the scanning frequency. Figure 6.15 shows the dependence of the optical path difference measurement error depending on the parameters $p = D/L_{\mathrm{C}}$ and $q = L_{\mathrm{C}}/\lambda$ for some standard values of the coherent zone length L_{C} (L_{C} = 10 μm), the scanning signal period T_{S} (T_{S} = 10 ms, f_{S} = 100 Hz) and effective noise power spectral density p_n ($p_n = 10^{-14}$ Hz^{-1}).

As can be seen from Figure 6.15, for the objectively selected values of the individual parameters of the sensor system, the error of measuring the difference of optical paths is somewhere around 10 pm. This is a very small value and of course represents only a theoretical value, while the value of a real measurement error can only be higher. There are many reasons for this, inter alia instability of the interferometer (drift and vibration), as well as the influence of ambient temperature (drift). The effects of interferometer instability as well as ambient temperature can be significantly reduced by proper construction and stabilization of the interferometer as well as its thermal insulation. In Figure 6.15, it can be observed that there are two slopes of the error curve giving the limits in which the measurement error value for different types of sources can move (different values of parameter $q = L_C/\lambda$). Based on the maximum and minimum sensitivity values of the sensor system, the following terms are given to define the defects of the measurement error function:

$$L_C\sqrt{\frac{2p_n}{T_S}\frac{\sqrt{N(p)}}{S_{max}(p,q)}} \le \sigma_x \le L_C\sqrt{\frac{2p_n}{T_S}\frac{\sqrt{N(p)}}{S_{min}(p,q)}}, \tag{6.81}$$

where the values for S_{max} and S_{min} are given by eq 6.26. In Figure 6.16, the limits of the optical path difference measurement error are given.

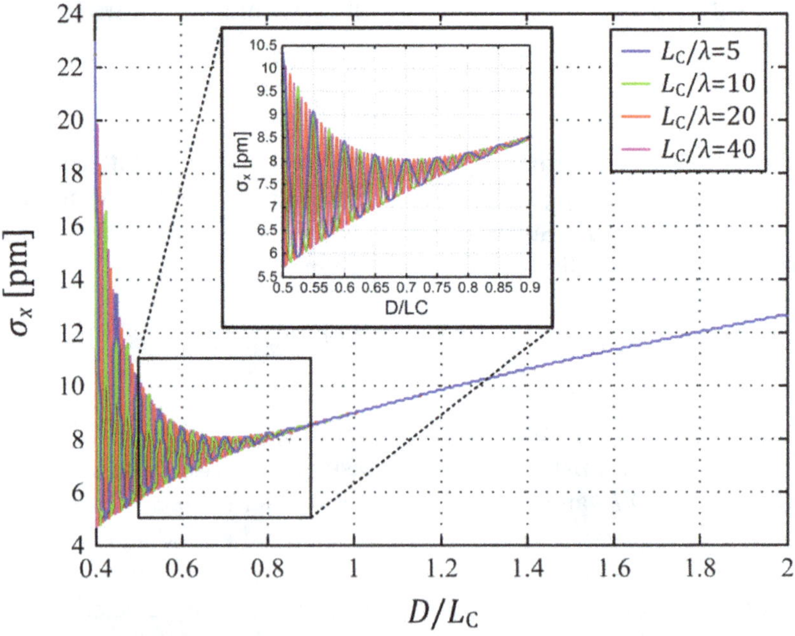

FIGURE 6.15 The dependence of the error of measuring the difference of optical paths σ_x depending on the parameters $p = D/L_C$ and $q = L_C/\lambda$ for some standard values of the size of the coherent zone $L_C = 10$ μm, scan signal periods T_S = ms, and the effective spectral density of the mean noise power $p_n = 10^{-14}$ Hz^{-1}.

As can be seen from Figure 6.16, there is an optimum value of the parameter $p_{opt} = (D/L_C)_{opt} \approx 0.71$ for which, in the worst case, a minimum measurement error of about $\sigma_x \approx 8$ pm is obtained. The value range of the parameter $p = D/L_C$ which in the worst case provides a measurement error of less than 10 pm is quite wide and ranges from $0.5 \leq D/L_C \leq 1.2$. Such a relatively wide range of parameter values $p = D/L_C$ gives an opportunity for relaxed operation while adjusting all parameters in the sensor system to achieve optimal operation of the complete system.

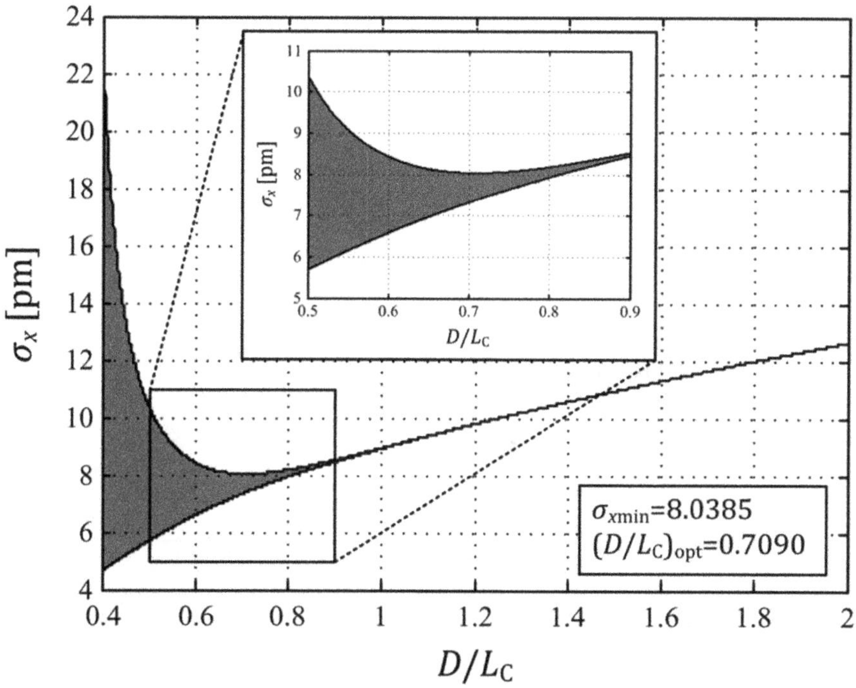

FIGURE 6.16 Limits within the values of the optical path difference measurement error may range.

Sections 6.2.1 and 6.2.2 provide a detailed mathematical analysis and presentation of basic characteristics, such as the functional dependence of the error signal on the optical path difference, the sensitivity, and measurement error of the modified centroid algorithm when a sawtooth voltage signal is given from the generator signal. Due to the periodicity and symmetry, the identical values of the sensitivity and the optical path difference measurement error would be obtained if triangular were used instead of the sawtooth voltage form. If the sawtooth voltage forms were fed to the inputs of the HVA from which output further the PA is fed, there would be some problems in the operation of the PA. The reason for this is the relatively large bandwidth required by the sawtooth or triangular voltage forms. Therefore, PAs having a relatively large dynamic range (typically several hundred micrometers) such as piezo bimorph actuators, on the other hand, have a relatively low resonance frequency (typically several hundred hertz), and some higher harmonics of the fundamental scan frequency may cause very irregular operation of the PA, so instead of the sawtooth scan of the coherence zone, we have a completely irregular scan, that is, the shape of the signal from the SG is not identical to the movement of the PA.

KEYWORDS

- **low-coherence fiber-optic sensor**
- **Michelson interferometer**
- **optical path difference**
- **receiving interferometer**
- **sensing interferometer**

REFERENCES

Berkcan, E.; Tiemann, J.; Brooksby, G. Sensors with centroid based common sensing and their multiplexing *Fiber Opt. Laser Sens.* **1992,** *1795* (10), 362–370.

Born, M.; Wolf, E. *Principles of Optics*, 5th ed.; Pergamon Press: Oxford, 1975.

Carre, P. Instalation et utilization du comparateur photoelectrique et interferantiel du Bureau International des Poids et Measures. *Metrologia* **1966,** *2*. (1), 13–23.

Chen, S.; Palmer, A. W.; Grattan, K. T. V.; Meggitt, B. T. Fringe Order Identification in Optical Fiber White-Light Interferometry Using Centroid Algorithm Method. *Electron. Lett.* **1992,** *28*, 553–555.

Cole, J. H.; Danver, B. A.; Bucaro, J. A. Syntetic-Heterodyne Interferometric Demodulation. *IEEE J. Quantum Elec.* **1982,** *QE-18* (4), 694–697.

Dandliker, R.; Zimmermann, E.; Frosio, G. Electronically Scanned White-Light Inteferometry: A Novel Noise-Resistant Signal Processing. *Opt. Lett.* **1992,** *17*, 679–681.

Dandridge, A.; Tveten, A. B.; Giallorenzi, T. G. Homodyne Demodulation Scheme for Fiber Optic Sensors Using Phase Generated Carrier. *IEEE J. Quantum Elec.* **1982,** *QE-18* (10), 1647–1653.

Danielson, B. L.; Whittenberg, C. D. Guided-Wave Reflectometry with Micrometer Resolution. *Appl. Opt.* **1987,** *26* (14), 2836–2842.

Diddams, S.; Diels, J. C. Dispersion Measurements with White-Light Interferometry. *J. Opt. Soc. Am. B* **1996,** *13* (6), 1120–1129.

Francon, M. *Optical Interferometry*; Academic Press: New York, NY, 1966.

Gerges, S.; Farahi, F.; Newson, T. P.; Jones, J. D. C.; Jackson, D. A. Interferometric Fibre-Optic Sensor Using a Short-Cohererence-Length Source. *Electonic Lett.* **1987,** *23* (21), 1110–1111.

Gerges, S.; Newson, T. P.; Jackson, D. A. Coherence Tuned Fiber Optic Sensing System with Self-Iinitialization Based on a Multimode Laser Diode. *Appl. Opt.* **1990,** *29*, 4473–4480.

Hariharan, P.; Malacara, D. Interference Interferometry and Interferometric Metrology. *SPIE* **1995,** *MS110*, 551–582.

Hariharan, P.; Malacara, D. Selected Papers on Interference, Interferometry and Interferometric Metrology. *SPIE Milestone* **1995,** *MS110.*

Hariharanm, P.; Oreb, B. F.; Eiju, T. Digital Phase-Shifting Interferometer: A Simple Error-Compensating Phase Calculation Algorithm. *Appl. Opt.* **1987,** *26*, 2504–2506.

Hauss, H. A. *Waves and Fields in Optoelectronics*; Prentice-Hall Inc., 1984.

Hecht, E. *Optics*, 2nd ed.; Addison-Wesley Publishing Company Inc., 1990.

Jensen, A.; Cour-Harbo, A. *Ripples in Mathematics: The Discrete Wavelet Transform*; Springer Verlag: Berlin Heidelberg, 2001.

Koch, U. R. Fiberoptic Displacement Sensor with 0.02 Mu-M Resolution by White-Light Interferometry. *Sens. Actuat. A* **1991,** *25*, 201–207.

Lai, S.; Yatagai, T. Use of the Fast Fourier Transform Method for Analyzing Linear and Equispaced Fizeau Fringes. *Appl. Opt.* **1994,** *33* (25), 5935–5940.

Lampard, D. Definitions of "Bandwidth" and "Time Duration" of Signals Which Are Connected By an Identity. *IRE Trans. Circ. Theor.* **1956,** *3*, pp. 286–288.

Larkin, K. G. Efficient Nonlinear Algorithm for Envelope Detection in White-Light Interferometry. *J. Opt. Soc. Am. A—Opt. Image Sci. Vis.* **1996,** *13*, 832–843.

Magyar, G.; Mandel, L. Interference Fringes Produced by Superposition of Two Independent Maser Light Beams. *Nature* **1963,** *198*, 255–256.

Mandel, L. *Optical Coherence and Quantum Optics*; Cambridge University Press, 1995.

Mandel, L. Quantum Theory of Interference Effects Produced by Inedependent Light Beams. *Physical Rewiew* **1964,** 134 (1A), A10–A15.

Marshall, R. H.; Ning, Y. N.; Palmer, A. W.; Meggit, B. T.; Grattan, K. T. V. A Novel Electronically Scanned White-Light Interferometer Using a Mach-Zehnder Approach. *J. Lightwave Technol.* **1996,** *14*, 397–402.

Matveev, A. N. *Optics*; Mir Publishers: Moscow, 1988.

McGarrity, A.; Jackson, D. A. The Frequency-Spectrum of a Phase-Modulated Interferometer Illuminated by a Gaussian Source. *J. Phys. A* **1996,** *80*, 4819–4830.

Meggit, T.; Boyle, W. J. O.; Grattan, K. T. V.; Baruch, A. E.; Palmer, A. W. Heterodyne Processing Scheme for Low Coherence Inerferometric Sensor Systems. *IEEE Proc. J.* **1991,** *138* (6), 393–395.

Mendlovic, D. Continuous Two-Dimensional On-Axis Optical Wavelet Transformer and Wavelet Processor with White-Light Illumination. *Appl. Opt.* **1998,** *37*, 1279–1282.

Meyer-Arendt, J. R. *Introduction to Classical and Modern Optics*; Prentice-Hall Inc., 1984.

Papoulis, A. *Probability, Random Variables, and Stochastics Processes* 3rd ed.; McGraw-Hill, Inc, 1991.

Park, J.; Li, X. Theoretical and Numerical Analysis of Superluminescent Diodes. *J. Lightwave Technol.* **2006,** *24* (6), 2473–2480.

Rao, Y. J.; Jackson, D. A. Universal Fiber-Optic Point Sensor System for Quasi-Static Absolute Measurements of Multiparameters Exploiting Low Coherence Interrogation. *J. Lightwave Technol.* **1996,** *14*, 592–600.

Romare, D.; Rizk, M. S.; Grattan, K. T. V.; Palmer, A. W. Superior LMS-Based Technique for White-Light Interferometric Systems. *IEEE Photon. Technol. Lett.* **1996,** *8* (1), 104–106.

Sandoz, P. Wavelet Transform as a Processing Tool in White-Light Interferometry. *Opt. Lett.* **1997,** *22*, 1065–1067.

Shabtay, G.; Mendlovic, D.; Zalevsky, Z. Optical Implementation of the Continuous Wavelet Transform. *Appl. Opt.* **1998,** *37*, 2964–2966.

Shough, D. Beyond Fringe Analysis, Interferometry VI: Techniques and Analysis. *Proc. SPIE* **1993,** *2003*, 208–223.

Takeda, H.; Kobayashi, S. Fourier-transform method of fringe-pattern analysis for computed-based topography and interferometry. *J. Opt. Soc. Am.* **1982,** 72, 156–160.

Wang, D. N. Multi-Wavelenght Combination Source with Fringe Pattern Transform Technique to Reduce the Equivalent Coherence Length in White Light Interferometry. *Sens. Actuat.* **2000,** *84*, 7–10.

Wang, D. N.; Grattan, K. T. V.; Palmer, A. W. Theoretical Analysis of the Optimum Wavelenght Combination in a Two-Wavelenght Combination Source for Optical Fiber Interferometric Sensing. *Sens. Actuat.* **2000,** *79*, 179–188.

Wang, N.; Ning, Y. N.; Grattan, K. T. V.; Palmer, A. W.; Weir, K. Optimized Multiwavelength Combination Sources for Interferometric Use. *Appl. Opt.* **1994,** *33*, 7326–7333.

Watkins, L. R.; Tan, S. M.; Barnes, T. H. Determination of Interferometer Phase Distrubitions By Use of Wavelets. *Opt. Lett.* **1999,** *13*, 905–907.

Wier, K.; Boyle, W. J. O.; Meggit, B. T.; Palmer, A. W.; Grattan, K. T. V. A Novel Adaptation of the Michelson Interferometer for the Measurement of Vibration. *J. Lightwave Tech.* **1992,** *10* (5), 700–703.

Weir, K.; Boyle, W. J. O.; Palmer, A. W.; Grattan, K. T. V.; Meggitt, B. T. Low Coherence Interferometric Fiber Optic Vibrometer Using Novel Optical Signal-Processing Scheme. *Electron. Lett.* **1991,** *27*, 1658–1660.

CHAPTER 7

Fiber-Optic Sensor Case Studies

ABSTRACT

The case studies of four different fiber-optic sensors are presented in this chapter. The first case represents a novel method for measuring the absolute position based on the white-light channeled spectrum. The second case deals with the rough surface height distribution measurement with the help of the low-coherence interferometry. The third case represents the novel principles of the wide-dynamic range low-coherence interferometry. Finally, the fourth case shows the optical coherence tomography technique with enhanced resolution.

7.1 ABSOLUTE POSITION MEASUREMENT WITH LOW-COHERENCE FIBER-OPTIC INTERFEROMETRY

Low-coherence interferometry in the form of the white-light channeled spectrum interferometry may serve as a base for a very simple absolute position measurement relaying on a simple signal processing of the captured spectral interferogram. This technique offers very accurate, high-resolution measurement with an error on the order of 10 nm in the measurement range of approximately 100 μm. Most of the two-beam interferometric configurations aimed for absolute position/displacement measurement are realized as the Michelson interferometer by using bulky optical components and custom-made spectrometers. Having separate and relatively long beam paths, this interferometric configuration is strongly affected by the changing surrounding environment conditions such as thermal drifts and/or vibrations, thus introducing additional errors in the measurement chain. To overcome the problems of the bulky optic interferometric setups, a very simple fiber optic–based absolute position measurement sensor using a commercially available spectrometer will be presented. The interferometer consists of the low-finesse Fabry–Perot interferometer formed between the fiber tip and the sensing mirror (SM) attached to the object which position is to be measured and which is positioned behind the fiber end at a relatively short distance. The air–glass interface reflection coefficient, located at the fiber end, and the coupling coefficient of the back-reflected light from the SM into the fiber end aperture are very low. Therefore, we assume that, in the case of such formed cavity, we have a simple two-beam interferometer. The first beam is obtained as a Fresnel reflection at the glass–air interface from the fiber tip and the second beam is obtained as a back-reflected light from the SM. Since both of the beams travel along the same sensing fiber (SF), any thermal and/or polarization drifts and/or vibrations that may arise along the SF are strongly suppressed. This interferometric sensing topology offers miniature sensing head (bear fiber tip) and strong disturbance rejection along the common path of the beams. Therefore, this measurement system provides the possibility of placing the signal processing unit at a relatively long distance from the sensing head.

Due to the occurred phase shift of the incident light after the reflection from the metallic mirror, the optical spectrum $S'(\omega)$, which is captured by the spectrometer, the following is valid:

$$S'(\omega) = S(\omega)\left[1 - V\cos(\Delta\tau\omega)\right], \tag{7.1}$$

where ω is the angular frequency of the optical signal, $S(\omega)$ is the optical spectrum of the used low-coherence light source (LCLS), V is the overall spectral fringe visibility for which $V \geq 0$ is valid, and it includes the effect of the response function of the spectrometer or equivalently the coherence length of the LCLS and also the effects of the geometry of the sensing interferometer, and $\Delta\tau$ is the time delay, which is given by:

$$\Delta\tau = \frac{2D}{c}, \tag{7.2}$$

where c is the speed of light in vacuum, D is the distance between the fiber tip and the SM. In Figure 7.1, the measured normalized channeled spectrum is depicted with the optical path difference of about 100 μm (distance between the fiber tip and the SM).

If we apply inverse Fourier transform on the captured optical power spectrum given in eq 7.1, we will obtain the following based on the Wiener–Khinchin theorem:

$$R'(\tau) = R(\tau) - \frac{V}{2}\left[R(\tau - \Delta\tau) + R(\tau + \Delta\tau)\right], \tag{7.3}$$

with $R'(\tau)$ and $R(\tau)$ being the autocorrelation functions of the captured channeled spectrum and LCLS spectrum, respectively. As one can see from eq 7.3, the autocorrelation function $R'(\tau)$ of the captured channeled spectrum consists of three components. The first component is the autocorrelation function of the LCLS spectrum. The second and the third terms represent time-delayed autocorrelation functions of the basic LCLS spectrum due to the modulation of the light spectrum in the sensing interferometer.

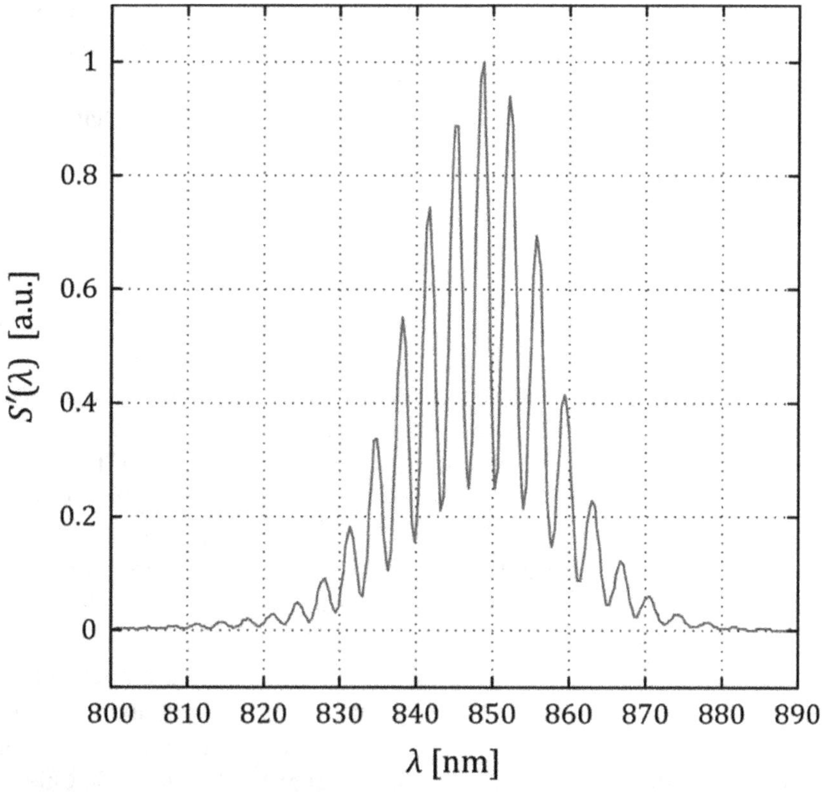

FIGURE 7.1 Measured normalized channeled spectrum for the optical path difference of about 100 μm.

The spectrum of the superluminescent diode (SLD) light takes an almost Gaussian shape. The inverse Fourier transform of the Gaussian-shaped signal has also Gaussian shape, so the autocorrelation functions $R'(\tau)$ and $R(\tau \pm \Delta\tau)$ are also approximately Gaussian functions. Since the inverse Fourier transform is an even function, we can take this function only for the positive values of the time delay τ ($\tau \geq 0$). If so, the normalized absolute value of the inverse Fourier transform of the spectrometer captured channeled spectrum can be approximated as a sum of two mutually delayed Gaussian functions in the following way:

$$|\rho(\tau)| \approx \exp\left[-\left(\frac{\tau}{\text{ä}\tau}\right)^2\right] + \frac{V}{2}\exp\left[-\left(\frac{\tau - \Delta\tau}{\text{ä}\tau}\right)^2\right], \tag{7.4}$$

where $\rho(\tau) = R'(0)/R'(\tau)$ is the normalized autocorrelation function ($\tau \geq 0$), $\delta\tau$ is the parameter that refers to $1/e$ width of the Gaussian function and it is directly proportional to the full width at half maximum (FWHM) coherence time τ_C of used LCLS. Approximately, we can take that correlation time (coherence time), the following is valid:

$$\tau_C = \frac{L_C}{c} \approx 2\sqrt{\ln 2} \cdot \delta\tau, \tag{7.5}$$

where L_C is the FWHM coherence length. To avoid overlapping of Gaussian functions from eq 7.4, it is necessary to ensure that the minimal possible distance from the fiber tip to the SM is larger than the coherence length of the used LCLS, so the coherence length represents at the same time the minimally possible measurable distance.

The SM absolute position measurement from the fiber tip can be performed by fitting the measured normalized autocorrelation function of the channeled spectrum with the derived function given in eq 7.4 and further by finding the time delay $\Delta\tau$ from the obtained fitted parameters of the function given in eq 7.4. From such obtained estimated time delay, the distance can be easily calculated according to the relation: $D = c\Delta\tau/2$. In Figure 7.2(a), it is presented the measured normalized autocorrelation function of the captured channeled spectrum (gray line) together with fitted data (black line). As it can be noticed from Figure 7.2(a) and (b), the fitting is very good with the goodness factor R^2 equal to $R^2 = 0.9997$. One can notice that the only discrepancies between the measured and fitted data are in the close proximity of the main peaks, marked as the "side-bands" that can be found near the larger peak in Figure 7.2(a) and near the smaller peak in Figure 7.2(b). The origin for these "side-bands" can be found in the slight asymmetry of the spectrum of the used LCLS.

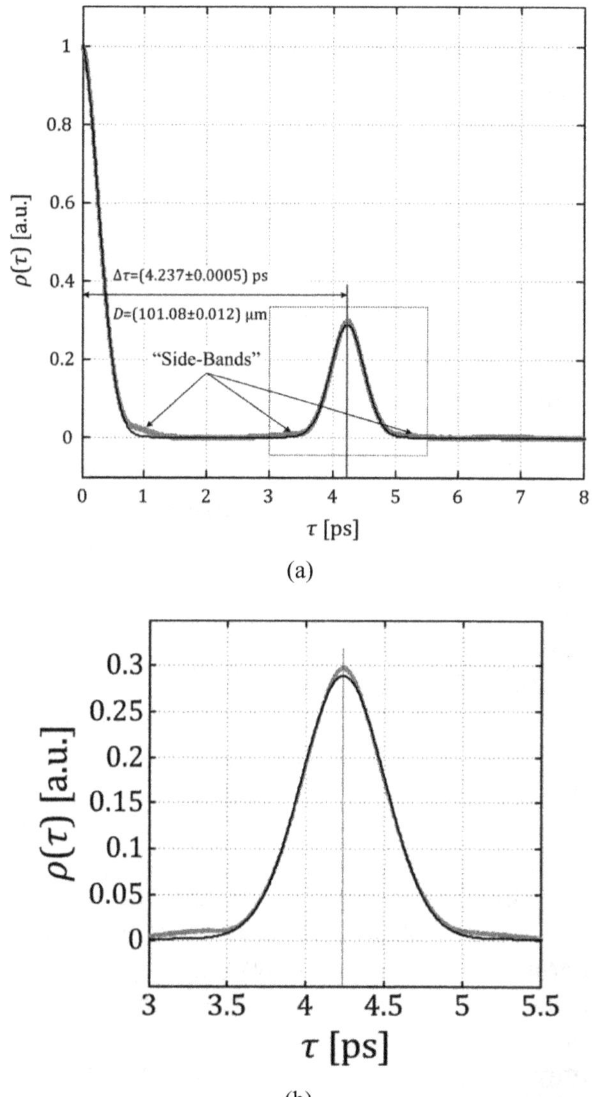

FIGURE 7.2 Normalized autocorrelation function $|\rho(\tau)|$ of the measured channeled spectrum: the gray line represents the measured normalized autocorrelation function and the black line represents the fitted data using the sum of two Gaussian functions.

Figure 7.3 represents the block schematic of the experimental setup. As an LCLS, a used pigtailed SLD SLD-381-MP2-DIL-SM-PD from

Superlum, Ltd., the SLD emits infrared light with the central wavelength λ_0 = 842.2 nm and with the spectral width $\Delta\lambda$ = 22.3 nm (FWHM) for the diode current of 140 mA and the junction temperature of 25°C. The coherence length of the SLD is given by:

$$L_C = k\frac{\lambda_0^2}{\Delta\lambda}, \tag{7.6}$$

where for the Gaussian spectrum shape, the following is valid:

$$k = \sqrt{\frac{2\ln 2}{\pi}} \approx 0.66, \tag{7.7}$$

so in the case of the abovementioned SLD, the coherence length is $L_C \approx$ 21 μm, which, at the same time, represents approximately the minimally possible measurable distance of about D_{min} 20 μm. SLD emits approximately P = 1.4 mW of optical power into a single-mode fiber at the same diode current and temperature as mentioned earlier. The commercial laser diode driver (LDD) multichannel laser diode control system PRO 800 with integrated temperature module TED 8xxx and current module LDC 8xxx from Profile Optische Systeme GmbH SLD is used for driving the SLD. SLD is connected to one of the input arms of the 2 × 2 fused silica single-mode fiber-optic directional coupler (FOC). The fiber-optic coupler is made of Corning Flexcore 850 5/125 μm type of the fiber and is made by Gould Technology, LLC. According to the accompanied data provided by the manufacturer, the insertion losses for both output arms are 2.8 and 3.4 dB at the wavelength 830 nm. The measured optical powers at two output arms after splitting are 710 μW (the total loss of 2.95 dB) and 560 μW (the total loss of 3.98 dB). To suppress parasitic back-reflection, one of the output arms, the one that with lower optical power, is immersed in the index matching gel (IMG). The other output arm representing the SF is directed toward the SM. To be able to test the sensing system capabilities the SM is firmly attached to the motorized micrometer positioning stage (MMPS) Z625B 25 mm Motorized Actuator from Thorlabs, Inc. The overall path of the motorized stage is approximately 25 mm and having the reduction gear backlash smaller than 8 μm. To measure the actual position of the motorized stage, the internal high-precision rotary Hall sensor-based encoder at the motor shaft side has been employed. The Hall sensor provides high-resolution position measurement with the error of about 40 nm. The C-843 DC-Motor Controller Card and Smart Move Operating

Software for C-843 Motor Controller PCI Card from Physik Instrumente (PI) GmbH & Co. KG and a personal computer (PC) are used for driving the motorized stage. The position of the SM is measured independently by the micrometer (MM) Digimatic Indicator ID-C125B from Mitutoyo Corp. In this way, the influence of the MMPS reduction gear backlash on the position measurement has been suppressed. The used micrometer has a position measurement range of 25.4 mm and a resolution of ±500 nm. Back-reflected light beams from the fiber end and the SM illuminate the optical spectrometer (OS). As OS, the S2000 Miniature Fiber Optic Spectrometer from Ocean Optics, Inc. is used. The spectral width to which the OS is sensitive is in the range from 333.35 to 1012.19 nm with 2048 resolvable points and with the average wavelength sampling period of about 0.33 nm. The second input arm of the FOC is connected to the OS. The spectrum of the back-reflected light is captured via microcontroller-based A/D card ADC1000-USB and OOIBase32™ spectrometer operating software from the same company and transferred to the PC via the USB interface. An average of 20 successive optical power spectrums captured with integration time of 10 ms is made for each channeled spectrum.

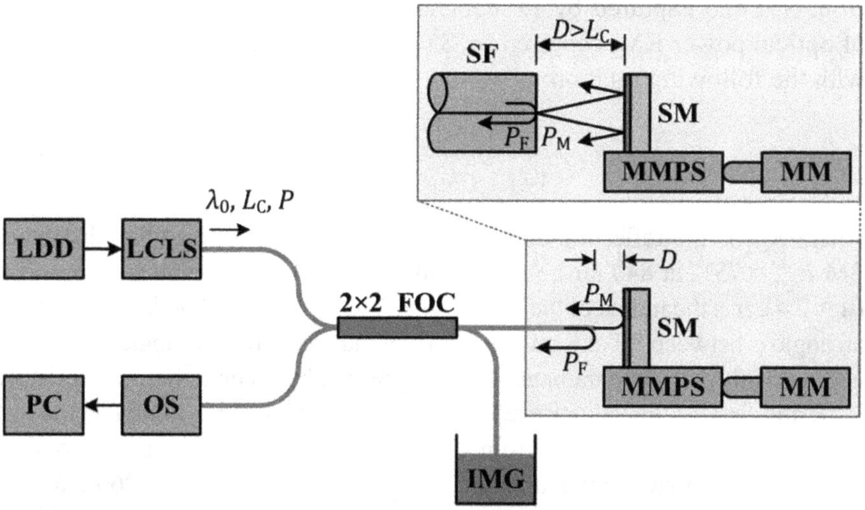

FIGURE 7.3 Block schematic of the experimental setup: LDD—laser diode driver, LCLS—low-coherence light source, PC—personal computer, OS—optical spectrometer, FOC—fiber-optic coupler, IMG—index matching gel, MMPS—motorized micrometer positioning stage, MM—micrometer, SF—sensing fiber, and SM—sensing mirror.

At the beginning of the measurement, the SM is positioned as close as possible to the fiber tip by using MMPS stage and optical microscope and resetting the micrometer for this position of the SM.

To measure the absolute position of the SM the position of the SM was changed with the help of the MMPS in the range from 50 to 250 μm with the step of 10 μm. The position of the SM was controlled with the help of MM.

The optical power of the OS captured back-reflected light depends on the position of the SM. The power of the back-reflected light, captured by the SF, consists of one fixed part (optical power) P_F obtained as a Fresnel reflection from the glass–air interface at the fiber end and one variable part (optical power) P_M obtained as a back-reflection from the SM. The fused silica has the refraction index of about $n_F \approx 1.45$ at the wavelength of 840 nm, so the reflection coefficient R_F at the glass–air interface is given by:

$$R_F = \left(\frac{n_F - 1}{n_F + 1} \right)^2 \approx 3.4\%, \tag{7.8}$$

thus giving the $P_F \approx 24$ μW. However, the optical power back-reflected from SM and captured by SF depends on the SM position. The fraction of optical power RM, captured by SM after the mirror reflection, is given with the following relation:

$$R_M = \frac{(1 - R_F)^2 R_{SM}}{1 + \left(\lambda_0 D \ln w / 2\pi a^2\right)^2} f(\varphi), \tag{7.9}$$

where R_{SM} is the reflection coefficient for SM (in the case of the aluminum SM $R_{SM} \approx 75\%$ at 840 nm), w is the single-mode fiber normalized frequency ($w \approx 2.4$), a is the radius of the fiber core ($a = 2.5$ μm), and $f(\varphi)$ is a function of an angle φ between SF axis and SM normal that describes the angle-dependent geometrical coupling parameter. If the alignment has been done, appropriately $\varphi \approx 0$ and $f(\varphi) \approx 1$ are valid. Based on these data and having in mind that the SM position changes in the range from $D_{min} = 50$ μm to $D_{max} = 250$ μm, the optical power P_M is changed in the range from $P_{Mmin} \approx 22$ μW to $P_{Mmax} \approx 260$ μW.

In Figure 7.4, it is presented the calculated overall spectral fringe visibility V. The fringe visibility is obtained from the parameters of the fitted curves. In Figure 7.4 are also presented the error bars giving information about all the possible values. The overall spectral fringe visibility is estimated as the double height of the time-delayed peak. Only the measured and not the calculated overall spectral fringe visibility is presented since the overall spectral fringe visibility depends on different parameters of the complete interferometric sensing setup. The effects of the geometry of the sensing interferometer dominate in the overall spectral fringe visibility term V, given in eq 7.1, and the theoretically estimated spectral fringe visibility V_G that depends only on the geometry of the sensing interferometer is given by:

$$V_G = \frac{2\sqrt{R_F R_M}}{R_F + R_M}, \tag{7.10}$$

where $V_G \approx V$ is valid. If we compare the experimental and calculated data obtained from eqs 7.9 and 7.10 for the overall spectral fringe visibility, one can find some discrepancies between these two results. The reason for this lies in the fact that some of the estimated values of the relevant parameters given earlier are just rough estimations and in the effect of the transfer function of the OS. Due to the nonideal cleaving of the SF and/or since there might be an inclination between SM and the SF and/or the possible presence of dirtiness at the SM surface, the relevant parameter values can be, also, significantly changed.

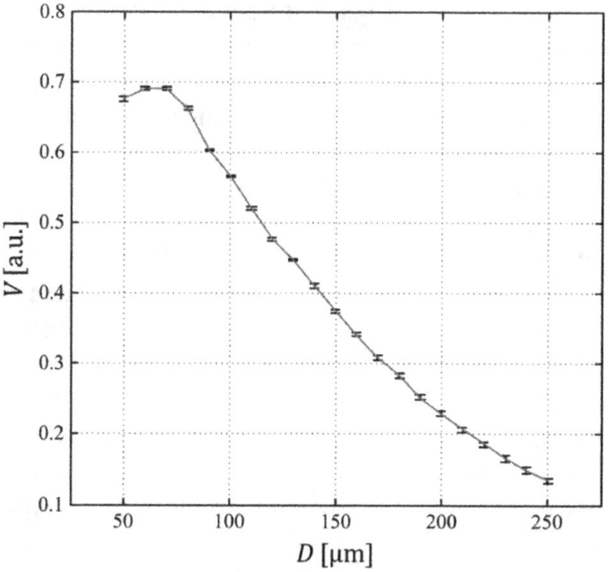

FIGURE 7.4 Measured spectral fringe visibility as a function of the distance measured by MM.

To determine the actual SM distance for each measurement point, we carry out the same procedure for absolute distance measurement as it was given earlier. We used the MathWorks™, Inc. MATLAB® Curve Fitting Toolbox™ for fitting the curve based on the calculated data. In Figure 7.5(a), it is presented the measured data (given with circles) vs. the position of the SM obtained from the MM. The linear fit based on the measured data set is presented with solid line on the same diagram. Since the goodness factor is $R^2 = 1$, the fitting is very good. Moreover, the obtained root-mean-squared error of RMSE = 440 nm is in a very good agreement with the measurement resolution of used MM (±500 nm), so the cause of most of the scattering errors lies in the low accuracy of the used MM. The dependence between the estimated SM distance D_E obtained by the presented measurement method and the distance D_S obtained from the MM stage is given by the following equation:

$$D_E = 1.009 D_S + 0.4397\ \mu\text{m}. \tag{7.11}$$

One can notice a slight discrepancy between the measured data and the data obtained from the MM stage. The cause of the discrepancy might lie the slight misalignment between the SF, MMPS, and MM (the slope is not exactly equal to 1 but it is equal to 1.009, for example, the axes of SF, MMPS, and MM are not perfectly collinear) where the small offset of approximately 440 nm is probably caused by the imperfect nulling of the MM at the very beginning of the measurement process, that is, at the beginning of the measurement, the distance between the fiber tip and SM was probably not equal to zero and also the MM resolution of about ±500 nm influenced the nulling. In the same diagram of Figure 7.5(a), it is presented the estimated measurement accuracy ΔD_A obtained as the difference between the estimated distance D_E and the distance measured by MM D_S, for example, $\Delta D_A = D_E - D_S$. Based on the estimated measurement accuracy, we obtained the worst-case measurement accuracy of about 2.5 μm. In Figure 7.5(b), it is presented the relative accuracy obtained as the ratio of measurement accuracy ΔD_A and distance D_S. For the average relative accuracy of about 1%, one can notice that this relative accuracy also originates from the imperfect misalignment of the complete measurement setup, because it is pretty close to the fitted line slope difference from an ideal case of about 0.9%.

The curve fitting tool as the result of the fitting process gives the 95% confidence boundaries (i.e., the 2σ boundaries) for the fitted parameters.

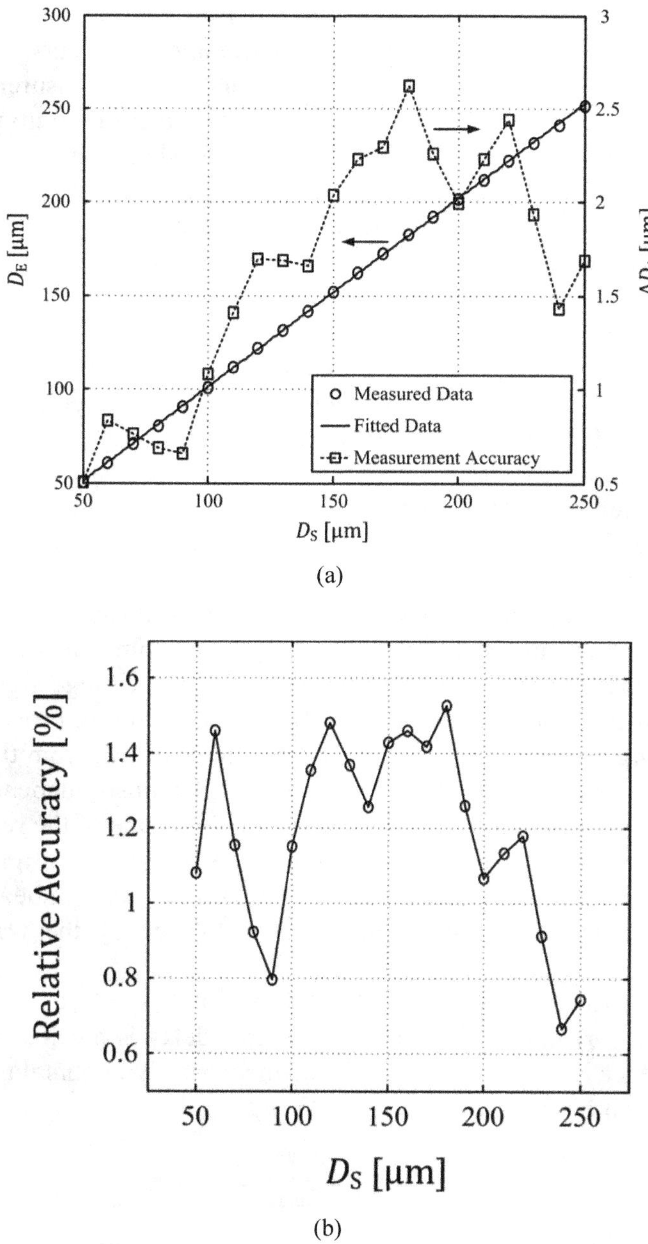

FIGURE 7.5 (a) Estimated distance of the sensing mirror vs. distance measured by MM: (○)—measured points, (-)—fitted data and (--□--)—measurement accuracy. (b) The relative accuracy vs. distance.

Based on the obtained confidence boundaries for the time delay, we obtained the 1σ errors for the SM distance measurements. In Figure 7.6(a), it is presented the 1σ estimated errors vs. the measurement SM distance. One can notice from Figure 7.6(a) that the minimum measurement error of less than 6 nm is obtained for the shorter distances (up to 110 μm but not for each point). Due to the small coupling coefficient of the back-reflected light and thus lower optical powers for larger distances (up to 250 μm), the measurement error increases rapidly. The maximum obtained measurement error is equal to approximately 120 nm for the distances in the measurement range of 240 – 250 μm. In the same diagram, it is also given the signal-to-noise ratio (SNR) measured as the ratio of the second peak height and its 1σ error (both values are obtained from the fitting process) vs. the SM distance. Significant decrease of the SNR can be noticed due to the decrease of the captured back-reflected light. In Figure 7.6(b), it is also presented the product of the SNR and the estimated error vs. the SM distance. One can notice that the corresponding product is pretty constant for the distances larger than 140 μm. The reason for slight changes of the measured product for shorter distances might lie in the fact that the estimated measurement error is rather small for short distances, and it can change significantly from measurement to measurement due to the rounding process of the curve fitting toolbox. The conclusion that can be drawn from the above-presented analysis is that the measurement error mostly depends on the SNR, and it is inversely proportional to the SNR. One of the reasons for the rapid estimated error jump for the larger SM distances might also lie in the fact that the maximum possible SM distance that can be measured with the presented measurement setup is limited by the wavelength sampling period of the used OS.

Taking into consideration the limitations due to the Nyquist–Shannon sampling theorem, eq 7.1, and since the time delay is much greater than the coherence time of the used LCS, the theoretically maximal time delay $\Delta\tau_{max}$ that can be sampled with used OS is given by:

$$\Delta\tau_{max} \leq \frac{\lambda_0^2}{2c\Delta\lambda_S}, \tag{7.12}$$

where $\Delta\lambda_S$ is the wavelength sampling period of the OS. According to eq 7.1 for the theoretically maximum possible distance D_{max}, which can be measured with the presented measurement system, the following is valid:

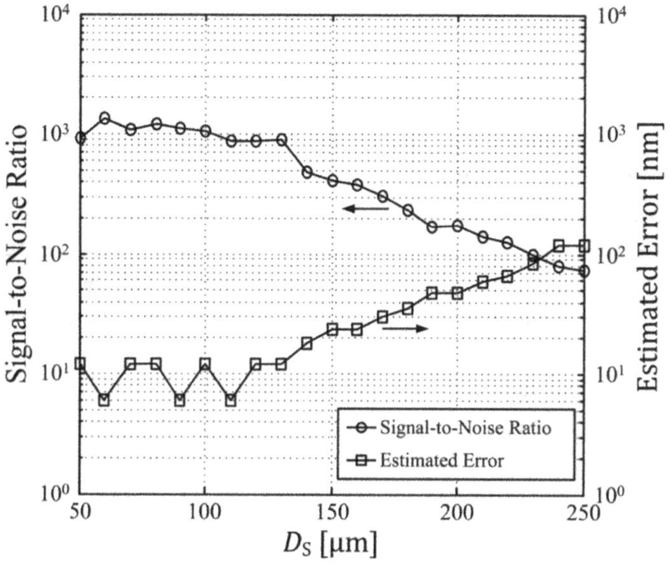

(a)

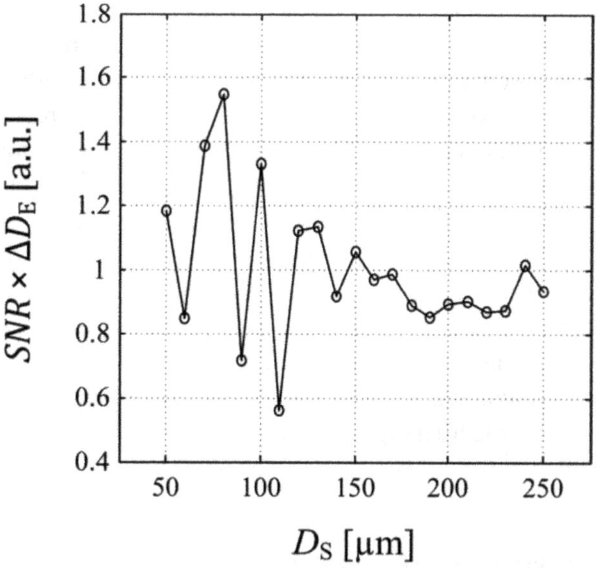

(b)

FIGURE 7.6 (a) Signal-to-noise ratio and estimated measurement error vs. distance: (-○-)—signal-to-noise ratio and (--□--)—estimated error. (b) Product of the SNR and the estimated error vs. the SM distance.

$$D_{max} \leq \frac{1}{4}\frac{\lambda_0^2}{\Delta\lambda_S}. \tag{7.13}$$

Having the LCSL with the central wavelength of approximately λ_0 = 842.2 nm and with the average wavelength sampling period of the OS of about $\Delta\lambda_S$ = 0.33 nm, we obtained the theoretically maximum possible measurable absolute distance of about D_{max} = 540 μm.

During the testing of the presented measurement sensing system, we also performed the measurement of the repeatability. Therefore, we positioned the SM at the distance of about 100 μm and captured 10 channeled spectrums by averaging 20 spectrums with the integration time of 5 ms. The current of the SLD is set to the value of 90 mA and the time between each spectrum capturing was 1 min. Based on the measured data, we obtained an average measured distance of 103.06 μm with an average drift of about 3 nm/mim and the root-mean-squared error of about 11 nm (after applying the linear fit on the measured position data set), so the repeatability of our measurement is approximately 11 nm. The measured repeatability is very close to the measurement error of 12 nm obtained for the same position of the SM.

Finally, we can say that the presented measurement sensing system has a very small sensing head (bare single-mode fiber with the diameter of only 125 μm). Moreover, the sensor has a very high resistance to the environmental influences, thus enabling us to use very long down-lead SF. The maximum possible measurement distance is limited by the wavelength sampling period of the OS, and it is limited to the maximal value of approximately D_{max} = 540 μm. On the other hand, for short distances, we have the minimal resolution of approximately ΔD_{min} = 6 nm. The theoretical dynamic range of the presented fiber-optic sensing system is defined as the ratio between the theoretically maximal possible measurable distance D_{max} and the minimal achievable resolution ΔD_{min} expressed in dB, so for the dynamic range of this fiber-optic sensing system defined in such a way, we have $DR = 20\log_{10}\left(D_{max} / \Delta D_{min}\right)$, for example, $DR \approx 99$ dB.

7.2 ROUGH SURFACE HEIGHT DISTRIBUTION MEASUREMENT WITH LOW-COHERENCE INTERFEROMETRY

Characterization of the rough surfaces (RSs) is of great importance in many fields of the high-precision material processing. The minimization of the production costs and significant reduction of the manufacturing time can be accomplished by using appropriate and convenient measurement techniques for the RS characterization in the automatically controlled machining systems. Such integrated measuring system will provide the possibility to achieve the desired roughness: not too smooth and not too rough in a relatively easy manner. RS characterization is also of the great importance in many other scientific and industrial fields such as contact mechanics, tribology, optics, medical industry, semiconductor industry, and machining industry.

RS characterization measurement techniques can be roughly divided into two broad categories: contact and contactless techniques. In the case of the microscopic-scale surface roughness measurements, a contact technique based on the stylus profiler using electronic signal amplification is the most popular due to their inherent high accuracy. To obtain RS height variation, the stylus of the stylus profiler is loaded onto the surface to be characterized and then moved across the surface at a constant velocity. The main drawbacks of the stylus profiler instruments for the in-process measurements are their need for the physical contact to the surface to be characterized and rather time-consuming measurement. However, there are on the other side numerous optical contactless techniques that are applicable to the in-process surface roughness measurement. The most common of these techniques are interferometry, laser speckle measurement, light scattering, confocal microscopy, ellipsometry, and other optical techniques. Optical techniques offer several advantages such as area-covering measurement, applicability to in-process measurement, and relatively fast measurement. The main drawback of optical methods is the difficulty to directly measure the widely accepted parameters for the RS characterization and also in the difficulty to transfer optically measured parameters into these RS parameters, such as centerline average roughness R_a or root-mean-squared average roughness R_q. Confocal microscopy as the output parameters gives the traditional RS parameters: the sensor follows a track in the same manner as a stylus during the measurement. In some other optical measurement techniques, indirect measurements can be made, and in some very restricted cases, the light-scattering parameter can be translated to R_a or R_q. Since being intensively used for a long period of time, the parameters, such as R_a and R_q, given in eqs 7.14 and 7.15, respectively, are standardized and widely accepted:

$$R_a = \lim_{S\to\infty}\left[\frac{1}{S}\int_S \left|\Delta z(x,y)\right| \mathrm{d}S\right] \text{ and} \tag{7.14}$$

$$R_q = \sqrt{\lim_{S\to\infty}\left\{\frac{1}{S}\int_S \left[\Delta z(x,y)\right]^2 \mathrm{d}S\right\}}, \tag{7.15}$$

where $\Delta z(x, y)$ is the RS profile departure from the line of average height at the position (x, y) on the RS, and S is the area of the RS that is to be characterized.

In this chapter, a very simple white light–based fiber-optic interferometric sensing unit for the RS height distribution measurement is presented. The sensor is realized as a two-beam interferometer where one of the beams illuminates the RS to be characterized and the other one illuminates the reference mirror (RM). By capturing both optical spectrums of the used white light source and of the interferometer channeled spectrum, the probability density function (PDF) can be found as the inverse Fourier transform of the ratio of two optical spectrums.

In Figure 7.7, it presented a simple sensing setup for the height distribution measurement of a RS that we will use in our theoretical analysis. The setup consists of a simple bulky white-light Michelson interferometer. A broadband light from the LCLS is spatially filtered and launched into the interferometer using achromatic collimating lens (L1). For the purpose of theoretical analysis, we will assume that, after passing the collimating lens, we have perfectly plane wave fronts. After the splitting on the beam splitter (BS), such perfectly plane waves impinge, in the first case, the RM and, in the second case, the RS, which height distribution is to be measured. After the back-reflections from the RM and RS, both of the beams are recombined in the BS and focused with the help of the second focusing also achromatic lens (L2) onto the slit of the optical spectrum analyzer (OSA). The OSA is connected with the personnel computer (PC) for data capturing and further analysis.

According to the Huygens–Kirchhoff principle for the electrical field $E_{\mathrm{CS}}(t)$, that we have at the input of the OSA we can write the following equation:

$$E_{\mathrm{CS}}(t) = E_{\mathrm{R}}(t) + E_{\mathrm{S}}(t) = E_{\mathrm{R}}(t) + \int_S \mathrm{d}E_{\mathrm{S}}(x, y, t), \tag{7.16}$$

where t is time, $E_{\mathrm{R}}(t)$ is the reference electrical field obtained as a back-reflection from the RM and focused onto the OSA entrance, $E_{\mathrm{S}}(t)$ is the overall electrical field obtained as a back-reflection from the RS and focused onto the OSA entrance, S is the illuminated area of the RS, x and y are the coordinates in the plane of the RS, and $dE_S(x, y, t)$ is elementary back-reflected electrical field obtained from the area $(x, x + dx; y, y + dy)$ of the RS and also focused onto the OSA entrance.

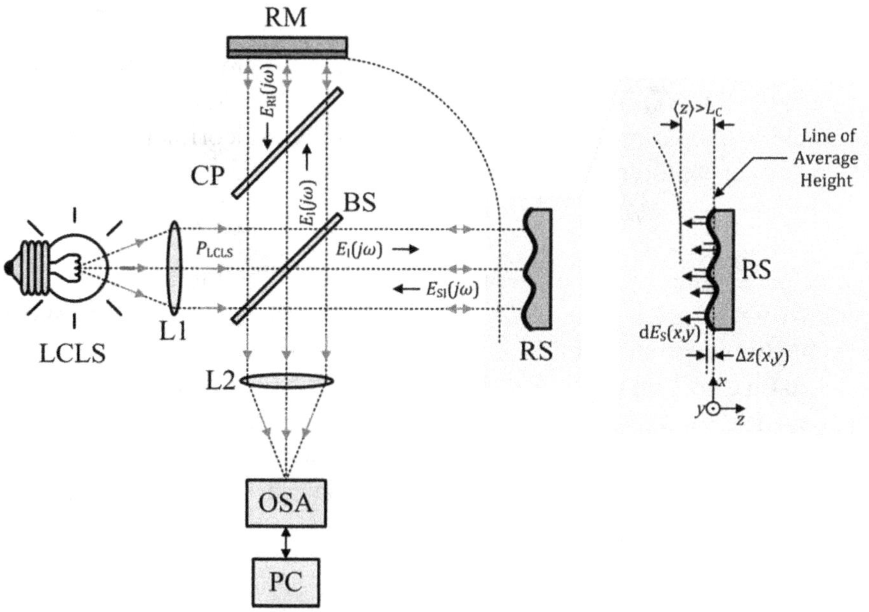

FIGURE 7.7 White-light Michelson interferometer for rough surface height distribution measurement.

Without losing the generality of the analysis, we will assume an ideal BS with the splitting ratio of 50%:50%, an ideal compensation plate (CP), and an ideal focusing lens, that is, each interferometer element doesn't change the incident light spectrum and doesn't introduce additional losses. This condition can be easily achieved when using the LCS with relatively narrow spectrum width, such as SLD. In this ideal case, for both electrical fields from eq 7.16, we can write the following:

$$E_R(t) = \frac{1}{\sqrt{2}} R_M A_F E_I(t) \text{ and} \tag{7.17}$$

$$dE_S(x, y, t) = E_I\left[t - \frac{2z(x, y)}{c}\right] dR_S(x, y), \tag{7.18}$$

where $E_I(t)$ is the incident electrical field after the BS, R_M is the reflection coefficient of the RM, and we take it as a real number (reflection coefficient can take the positive values for dielectric mirrors or the negative values for metallic mirrors), A_F is the gain factor obtained due to the focusing of the RM back-reflected light and assuming homogenous irradiance in the illuminated area it can be approximately estimated by $A_F \approx \sqrt{S/S_S}$, where S_S is the OSA slit area, $z(x,y)$ is the optical path difference at the position (x,y) on the RS, c is speed of light, and $dR_S(x,y)$ is the elementary coupling coefficient at the position (x,y) on the RS. For the autocorrelation function $R_{CS}(\tau)$ of the captured electrical field $E_{CS}(t)$, we have:

$$R_{CS}(\tau) = \langle E_{CS}(t) E_{CS}(t+\tau) \rangle, \tag{7.19}$$

where $\langle \bullet \rangle$ denotes the time average if we assume $E_{CS}(t)$ to be ergodic and stationary stochastic process. We will also assume that our OSA consists of a monochromator and the linear CCD array. Optical detectors in the OSA are sensitive only on the time average value of the light power falling onto the detector, so OSA measures average power spectral density (PSD) of the captured optical signal. According to the Wiener–Khintchine theorem, for the Fourier transform of the autocorrelation function given in eq 7.19, we have:

$$P_{CS}(\omega) = \frac{S_S}{Z_0} \mathcal{F}\{R_{CS}(\tau)\} = \frac{S_S}{Z_0} |E_{CS}(j\omega)|^2, \tag{7.20}$$

where $P_{CS}(\omega)$ is the OSA overall captured PSD, $\mathcal{F}\{\bullet\}$ denotes the Fourier transform operator, and Z_0 is the vacuum impedance. For the Fourier transform $E_{CS}(j\omega)$ of the captured electrical field, we have:

$$E_{CS}(j\omega) = E_I(j\omega) \left\{ \frac{1}{\sqrt{2}} R_M A_F + \int_S \exp\left[-j\omega \frac{2z(x,y)}{c} \right] dR_S(x,y) \right\}. \tag{7.21}$$

For the OSA captured optical PSD, we have according to eqs 7.20 and 7.21:

$$P_{CS}(\omega) = P_I(\omega) \left| \frac{R_M}{\sqrt{2}} + \frac{1}{A_F} \int_S \exp\left[-j\omega \frac{2z(x,y)}{c} \right] dR_S(x,y) \right|^2, \tag{7.22}$$

where $P_I(\omega) = S|E_I(j\omega)|^2 / Z_0$ is the incident optical PSD. Light, with the optical PSD given in eq 7.22, impinges the detectors in the OSA. For each detector in the linear CCD array, we can write that electrical voltage signal

$V_{CS}i$, at the output of *i*th detector (before A/D converter in the OSA) and neglecting the influence of the detector dark current due to the possibility of its cancellation, is given with:

$$V_{CSi} = G\frac{q\eta_D(\omega_i)}{\hbar\omega_i}\frac{T_{CS}}{C_{Di}}P_{CS}(\omega_i)\Delta\omega_{OSA}(\omega_i), \tag{7.23}$$

where G is the voltage gain of the readout electronics and it is constant for every detector in the CCD array, because due to the multiplexing process, always the same voltage amplifier and signal processing electronics are used for each detector, q is the elementary charge, $\eta_D(\omega_i)$ is the detector quantum efficiency and it is, in general case, wavelength dependent and is the same for the same wavelength for each detector, $\hbar$ is the reduced Planck constant, T_{CS} is the integration time during this measurement, C_{Di} is the *i*th detector capacitance, and $\Delta\omega_{OSA}(\omega_i)$ is the OSA angular frequency resolution and in general case due to the monochromator nonlinearity is frequency dependent. In combination with the monochromator, *i*th detector in the CCD array is illuminated with the light with the angular frequency in the range from ω_i to $\omega_i + \Delta\omega_{OSA}(\omega_i)$. In general case, we can write $\omega_i = \omega$, so we have $V_{CSi} = V_{CS}(\omega)$ and $C_{Di} = C_D(\omega)$. Finally, eq 7.23 yields to:

$$V_{CS}(\omega) = \frac{qG}{\hbar}\frac{\eta_D(\omega)\Delta\omega_{OSA}(\omega)}{\omega C_D(\omega)}T_{CS}P_{CS}(\omega). \tag{7.24}$$

As we can see from eq. 7.24, conversion factor is frequency dependent. Of course, the influence of the wavelength on the conversion factor can be eliminated by the OSA calibration. Similarly, we can measure the optical PSD of the light reflected from the RM and coupled into the OSA by removing the RS, thus eliminating the light coming from its surface. In this case, we have captured at the input of the OSA, the following electrical voltage signal $V_R(\omega)$:

$$V_R(\omega) = \frac{qG}{\hbar}\frac{\eta_D(\omega)\Delta\omega_{OSA}(\omega)}{\omega C_D(\omega)}T_R P_R(\omega), \tag{7.25}$$

where T_R is the integration time during this measurement and $P_R(\omega) = R_M^2 P_I(\omega)/2$ is the OSA measured optical PSD of the RM back-reflected light. Taking into consideration eqs 7.22, 7.23, and 7.25, we have:

$$V_{CS}(\omega) = \frac{T_{CS}}{T_R}V_R(\omega)\left|1 + \frac{\sqrt{2}}{R_M A_F}\int_S \exp\left[-j\omega\frac{2z(x,y)}{c}\right]dR_S(x,y)\right|^2. \tag{7.26}$$

As we can see from eq 7.26, by measuring both channeled spectrum and optical PSD of the RM back-reflected light, we eliminated the wavelength influence of the OSA parameters on the measurement of the surface roughness, that is, we suppressed the influence of the OSA multiplicative noise. This is important, because we can also use the OSA, which sensitivity is not calibrated with the wavelengths.

According to eq 7.22 and Huygens–Kirchhoff's principle, for the elementary coupling coefficient, we can write $\mathrm{d}R_S(x,y) = \rho(x,y)\,\mathrm{d}x\mathrm{d}y = \rho(x,y)\mathrm{d}S$ where $\rho(x,y)$ is a random function of the position (x,y) in the RS plane and depends on the geometrical parameters of the RS and on the reflection coefficient of the material the RS is made of. We assume that every elementary surface acts like a small mirror with the reflection coefficient equal to the reflection coefficient of the RS material and every back-reflected beam can be treated using the geometrical optics principle. At the input of the OSA, we have a small slit that serves as a spatial filter and because of that, every back-reflected beam that is inclined regarding the incident beam axis, miss the small opening at the input of the OSA. For that reason, we have coupled into the OSA only those reflections that originate from the surface areas where we have the following conditions fulfilled: $\partial z^2(x,y)/\partial x\partial y \approx 0$, that is, from the extreme (minima and maxima) and saddle points of the RS where the surface slope is approximately equal to zero. In this way, we eliminated the influence of the different RS slopes on the reflection coefficient as it is the case for dielectric materials, that is, for dielectric materials, reflection coefficient depends on the angle of incidence. So, we can define the function $\rho(x,y)$ in the following way:

$$\rho(x,y) \approx \begin{cases} \rho, & \dfrac{\partial z^2(x,y)}{\partial x \partial y} \approx 0 \\ 0, & \dfrac{\partial z^2(x,y)}{\partial x \partial y} \neq 0 \end{cases}, \qquad (7.27)$$

where ρ is the real number constant that depends on the reflection coefficient of the RS material for the normal incidence and on the macroscopic geometry of the complete optical setup. As we have coupled into the OSA back-reflected light only from a small portion of the RS, we can make a solid assumption that optical power of the back-reflected light from the RS is much lower than the optical power back-reflected from the RM. Taking into consideration all assumptions we made, we neglected the squared term of the integral term in eq 7.22, so for the captured optical PSD we can write:

$$V_{\mathrm{CS}}(\omega)=\frac{T_{\mathrm{CS}}}{T_{\mathrm{R}}}V_{\mathrm{R}}(\omega)\left\{1+\frac{2\sqrt{2}}{R_{\mathrm{M}}A_{\mathrm{F}}}\int_{S}\cos\left[\frac{2\omega}{c}z(x,y)\right]\rho(x,y)\mathrm{d}S\right\}. \quad (7.28)$$

In this way, by coupling only small amount of the back-reflected light from the RS, we linearize complete relation between measured channeled spectrums vs. RS height distribution, thus enabling, as we will see later on, finding the RS heights distribution in a closed form. As we do not have any contribution from the points of the RS, which have slopes different form zero (inclined with the respect to the RS), we can take into consideration new stochastic variable $\tilde{z}(x,y)$ obtained from $z(x,y)$ taking only those points where the following condition $\partial z^2(x,y)/\partial x\partial y \approx 0$ is fulfilled, that is:

$$\tilde{z}(x,y)=z(x,y)\Big|_{\frac{\partial z^2(x,y)}{\partial x\partial y}\approx 0}. \quad (7.29)$$

In this case, eq 7.28 can be written as:

$$V_{\mathrm{CS}}(\omega)\approx V_{\mathrm{R}}(\omega)\left\{1+\frac{2\sqrt{2}\rho}{R_{\mathrm{M}}A_{\mathrm{F}}}\int_{S}\cos\left[\frac{2\omega}{c}\tilde{z}(x,y)\right]\mathrm{d}S\right\}. \quad (7.30)$$

where we took for the integration times $T_{\mathrm{CS}} = T_{\mathrm{R}} = T_{\mathrm{I}}$ without losing the generality of the analysis. Equation 7.30 can be rearranged in the following way:

$$V_{\mathrm{CS}}(\omega)\approx V_{\mathrm{R}}(\omega)\left\{1+\frac{2\sqrt{2}\rho}{R_{\mathrm{M}}A_{\mathrm{F}}}S\left\{\frac{1}{S}\int_{S}\cos\left[\frac{2\omega}{c}\tilde{z}(x,y)\right]\mathrm{d}S\right\}\right\}. \quad (7.31)$$

If the area of the illuminated part of the RS S is much larger than the square of the RS heights correlation length, we can approximately write:

$$V_{\mathrm{CS}}(\omega)\approx V_{\mathrm{R}}(\omega)\left\{1+\frac{2\sqrt{2}\rho}{R_{\mathrm{M}}A_{\mathrm{F}}}S\cdot\lim_{S\to\infty}\left\{\frac{1}{S}\int_{S}\cos\left[\frac{2\omega}{c}\tilde{z}(x,y)\right]\mathrm{d}S\right\}\right\}, \quad (7.32)$$

or equivalently:

$$V_{\mathrm{CS}}(\omega)\approx V_{\mathrm{R}}(\omega)\left\{1+V\cos\left[\frac{2\omega}{c}\tilde{z}(x,y)\right]\right\}, \quad (7.33)$$

where $V = 2\sqrt{2}\rho S / R_{\mathrm{M}}A_{\mathrm{F}}$ is the spectral interferometer fringe visibility and $\langle\bullet\rangle$ denotes the surface average value of the stochastic variable function on the RS illuminated area. If we assume that stochastic process $\tilde{z}(x,y)$ is ergodic and stationary, we can assume that average value of the stochastic variable function on the RS illuminated area is equal to the ensemble average value of the stochastic variable function, so eq 7.33 becomes:

$$V_{\mathrm{CS}}(\omega) \approx V_{\mathrm{R}}(\omega)\left[1+V\int_{-\infty}^{+\infty}\cos\left(\frac{2\omega}{c}\tilde{z}\right)p_{\tilde{\xi}}(\tilde{z})\mathrm{d}\tilde{z}\right], \quad (7.34)$$

where $p_{\tilde{\xi}}(\tilde{z})$ denotes the PDF of the stochastic variable $\tilde{z}(x,y)$.

According, again, to the Wiener–Khintchine theorem, for the autocorrelation function $R_{\mathrm{CS}}(\tau)$ of the OSA captured equivalent electrical field (we take the term equivalent, because this equivalent electrical field differs from the real electrical field only for a constant factor due to the OSA data capturing process), we have:

$$R_{\mathrm{CS}}(\tau) = \frac{1}{2\pi}\int_{-\infty}^{+\infty}V_{\mathrm{CS}}(\omega)\exp(j\omega\tau)\mathrm{d}\omega, \quad (7.35)$$

or equivalently:

$$R_{\mathrm{CS}}(\tau) \approx R_{\mathrm{R}}(\tau) + \frac{V}{2\pi}\int_{-\infty}^{+\infty}V_{\mathrm{R}}(\omega)\left[\int_{-\infty}^{+\infty}\cos\left(\frac{2\omega}{c}\tilde{z}\right)p_{\tilde{\xi}}(\tilde{z})\mathrm{d}\tilde{z}\right]\exp(j\omega\tau)\mathrm{d}\omega \quad (7.36)$$

where $R_{\mathrm{R}}(\tau)$ is the autocorrelation function of the equivalent RM back-reflected electrical field. If we change the order of integration in eq. 7.36, we have:

$$R_{\mathrm{CS}}(\tau) \approx R_{\mathrm{R}}(\tau) + \frac{V}{2\pi}\int_{-\infty}^{+\infty}\left[\int_{-\infty}^{+\infty}V_{\mathrm{R}}(\omega)\cos\left(\frac{2\tilde{z}}{\mathrm{c}}\omega\right)\exp(j\omega\tau)\mathrm{d}\omega\right]p_{\tilde{\xi}}(\tilde{z})\mathrm{d}\tilde{z} \quad (7.37)$$

or equivalently, we have:

$$R_{\mathrm{CS}}(\tau) \approx R_{\mathrm{R}}(\tau) + \frac{V}{2}\int_{-\infty}^{+\infty}\left[R_{\mathrm{R}}\left(\tau-\frac{2\tilde{z}}{c}\right)+R_{\mathrm{R}}\left(\tau+\frac{2\tilde{z}}{c}\right)\right]p_{\tilde{\xi}}(\tilde{z})\mathrm{d}\tilde{z}. \quad (7.38)$$

Now if we take the following substitution: $t = 2\tilde{z}/c$ and taking into consideration that for the PDFs is valid: $\left|p_{\tilde{\xi}}(\tilde{z})\mathrm{d}\tilde{z}\right| = \left|p_{\tau}(t)\mathrm{d}t\right|$ (PDF always takes positive values), where $p_{\tau}(t)$ represents the PDF of the new stochastic variable t. Taking these substitutions into consideration, eq 7.38 becomes:

$$R_{\mathrm{CS}}(\tau) \approx R_{\mathrm{R}}(\tau) + \frac{V}{2}\int_{-\infty}^{+\infty}\left[R_{\mathrm{R}}(\tau-t)+R_{\mathrm{R}}(\tau+t)\right]p_{\tau}(t)\mathrm{d}t. \quad (7.39)$$

This is an integral equation that is very similar to the homogenous Fredholm integral equation of the first kind. The only difference is the second member under the integral. As the autocorrelation function is an even function, we can write $R_R(\tau + t) = R_R(-\tau - t)$ and eq 7.39 becomes:

$$R_{CS}(\tau) \approx R_R(\tau) + \frac{V}{2}\int_{-\infty}^{+\infty} R_R(\tau - t)\, p_\tau(t)\,dt + \frac{V}{2}\int_{-\infty}^{+\infty} R_R(-\tau - t)\, p_\tau(t)\,dt, \quad (7.40)$$

which also gives the following:

$$R_{CS}(\tau) \approx R_R(\tau) + \frac{V}{2}(R_R * p_\tau)(\tau) + \frac{V}{2}(R_R * p_\tau)(-\tau), \quad (7.41)$$

where the asterisk sign (*) denotes the functions convolution operator. According, again, to the Wiener–Khintchine theorem, if we made the Fourier transforms of both sides in eq 7.41 and taking into consideration that optical PSD is also an even function, that is, $V_R(\omega) = V_R(-\omega)$, we obtained:

$$V_{CS}(\omega) - V_R(\omega) \approx \frac{V}{2} V_R(\omega)\left[P_\omega(j\omega) + P_\omega(-j\omega)\right], \quad (7.42)$$

where $P_\omega(j\omega)$ is the Fourier transform of the PDF $p_\tau(t)$. Term on the left side of eq 7.42 represents only the "high frequency" part $V_{CS}^{HF}(\omega)$ of the OSA captured channeled spectrum $V_{CS}(\omega)$. It is better to find the "high frequency" part of the OSA captured channeled spectrum than subtract two optical PSDs, because between these two measurements, some of the parameters, such as LCLS power, can be changed, which can lead to the incorrect result. So finally, we have the following:

$$V_{CS}^{HF}(\omega) \approx \frac{1}{C} V_R(\omega)\left[P_\omega(j\omega) + P_\omega(-j\omega)\right], \quad (7.43)$$

where C is the real constant that will be found to fulfill the condition: $\int_{-\infty}^{+\infty} p_\tau(t)\,dt = 1$. In general, this constant is equal to $C = 2/V$ but due to lack of accurate values of the fringe visibility finally this parameter will be found according the abovementioned integral equation. If we rearrange eq 7.43, we have:

$$P_\omega(j\omega) + P_\omega(-j\omega) \approx C\frac{V_{CS}^{HF}(\omega)}{V_R(\omega)}. \quad (7.44)$$

If we make an inverse Fourier transforms of the both sides of eq 7.44, we have:

$$p_\tau(t) + p_\tau(-t) \approx C \cdot \mathcal{F}^{-1}\left\{\frac{V_{CS}^{HF}(\omega)}{V_R(\omega)}\right\}, \quad (7.45)$$

where $\mathcal{F}^{-1}\{\bullet\}$ denotes the inverse Fourier transform. As the PDF $p_\tau(t)$ is concentrated in the close proximity of the point $t = 2\langle\tilde{z}\rangle/c$ and as in general case, this function is nor even nor odd function, in eq 7.45, we have the sum of $p_\tau(t) + p_\tau(-t)$ to artificially made an even function, because inverse Fourier transform of an even real function is also an even real function. Finally, for the PDF $p_{\tilde{\zeta}}(\tilde{z})$, we can write:

$$p_{\tilde{\zeta}}(\tilde{z}) = \left|p_\tau(t)\right|\Big|_{t=\frac{2\tilde{z}}{c},t>0} \approx C\left|\mathcal{F}^{-1}\left\{\frac{V_{\mathrm{CS}}^{\mathrm{HF}}(\omega)}{V_{\mathrm{R}}(\omega)}\right\}\right|\Bigg|_{t=(2\langle\tilde{z}\rangle/c),t>0}, \tag{7.46}$$

where we have taken the absolute value of the Fourier transform, because the PDF always takes positive real values and the constant C assumed to be positive.

As we mentioned elsewhere, this is the PDF of the stochastic variable $\tilde{z}$ but not of the stochastic variable z we are interested in. To find the PDF function $p_\zeta(z)$ of the RS heights, in the following section will be given the mathematical analysis, which will give us the relation between these two PDFs.

In eq 7.46, it is given the PDF of the extreme and saddle point heights onto the RS. However, we are not interested in the PDF of the extreme and saddle point heights but in the PDF of the overall heights onto the RS. In this section, it will be presented the mathematical analysis to find the mathematical relation between these two PDFs. The probability $\mathrm{d}P_{\Delta\dot{z}} = 0$ that we have extreme or saddle point in the rectangular $(x, x+\mathrm{d}x; \Delta z, \Delta z + \mathrm{d}(\Delta z))$ is given by:

$$\mathrm{d}P_{\Delta\dot{z}=0} = -\mathrm{d}x\mathrm{d}(\Delta z)\int_{-\infty}^{+\infty} p_{\Delta\zeta}(\Delta z, \Delta\dot{z}=0, \Delta\ddot{z})\Delta\ddot{z}\mathrm{d}(\Delta\ddot{z}), \tag{7.47}$$

where Δz is the RS height, where the following condition is valid $\langle\Delta z\rangle = 0$, $\Delta\dot{z} = \mathrm{d}(\Delta z)/\mathrm{d}x$ and $\Delta\ddot{z} = \mathrm{d}^2(\Delta z)/\mathrm{d}x^2$ are the first and the second derivative of RS height, respectively, and $p_{\Delta\zeta}(\Delta z, \Delta\dot{z}=0, \Delta\ddot{z})$ is the joint PDF. In our analysis, we consider only dependence of the stochastic variable Δz upon x coordinate. This is mathematically justified because we made an assumption that stochastic variable Δz is stationary. For the joint PDF $p_{\Delta\zeta}(\Delta z, \Delta\dot{z}=0, \Delta\ddot{z})$, where we assumed the PDF of stochastic variable Δz to be Gaussian with a zero mean, we have:

$$p_{\Delta\zeta}(\Delta z, \Delta\dot{z}=0, \Delta\ddot{z}) = (2\pi)^{-3/2}|M|^{-1/2}\exp(\alpha), \tag{7.48}$$

where

$$\alpha = -\frac{1}{2|M|}\left[M_{11}(\Delta z)^2 + M_{33}(\Delta \ddot{z})^2 + 2M_{13}\Delta z\Delta \ddot{z}\right], \tag{7.49}$$

and where we have the following relations:

$\langle(\Delta z)^2\rangle = \sigma_{\Delta\varsigma}^2$, $\langle(\Delta \dot{z})^2\rangle = \dot{\sigma}_{\Delta\varsigma}^2$, $\langle\Delta z\Delta \dot{z}\rangle = 0$, $\langle\Delta \dot{z}\Delta \ddot{z}\rangle = 0$, $\langle\Delta z\Delta \ddot{z}\rangle = -\dot{\sigma}_{\Delta\varsigma}^2$, $\langle(\Delta \ddot{z})^2\rangle = \ddot{\sigma}_{\Delta\varsigma}^2$, $|M| = \dot{\sigma}_{\Delta\varsigma}^2\left(\sigma_{\Delta\varsigma}^2\ddot{\sigma}_{\Delta\varsigma}^2 - \dot{\sigma}_{\Delta\varsigma}^4\right)$, $M_{11} = \dot{\sigma}_{\Delta\varsigma}^2\ddot{\sigma}_{\Delta\varsigma}^2$, $M_{13} = \dot{\sigma}_{\Delta\varsigma}^4$, and $M_{33} = \sigma_{\Delta\varsigma}^2\dot{\sigma}_{\Delta\varsigma}^2$.

For the joint PDF $p_{\Delta\varsigma}(\Delta z, \Delta \dot{z} = 0)$ of the extreme and saddle points, we have:

$$p_{\Delta\varsigma}(\Delta z, \Delta \dot{z} = 0) = \int_{-\infty}^{+\infty} p_{\Delta\varsigma}(\Delta z, \Delta \dot{z} = 0, \Delta \ddot{z})\,\mathrm{d}(\Delta \ddot{z}), \tag{7.50}$$

which further yields to:

$$p_{\Delta\varsigma}(\Delta z, \Delta \dot{z} = 0) = (2\pi)^{-3/2}|M|^{-1/2}\exp\left[-\frac{1}{2|M|}\left(M_{11} - \frac{M_{13}^2}{M_{33}}\right)(\Delta z)^2\right]$$
$$\int_{-\infty}^{+\infty}\exp\left[-\frac{M_{33}}{2|M|}\left(\Delta \ddot{z} + \frac{M_{13}}{M_{33}}\Delta z\right)^2\right]\mathrm{d}(\Delta \ddot{z}), \tag{7.51}$$

After solving eq 7.51, we obtained for the joint PDF $p_{\Delta\varsigma}(\Delta z, \Delta \dot{z} = 0)$ the following equation:

$$p_{\Delta\varsigma}(\Delta z, \Delta \dot{z} = 0) = \frac{1}{2\pi\sigma_{\Delta\varsigma}\dot{\sigma}_{\Delta\varsigma}}\exp\left[-\frac{(\Delta z)^2}{2\sigma_{\Delta\varsigma}^2}\right], \tag{7.52}$$

where $\sigma_{\Delta\varsigma}$ is the RS height standard deviation, that is,: $\sigma_{\Delta\varsigma} = R_q$. According to Bayes' theorem, for the conditional PDF $p_{\Delta\varsigma}(\Delta z|\Delta \dot{z} = 0)$, that is, for the PDF of the extreme and saddle point heights $p_{\Delta\varsigma}(\Delta \tilde{z})$ of the RS, we have:

$$p_{\Delta\varsigma}(\Delta z|\Delta \dot{z} = 0) = p_{\Delta\tilde{\varsigma}}(\Delta \tilde{z}) = \frac{p_{\Delta\varsigma}(\Delta z, \Delta \dot{z} = 0)}{p_{\Delta\varsigma}(\Delta \dot{z} = 0)}, \tag{7.53}$$

which further leads to:

$$p_{\Delta\varsigma}(\Delta z|\Delta \dot{z} = 0) = \frac{1}{\sqrt{2\pi}\sigma_{\Delta\varsigma}}\exp\left[-\frac{(\Delta z)^2}{2\sigma_{\Delta\varsigma}^2}\right] = p_{\Delta\varsigma}(\Delta z), \tag{7.54}$$

where $\Delta\tilde{z}$ is the RS extreme and saddle point height, where the following condition is valid $\langle\Delta\tilde{z}\rangle = 0$. The complete analysis conducted earlier takes into consideration stochastic variables with Gaussian distribution with zero mean. However, optical path differences, in general case, do not have zero mean, but some real value different from zero. In this case, we have also the same distribution for both cases, because the mean value different from zero just makes the shift of the zero-mean Gaussian distribution, that is, $\Delta z = z - \langle z\rangle$. Finally, according to eq 7.54, for the Gaussian distribution of the RS heights, we have the same PDF for the overall height distribution as for the extreme and saddle point height distribution. At the end, if we assume approximately Gaussian distribution for the RS heights, we can write:

$$p_{\varsigma}(z) \approx C\left|\mathcal{F}^{-1}\left\{\frac{V_{\text{CS}}^{\text{HF}}(\omega)}{V_{\text{R}}(\omega)}\right\}\right|_{t=2z/c,t>0}. \tag{7.55}$$

Due to the mathematical operation of division in eq 7.55 for finding the PDF of the RS heights, which is sensitive to the optical PSD measurement uncertainty, that is, the random noise, especially in the parts where we have optical PSD tails (relatively low level of the captured optical signal), we will perform the mathematical analysis to find the measurement uncertainty of the PDF. For the measured ratio between two optical PSD, we have:

$$\frac{\hat{V}_{\text{CS}}^{\text{HF}}(\omega)}{\hat{V}_{\text{R}}(\omega)} = \frac{V_{\text{CS}}^{\text{HF}}(\omega) + N_{\text{CS}}(\omega)}{V_{\text{R}}(\omega) + N_{\text{R}}(\omega)} = S(\omega) + N(\omega), \tag{7.56}$$

where we have:

$$S(\omega) = \frac{V_{\text{CS}}^{\text{HF}}(\omega)}{V_{\text{R}}(\omega)}, \tag{7.57}$$

and

$$N(\omega) \approx \frac{N_{\text{CS}}(\omega)}{V_{\text{R}}(\omega)} - \frac{V_{\text{CS}}^{\text{HF}}(\omega)N_{\text{R}}(\omega)}{V_{\text{R}}^{2}(\omega)}, \tag{7.58}$$

where $\hat{V}_{\text{CS}}^{\text{HF}}(\omega)$ and $\hat{V}_{\text{R}}(\omega)$ are measured optical PSDs, $V_{\text{CS}}^{\text{HF}}(\omega)$ and $V_{\text{R}}(\omega)$ are accurate values of the optical PSDs, $N_{\text{CS}}(\omega)$ and $N_{\text{R}}(\omega)$ are random noises due to the measurement process, $S(\omega)$ is the equivalent useful signal, and $N(\omega)$ is the equivalent random noise. We also assumed to be $N_{\text{CS}}(\omega) \ll V_{\text{CS}}^{\text{HF}}(\omega)$ and $N_{\text{R}}(\omega) \ll V_{\text{R}}(\omega)$, that is, the overall SNR of the used

OSA is large enough. By combining eqs 7.56, 7.57, 7.58, and 7.55, we have:

$$p_{\varsigma}(z) \approx C\left|s(t)+n(t)\right|\Big|_{t=\frac{2z}{c},t>0}. \tag{7.59}$$

where it is fulfilled:

$$s(t)=\frac{1}{2\pi}\int_{-\infty}^{+\infty} S(\omega)\exp(j\omega t)\mathrm{d}\omega. \tag{7.60}$$

and

$$n(t)=\frac{1}{2\pi}\int_{-\infty}^{+\infty} N(\omega)\exp(j\omega t)\mathrm{d}\omega. \tag{7.61}$$

and where *s*(*t*) is the equivalent useful signal and *n*(*t*) is the equivalent random noise in time domain, where $|S(\omega)|^2$ and $|N(\omega)|^2$ are their energy spectral densities, respectively. For the SNR SNR_p of the measured PDF according to eqs 7.59, 7.60, and 7.61, we have:

$$\mathrm{SNR_P} \approx \frac{\max\left\{\left|\int_{-\infty}^{+\infty} S(\omega)\exp(j\omega t)\mathrm{d}\omega\right|^2\right\}}{\langle\left|\int_{-\infty}^{+\infty} N(\omega)\exp(j\omega \mathrm{t})\mathrm{d}\omega\right|^2\rangle s}. \tag{7.62}$$

As the ratio $R_{\mathrm{RS}}(\omega)=V_{\mathrm{CS}}^{\mathrm{HF}}(\omega)/V_{\mathrm{R}}(\omega)$ is independent on the shape of the used LCLS optical PSD but only on the RS characteristics, eq 7.62 becomes:

$$\mathrm{SNR_P} \approx \frac{\max\left\{\left|\int_{-\infty}^{+\infty} R_{\mathrm{RS}}(\omega)\exp(j\omega t)\mathrm{d}\omega\right|^2\right\}}{\langle\left|\int_{-\infty}^{+\infty} \frac{N_{\mathrm{CS}}(\omega)-R_{\mathrm{RS}}(\omega)N_{\mathrm{R}}(\omega)}{V_{\mathrm{R}}(\omega)}\exp(j\omega t)\mathrm{d}\omega\right|^2\rangle}, \tag{7.63}$$

The term in the denominator of eq 7.63 represents the equivalent random noise signal *n*(*t*) in time domain for which is valid:

$$n(t)=\frac{1}{2\pi}\int_{-\infty}^{+\infty} \frac{N_{\mathrm{CS}}(\omega)-R_{\mathrm{RS}}(\omega)N_{\mathrm{R}}(\omega)}{V_{\mathrm{R}}(\omega)}\exp(j\omega t)\mathrm{d}\omega. \tag{7.64}$$

For random noise in the optical PSD measurement is valid, the following relations: $\langle N_{\mathrm{CS}}(\omega)\rangle = 0$ and $\langle N_{\mathrm{R}}(\omega)\rangle = 0$, that is, they have zero mean value and they are mutually independent stochastic processes, so we have $\langle N_{\mathrm{CS}}(\omega)N_{\mathrm{R}}(\omega)\rangle = 0$, because they originated from two independent measurements. We neglect the influence of other measurement errors, for example, fix pattern noise, because it is simple to eliminate this kind of measurement errors due to their predictability. However, we take into consideration only the influence of the stochastic errors like random noise, because they cannot be easily eliminated due to its randomness. The term under the integral in the denominator of eq 7.64 represents the Fourier transform of the overall equivalent random noise $N(\omega) = \mathcal{F}\{n(t)\}$, so we can write:

$$N(\omega)=\frac{N_{\mathrm{CS}}(\omega)-R_{\mathrm{RS}}(\omega)N_{\mathrm{R}}(\omega)}{V_{\mathrm{R}}(\omega)}. \tag{7.65}$$

According to Parseval's theorem, we have:

$$\int_{-\infty}^{+\infty} n^2(t)\mathrm{d}t=\frac{1}{2\pi}\int_{-\infty}^{+\infty}\left|N(\omega)\right|^2\mathrm{d}\omega, \tag{7.66}$$

Taking the ensemble average of both sides in eq 7.66, we have:

$$\int_{-\infty}^{+\infty} n^2(t)\mathrm{d}t=\langle\frac{1}{2\pi}\int_{-\infty}^{+\infty}\left|N(\omega)\right|^2\mathrm{d}\omega\rangle, \tag{7.67}$$

or equivalently:

$$\int_{-\infty}^{+\infty} n^2(t)\mathrm{d}t=\frac{1}{2\pi}\int_{-\infty}^{+\infty}\left|N(\omega)\right|^2\mathrm{d}\omega, \tag{7.68}$$

where we put the ensemble averaging operator under the integral by changing the order of integration. Due to the existence of the Fourier transform of the signal $n(t)$, we have fulfilled the following condition: $\int_{-\infty}^{+\infty}\left|n(t)\right|\mathrm{d}t<+\infty$. This is only possible if we have time-limited signal $n(t)$. Any real OSA measures the optical PSD only in some limited range of the wavelengths, so the Fourier transform $N(\omega)$ of the signal $n(t)$ is frequency-limited. Therefore, for the time duration of the band-limited signal, we approximately have $\Delta t_{\mathrm{D}}\Delta\omega_{\mathrm{B}}\approx 1$, where Δt_{D} is time duration of the signal $n(t)$ and $\Delta\omega_{\mathrm{B}}=2\pi c\left(1/\lambda_{\mathrm{RL}}-1/\lambda_{\mathrm{RH}}\right)$ is the frequency range of the OSA, where λ_{RL} and λ_{RH} are the lower and the higher wavelength range, respectively. Taking all this into consideration and also by assuming: $\int_{-\infty}^{+\infty} n^2(t)\mathrm{d}t\approx n^2(t)\Delta t_{\mathrm{D}}$, eq 7.68 becomes:

$$\langle n^2(t)\rangle\approx\frac{1}{2\pi\Delta t_{\mathrm{D}}}\int_{-\infty}^{+\infty}\langle\left|N(\omega)\right|^2\rangle\mathrm{d}\omega. \tag{7.69}$$

By combining eqs 7.65 and 7.69, for the equivalent random noise variance, we have:

$$\langle n^2(t)\rangle \approx \frac{\Delta\omega_B}{2\pi}\int_{-\infty}^{+\infty}\langle\frac{N_{CS}^2(\omega)+R_{RS}^2(\omega)N_R^2(\omega)}{V_R^2(\omega)}\rangle\,d\omega, \tag{7.70}$$

where we take into consideration that two random noise processes are independent. As we have very similar measurement conditions during the measurement of the RM back-reflected light optical PSD and during the measurement of channeled spectrum (the same integration time $T_{CS} = T_R = T_I$), we approximately have $\langle N_{CS}^2(\omega)\rangle \approx \langle N_R^2(\omega)\rangle$ and if we assume that captured optical signal is much higher than random noise, we can say that we fulfilled the condition $R_{RS}^2(\omega) >> \langle N_{CS}^2(\omega)\rangle, \langle N_R^2(\omega)\rangle$. In this case, eq 7.70 becomes:

$$\langle n^2(t)\rangle \approx \frac{\Delta\omega_B}{2\pi}\int_{-\infty}^{+\infty}\langle\frac{R_{RS}^2(\omega)}{V_R^2(\omega)}N_R^2(\omega)\rangle\,d\omega. \tag{7.71}$$

As we take the random optical PSD noise pattern $N_R(\omega)$ only in one measurement, we need to make the ensemble averaging of the measurement to finally obtain the equivalent random noise average power to be:

$$\langle n^2(t)\rangle \approx \frac{\Delta\omega_B}{2\pi}\int_{-\infty}^{+\infty}\frac{R_{RS}^2(\omega)}{V_R^2(\omega)}\langle N_R^2(\omega)\rangle d\omega, \tag{7.72}$$

which further yields to

$$\langle n^2(t)\rangle \approx \frac{\Delta\omega_B}{2\pi}\langle N_R^2(\omega)\rangle\int_{-\infty}^{+\infty}\frac{R_{RS}^2(\omega)}{V_R^2(\omega)}d\omega, \tag{7.73}$$

and where we finally got:

$$\langle n^2(t)\rangle \approx \frac{\Delta\omega_B}{2\pi}\sigma_{OSA}^2\int_{-\infty}^{+\infty}\frac{R_{RS}^2(\omega)}{V_R^2(\omega)}d\omega, \tag{7.74}$$

where we assumed $N_R(\omega)$ to be stationary stochastic process, σ_{OSA} is the OSA detectors random voltage noise standard deviation and it was assumed to be constant for each detector of the linear CCD array (stationary stochastic process, that is, σ_{OSA} is not a function of frequency). The condition of stationarity is fulfilled when the detectors voltage noise almost doesn't depend on the detectors captured optical signal power level but depends only on the detectors dark current and readout electronics noise. We have this situation when we have low optical power levels, that is, when we work far away from the quantum limit regime. Low optical signal levels can significantly degrade SNR and because of that there is a good reason for choosing the optimal interferometer parameters that will maximize SNR. Hence, we performed this mathematical analysis to maximize SNR in RS heights distribution measurement. In the opposite case, when we almost reach quantum limit regime, we usually have high SNR that enables us much-relaxed choice of interferometer parameters, so this situation is not so interesting for the analysis and optimization. Finally, for the SNR, we have:

$$SNR_{\mathrm{P}} \approx \frac{2\pi}{\Delta\omega_{\mathrm{B}}\sigma_{\mathrm{OSA}}^{2}} \frac{\max\left\{\left|\int_{-\infty}^{+\infty} R_{\mathrm{RS}}(\omega)\exp(j\omega t)\mathrm{d}\omega\right|^{2}\right\}}{\int_{-\infty}^{+\infty}\left(R_{\mathrm{RS}}^{2}(\omega)/V_{\mathrm{R}}^{2}(\omega)\right)\mathrm{d}\omega}. \tag{7.75}$$

As the term given in the nominator of eq 7.75 represents the peak value of the PDF $p_{\tau}(t)$, it has its maximal value for the time approximately equal to $t = t_0 = 2\langle z\rangle/c$, where $\langle z\rangle$ is the mean optical path difference of the RS. In this case, eq 7.75 becomes:

$$SNR_{\mathrm{P}} \approx \frac{2\pi}{\Delta\omega_{\mathrm{B}}\sigma_{\mathrm{OSA}}^{2}} \frac{\left|\int_{-\infty}^{+\infty} R_{\mathrm{RS}}(\omega)\exp(j\omega t_0)\mathrm{d}\omega\right|^{2}}{\int_{-\infty}^{+\infty}\left(R_{\mathrm{RS}}^{2}(\omega)/V_{\mathrm{R}}^{2}(\omega)\right)\mathrm{d}\omega}. \tag{7.76}$$

The ratio $R_{\mathrm{RS}}(\omega)$ is an even function, because it represents "high frequency" part of the ratio of two optical PSDs. For that reason, imaginary term in the nominator vanishes due to the integration process, so we have:

$$SNR_{\mathrm{P}} \approx \frac{2\pi}{\Delta\omega_{\mathrm{B}}\sigma_{\mathrm{OSA}}^{2}} \frac{\left[\int_{-\infty}^{+\infty} R_{\mathrm{RS}}(\omega)\exp(j\omega t_0)\mathrm{d}\omega\right]^{2}}{\int_{-\infty}^{+\infty}\left(R_{\mathrm{RS}}^{2}(\omega)/V_{\mathrm{R}}^{2}(\omega)\right)\mathrm{d}\omega}. \tag{7.77}$$

Using the Cauchy–Schwarz–Bunyakovsky inequality and conducting the same analysis as in the case of finding the transfer function of the matched filter, the maximal SNR is achieved if the used optimal RM back-reflected light optical PSD, $V_{\mathrm{R}}^{\mathrm{opt}}(\omega)$, fulfills the following condition:

$$V_{\mathrm{R}}^{\mathrm{opt}}(\omega) = K\sqrt{\frac{R_{\mathrm{RS}}(\omega)}{\cos(\omega t_0)}}, \tag{7.78}$$

where K is an arbitrary real and positive constant. The analysis is given in more detail afterward.

For the optical path differences between the RS and RM, we can take that $z = \langle z \rangle + \Delta z$ is valid, where $\langle z \rangle$ is an average optical path difference and Δz is the surface roughness. Taking this into consideration and taking that both stochastic variables $\tilde{z}$ and z have the same distribution, we can rearrange eq 7.33 in the following way:

$$R_{\text{RS}}(\omega) \approx V\langle \cos\left[\frac{2\omega}{c}\left(\langle z \rangle + \Delta z\right)\right] \rangle, \tag{7.79}$$

which can be rearranged into:

$$R_{\text{RS}}(\omega) \approx V\left[\cos\left(\frac{2\omega}{c}\langle z \rangle\right)\langle \cos\left(\frac{2\omega}{c}\Delta z\right)\rangle - \sin\left(\frac{2\omega}{c}\langle z \rangle\right)\langle \sin\left(\frac{2\omega}{c}\Delta z\right)\rangle\right], \tag{7.80}$$

In the previous analysis, we assumed PDF of the optical path difference z to be approximately Gaussian with the mean value different from zero. In this case, the PDF of the surface roughness Δz is also Gaussian but with the zero mean. If so, the surface roughness PDF can be assumed approximately to be an even function, and in this case, the following condition $\langle \cos(2\omega\,\Delta z/c) \rangle \gg \langle \sin(2\omega\,\Delta z/c) \rangle$ is fulfilled. Hence according to eq 7.80 for the ratio $R_{\text{RS}}(\omega)$, we can write the following:

$$R_{\text{RS}}(\omega) \approx V\langle \cos\left(\frac{2\omega}{c}\Delta z\right)\rangle \cos\left(\frac{2\omega}{c}\langle z \rangle\right), \tag{7.81}$$

or equivalently:

$$R_{\text{RS}}(\omega) \approx A_{\text{RS}}(\omega)\cos(\omega t_0), \tag{7.82}$$

where $A_{\text{RS}}(\omega)$ is proportional to the Fourier transform of the surface roughness PDF $p_{\Delta\tau}(\Delta t)$ in time domain and $\Delta t = 2\Delta z/c$ and $t_0 = 2\langle z \rangle/c$ is valid. If we assume surface roughness PDF to be approximately Gaussian with the surface roughness R_q (standard deviation of the surface heights $\sigma_{\Delta\zeta} = R_q$), we have:

$$A_{\text{RS}}(\omega) \propto \exp\left[-2R_q^2\left(\frac{\omega}{c}\right)^2\right]. \tag{7.83}$$

Taking into consideration eqs 7.78, 7.82, and 7.83 for the optimal RM back-reflected light optical PSD, we have:

$$V_{\mathrm{R}}^{\mathrm{opt}}(\omega) = K\sqrt{\frac{R_{\mathrm{RS}}(\omega)}{\cos(\omega t_0)}} \propto \exp\left[-R_q^2\left(\frac{\omega}{c}\right)^2\right]. \tag{7.84}$$

or equivalently:

$$V_{\mathrm{R}}^{\mathrm{opt}}(\lambda) \propto \exp\left[-\left(\frac{R_q}{\lambda}\right)^2\right]. \tag{7.85}$$

As from eq 7.25, we know that $V_{\mathrm{R}}(\omega)$ is linearly proportional to the LCLS optical PSD $V_{\mathrm{R}}(\omega) = \Re(\omega)P_{\mathrm{R}}(\omega)$, where $\Re(\omega) \approx$ const. is the OSA conversion factor and is constant for calibrated OSA and $P_{\mathrm{R}}(\omega) = R_{\mathrm{M}}^2(\omega)P_{\mathrm{I}}(\omega)/2$ and $P_{\mathrm{I}}(\omega) = P_{\mathrm{LCLS}}(\omega)/2$ is valid, where $P_{\mathrm{LCLS}}(\omega)$ is the used LCLS optical PSD, so for the optimal LCLS PSD $P_{\mathrm{LCLS}}^{\mathrm{opt}}(\lambda)$ that will provide maximal possible SNR in RS height distribution measurement, we have:

$$P_{\mathrm{LCLS}}^{\mathrm{opt}}(\lambda) \propto \exp\left[-\left(\frac{R_q}{\lambda}\right)^2\right]. \tag{7.86}$$

In the general case of the optimal LCLS PSD shape, for the maximum measurement SNR $SNR_{\mathrm{P}}^{\max}$, we approximately have:

$$SNR_{\mathrm{P}} \approx \frac{2\pi}{\Delta\omega_{\mathrm{B}}\sigma_{\mathrm{OSA}}^2}\int_{-\infty}^{+\infty}\left[V_{\mathrm{R}}^{\mathrm{opt}}(\omega)\right]^2 \mathrm{d}\omega, \tag{7.87}$$

where we assumed that time delay t_0 is much higher than LCLS coherence time (LCLS coherence length is much shorter than mean optical path difference), that is, the following condition is fulfilled: $\int_{-\infty}^{+\infty}\left[V_{\mathrm{R}}^{\mathrm{opt}}(\omega)\right]^2 \mathrm{d}\omega >> \int_{-\infty}^{+\infty}\left[V_{\mathrm{R}}^{\mathrm{opt}}(\omega)\right]^2 \cos(2\omega t_0)\mathrm{d}\omega$. According to the Chebyshev integral inequality and assuming finite angular frequency bandwidth of the LCLS, $\Delta\omega_{\mathrm{LCLS}} = \omega_{\mathrm{LCLS}}^{\mathrm{H}} - \omega_{\mathrm{LCLS}}^{\mathrm{L}} \approx 2\pi c\Delta\lambda_{\mathrm{LCLS}}/\lambda_{\mathrm{LCLS0}}^2$, where $\omega_{\mathrm{LCLS}}^{\mathrm{H}}$ is the upper and $\omega_{\mathrm{LCLS}}^{\mathrm{L}}$ is the lower angular frequency range of the used LCLS spectrum, respectively, λ_{LCLS0} is the LCLS central wavelength and $\Delta\lambda_{\mathrm{LCLS}}$ is the LCLS spectrum width of the used LCLS, we have:

$$\Delta\omega_{\mathrm{LCLS}}\int_{-\infty}^{+\infty}\left[V_{\mathrm{R}}^{\mathrm{opt}}(\omega)\right]^2 \mathrm{d}\omega \geq \left[\int_{-\infty}^{+\infty} V_{\mathrm{R}}^{\mathrm{opt}}(\omega)\mathrm{d}\omega\right]^2 = \Re^2(T_{\mathrm{I}})P_{\mathrm{R}}^2, \tag{7.88}$$

where P_{R} is the overall optical power of the light back-reflected from the RM and $\Re$ is the conversion factor that depends on the OSA parameters and on the integration time T_{I}, that is, $\Re = \Re(T_{\mathrm{I}})$. By combining eqs 7.87 and 7.88 for the SNR, we obtained:

$$SNR_{\mathrm{P}}^{\max} \geq \frac{R_{\mathrm{M}}^{4}}{32c} \frac{\lambda_{\mathrm{LCLS0}}^{2}}{\Delta\lambda_{\mathrm{LCLS}}} \frac{\Re^{2}(T_{\mathrm{I}})}{\Delta\omega_{\mathrm{B}}\sigma_{\mathrm{OSA}}^{2}} P_{\mathrm{LCLS}}^{2}(\omega), \tag{7.89}$$

where the OSA conversion factor $\Re(T_{\mathrm{I}})$ is given with:

$$\Re(T_{\mathrm{I}}) = G\frac{q\eta_{\mathrm{D}}}{\hbar C_{\mathrm{D}}} \frac{\Delta\omega_{\mathrm{OSA}}}{\omega} T_{\mathrm{I}} \approx G\frac{q\eta_{\mathrm{D}}}{\hbar C_{\mathrm{D}}} \frac{\Delta\lambda_{\mathrm{OSA}}}{\lambda_{\mathrm{LCLS0}}} T_{\mathrm{I}}, \tag{7.90}$$

where $\Delta\lambda_{\mathrm{OSA}}$ is the OSA wavelength resolution and we assumed that LCLS wavelengths are concentrated in the close proximity to the central wavelength.

The OSA detectors random voltage noise standard deviation is given for the linear CCD array as:

$$\sigma_{\mathrm{OSA}i}^{2} = G^{2}\left[\frac{qT_{\mathrm{I}}}{C_{\mathrm{Di}}^{2}}\left(I_{\mathrm{D}i} + \frac{q\eta_{\mathrm{D}i}}{\hbar} \frac{\Delta\lambda_{\mathrm{OSA}}}{\lambda_{\mathrm{LCLS0}}} P_{\mathrm{R}i}\right) + e_{\mathrm{n}}^{2}\right], \tag{7.91}$$

where $\sigma_{\mathrm{OSAi}}^{2}$ is the *i*th detector random voltage noise variance that is dependent on the integration time, that is, $\sigma_{\mathrm{OSAi}}^{2} = \sigma_{\mathrm{OSAi}}^{2}(T_{\mathrm{I}})$, I_{Di} is the *i*th detector dark current, η_{Di} is the *i*th detector quantum efficiency, P_{Ri} is the *i*th detector measured optical PSD, and e_{n}, and is the readout electronics voltage noise, where we assumed that fixed pattern noise, introduced by the different detector dark currents, can be subtracted from the useful signal during the calibration process. Further, we will assume that readout electronics voltage noise is much higher than the voltage noise induced by optically and thermally (dark current) randomly generated electron–hole pairs in each detector, that is, we are working far away from the quantum limit. This condition is usually fulfilled when we have low-resolution quantization of the OSA captured data, which is usually the case for the OSAs that are available on the market. Due to the assumption we made and according to eq 7.91 for the detector random voltage noise variance, we have $\sigma_{\mathrm{OSAi}}^{2} \approx G^{2}\langle e_{\mathrm{n}}^{2}\rangle$. Finally, by combining eqs 7.89, 7.90, and 7.91 for the maximal SNR of the RS height distribution measurement, we have:

$$SNR_{\mathrm{P}}^{\max} \geq \frac{\pi R_{\mathrm{M}}^{4}}{16} \frac{\lambda_{\mathrm{RH}}\lambda_{\mathrm{RL}}\Delta\lambda_{\mathrm{OSA}}^{2} P_{\mathrm{LCLS}}^{2}(\omega)}{(\lambda_{\mathrm{RH}} - \lambda_{\mathrm{RL}})\Delta\lambda_{\mathrm{LCLS}}} \left(\frac{q\eta_{\mathrm{D}}T_{\mathrm{I}}}{hcC_{\mathrm{D}}\sqrt{\langle e_{\mathrm{n}}^{2}\rangle}}\right)^{2}, \tag{7.92}$$

where *h* is the Planck constant. Relation given in eq 7.92 can give us a fast estimate of the worst-case scenario when we use an optimal PSD shape of the LCLS for the surface roughness height distribution measurement.

To experimentally verify the result obtained by the mathematical analysis given in the previous section, we performed an experimental characterization of the RS with the known surface roughness. The complete mathematical analysis is performed assuming bulky Michelson interferometer as the sensing unit. Due to the separate and relatively long beam paths, such configuration strongly suffers from environmental influences such as thermal drifts and/or vibrations, thus introducing additional errors in the measurement process. In our experiment, we use a very simple fiber optic–based interferometric sensing unit using commercially available OS. Our two-beam interferometer consists of the low-finesse Fabry–Perot interferometer obtained between the fiber end and the RS to be characterized positioned behind the fiber tip at a relatively short distance. The first beam is obtained as a Fresnel reflection at the glass–air interface form the fiber tip and the second beam is obtained as a back-reflected light form the RS and coupled into the SF. In the previous section, we assumed that RM is an ideal mirror. Any real mirror has finite roughness that will influence the measurement accuracy and introduce additional errors. Also, slightly inclined RM regarding the incident light beam will introduce an additional error due to the large area illuminated with the incident light beam and different optical paths that are obtained from different parts of the RM. For that reason, by using bulky Michelson interferometer, we need to perform very good alignment of all optical parts of the interferometer to obtain minimal measurement error. By choosing the fiber end as the RM, we will suppress all types of additional errors that can originate in the nonideal bulky RM, thus enabling much-relaxed alignment of the complete optical setup. Small surface area of the cleaved single-mode fiber core will strictly define the reference phase front of the reference beam. At the same time, single-mode fiber serves as a modal/spatial filter of the light that is back-reflected from the RS and coupled into the SF. In this way, only light that is back-reflected from those points on the RS surface with the slopes approximately equal to zero (i.e., extreme and saddle points) is coupled back into the SF. As a matter of fact, all beams of the RS back-reflected light with an inclination angle regarding the SF axis that is smaller than acceptance angle of the used single-mode fiber will be coupled into the SF but with additional losses. In our experimental setup, we used optical fiber with the numerical aperture of approximately $NA \approx 0.12$, which gives us a relatively small acceptance angle θ_A of about: $\theta_A = \arcsin(NA) \approx 6.9°$, and due to this, we can take as a solid assumption that we have back-coupled into the SF only light that is originated from the extreme and saddle points of the RS. In addition to this, both of the beams travel along the same SF to suppress any thermal and/or polarization drifts and/or vibrations that can occur along the SF. Such sensing configurations offers very small sensing head (bear fiber tip) and high disturbances rejection along the beam's path, thus providing possibility of placing the signal processing unit at a relatively long distance from the sensing head.

In Figure 7.8, it is presented the block schematics of the experimental setup for the RS heights distribution measurement. As an LCLS, we used pigtailed SLD SLD-381-MP2-DIL-SM-PD from Superlum, Ltd. The SLD emits infrared light with the central wavelength of λ_{LCLS0} = 854.4 nm and spectral bandwidth of $\Delta\lambda_{LCLS}$ = 24.6 nm (FWHM) at the diode current of 20 mA and the junction temperature of 25°C. The coherence length of the SLD L_C is equal to $L_C = k\lambda_{LCLS0}^2 / \Delta\lambda_{LCLS}$, where $k = \sqrt{2\ln 2 / \pi} \approx 0.66$ for the Gaussian spectrum shape and the coherence length for the given spectral width and central wavelength is approximately equal to LC $\approx$ 20 μm. The SLD is driven by the commercial LDD Multichannel laser diode control system PRO 800 with integrated Temperature module TED 8xxx and current module LDC 8xxx from Profile Optische Systeme GmbH. The pigtailed SLD is connected with the universal optical fiber splice Fiber-lokTM II 2529 from 3M Telecom Systems to one of the input arms of the 2 × 2 fused silica single-mode FOC. The fiber-optic coupler is made of Corning Flexcore 850 5/125-μm type of the fiber and is produced by Gould Technology, LLC. The insertion losses for both output arms are 2.8 and 3.4 dB at 830 nm. One of the output arms, which carries smaller amount of the optical power, is immersed into the IMG to suppress spurious back-reflection. Another output arm, assigned as the SF, is directed toward the RS, which height distribution is to be measured. The RS is firmly fixed to the MMPS Z625B 25-mm Motorized Actuator from Thorlabs, Inc for fine RS positioning regarding the SF fiber end. To estimate the position of the RS, we mounted the micrometer Digimatic Indicator ID-C125B form Mitutoyo Corp with the accuracy of ±500 nm. We positioned the RS from the SF fiber end at the distance of approximately 80 μm that can provide sufficiently high RS back-reflected light level and sufficient spectral interferometric pattern fringe visibility. Back-reflected light beams from the SF fiber end and the RS are used for illumination of the OS S2000 Miniature Fiber Optic Spectrometer form Ocean Optics, Inc. OS is sensitive for the wavelengths in the range from λ_{RL} = 333.35 nm to λ_{RH} = 1012.19 nm with

2048 resolvable points and with the average resolution of about $\Delta\lambda_{OSA}$ = 0.33 nm. This OS is connected at the second input arm of the fiber-optic coupler. Optical PSD is captured via microcontroller-based A/D card ADC1000-USB and OOIBase32™ spectrometer operating software from the same company and transferred to the PC via USB interface. Channeled spectrum, captured during the measurement, is an average of 20 successive optical PSDs with integration time of $T_I = 6$ ms.

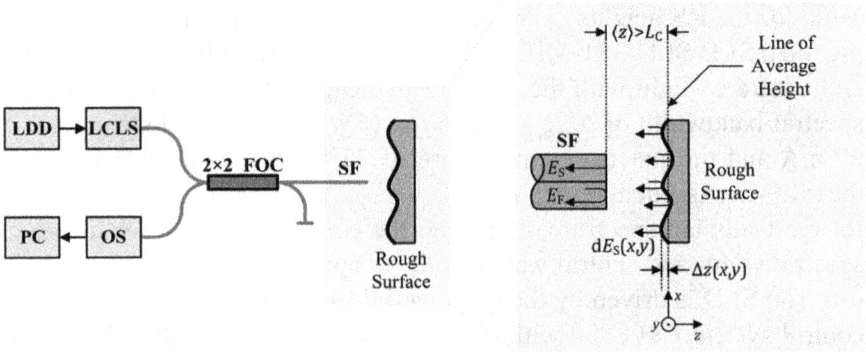

FIGURE 7.8 Block schematic of the experimental setup: LDD—laser diode driver, LCLS—low-coherence light source, PC—personal computer, OS—optical spectrometer, FOC—fiber-optic coupler, SF—sensing fiber.

In our experimental setup as the RS, we used Silicon Carbide (SiC) fiber lapping tape (Part. No. M2124-X, 4" No PSA) from ULTRA TEC MANUFACTURING, INC. The average grain size of the SiC particles onto the lapping tape is 9.0 μm, thus providing the RS with surface roughness approximately equal to the half of the grain size, that is, $R_q \approx 4.5$ μm (the derivation of the surface roughness is given in more details afterward). In Figure 7.9, it is given the microscope image of the RS with, by a circle marked, approximate dimension of the light spot onto the RS. As we used single-mode fiber with the core radius of $a = 5$ μm and numerical aperture of $NA \approx 0.12$, the light spot radius r_S at the distance of approximately $d \approx 80$ μm is given with $r_S \approx a + d \cdot NA / \sqrt{1 - NA^2}$, that is, $r_S \approx 15$ μm. This is rather small radius onto the RS in comparison with the average grain size, but it is still greater than average grain size and sufficient for surface roughness estimation that we need for our proof of concept of the measurement principle given in the previous section.

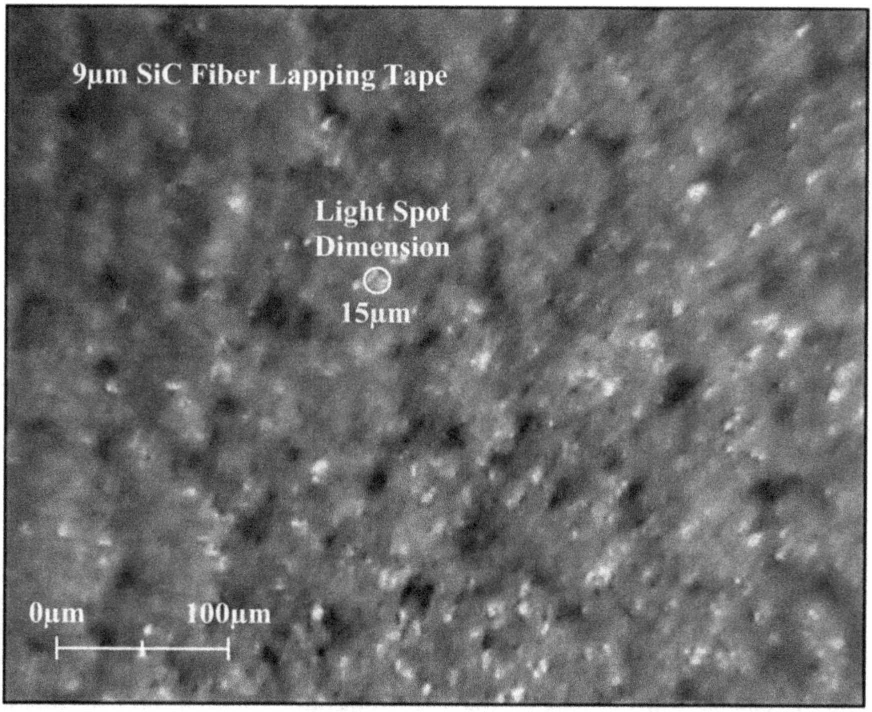

FIGURE 7.9 Microscope image of the RS with, by a circle marked, approximate dimension of the light spot onto the RS.

In our experimental setup as the OSA, we used commercially available OS. This spectrometer has an analog-to-digital (A/D) converter with a 12-bit resolution. According to eq 7.85, for the maximal SNR in the measurement of the surface roughness PDF, we approximately have:

$$SNR_{\mathrm{p}} \approx \frac{\pi\Delta\omega_{\mathrm{LCLS}}}{\Delta\omega_{\mathrm{B}}}\frac{\langle\left[V_{\mathrm{R}}^{\mathrm{opt}}(\omega)\right]^2\rangle}{\sigma_{\mathrm{OSA}}^2}, \tag{7.93}$$

where $\langle\left[V_{\mathrm{R}}^{\mathrm{opt}}(\omega)\right]^2\rangle$ is the average value of squared captured optical PSD. For the A/D converter quantization noise, we have $\sigma_{\mathrm{OSA}i}^2 = LSB^2/12$, where *LSB* is the last significant bit value. If we launch sufficient optical power from the LCLS into the fiber-optic coupler and choose integration time to be large enough to enable that captured signal is large enough, that is, almost at the maximal possible level, we approximately have

$\langle [V_R^{opt}(\omega)]^2 \rangle \approx FSR^2/4$, where FSR = $(2^n - 1)$ LSB $\approx 2^n$ LSB is the A/D converter full-scale range and $n = 12$ is the A/D converter resolution. By substituting all these values and values for $\Delta\omega_B$ and $\Delta\omega_{LCLS}$ into eq 7.94, we have:

$$\mathrm{SNR_P} \approx 3\pi \cdot 2^n \frac{\Delta\lambda_{LCLS}\lambda_{RH}\lambda_{RL}}{\lambda_{LCLS0}^2(\lambda_{RH} - \lambda_{RL})}. \tag{7.94}$$

By substituting all values, for the maximal SNR, we approximately have $\mathrm{SNR_P} \approx 2.65 \times 10^6$ and if we express SNR in decibels, we have $\mathrm{SNR_P^{max}} \approx 64\mathrm{dB}$. This is maximally possible obtainable SNR in measurement surface roughness with the parameters that we have in our experimental setup. Real SNR that we have in our measurement must be lower than this estimated maximal value of SNR.

At the beginning of the measurement process, we removed the RS to record the SLD spectrum obtained as the Fresnel reflection from the fiber end. Afterward, we placed the RS at the distance of approximately 80 μm from the SF fiber end. In Figure 7.10, there are presented both of the OSA captured optical PSDs. With the solid gray line is presented the recorded SLD spectrum $V_R(\lambda)$ and with dashed black line is presented the recorded channeled spectrum $V_{CS}(\lambda)$. At the same diagram, with solid black line, is presented the "high frequency" part of the recorded channeled spectrum $V_{CS}^{HF}(\lambda)$.

As we can see from Figure 7.10, spectral interferometric pattern fringe visibility is rather small and is approximately equal to $V \approx 0.1$. The reason for this lies in the fact that optical power of the RS back-reflected light and coupled into the SF is much lower than optical power obtained as a fiber end back-reflection. This is in very good agreement with our assumption at the beginning of the analysis, that only reflection from a fraction of the RS is back-coupled into the SF. This enables us to apply eq 7.55 to find the RS heights PDF.

For the calculation of the RS heights PDF, we need to find the ratio between the "high frequency" part of the recorded channeled spectrum and SLD optical PSD. In Figure 7.11(a), there are presented both of the captured optical PSDs of interest, where we normalized the "high frequency" part of the recorded channeled spectrum to have a closer look at this PSD. Also, in Figure 7.11(b), it is given the ratio between the normalized "high frequency" part of the recorded channeled spectrum and SLD optical PSD.

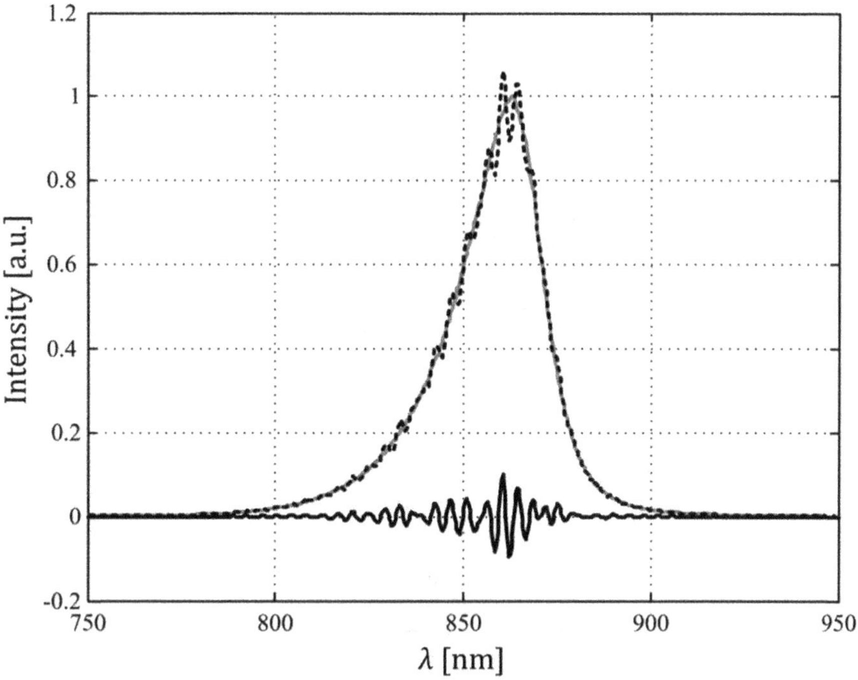

FIGURE 7.10 OSA captured optical PSDs. Solid gray line (—) presents the recorded SLD spectrum, dashed black line (- -) presents the recorded channeled spectrum, and solid black line (—) presents the "high frequency" part of the recorded channeled spectrum.

We depicted this ratio of the PSDs only in the wavelength range from 750 to 950 nm. The reason for this is that due to the low optical power levels at the tails of the SLD spectrum and relatively low resolution of the used A/D converter of the OS, we have large errors. This can be seen at the boundaries of the diagram given in Figure 7.11(b), where we have sharp peaks with discrete values obtained by low-resolution quantization of low-level signals.

Finally, the RS heights PDF can be obtained by making the inverse Fourier transform of the ratio between the normalized "high frequency" part of the recorded channeled spectrum and SLD optical PSD and normalizing such obtained function to fulfill the condition that integral over all values of the RS heights gives the value of one. In our analysis, we neglect the part of the spectrum out of the range given in Figure 7.11(b). We take that the ratio of these two PSDs has zero value for the wavelength out of

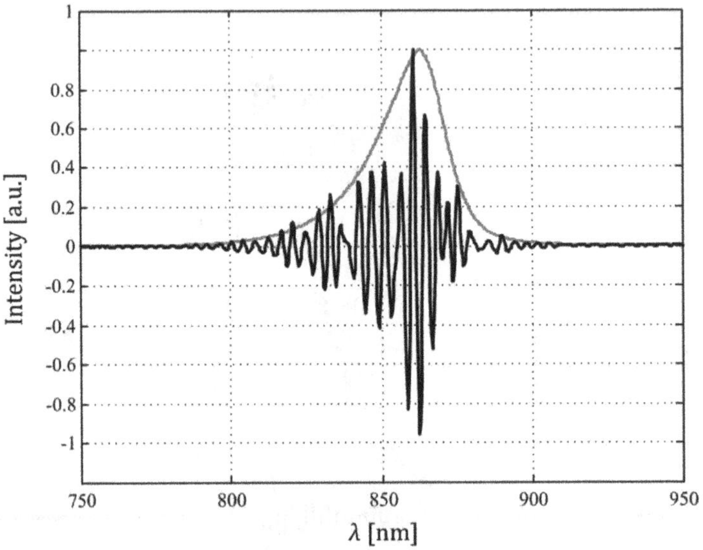

(a)

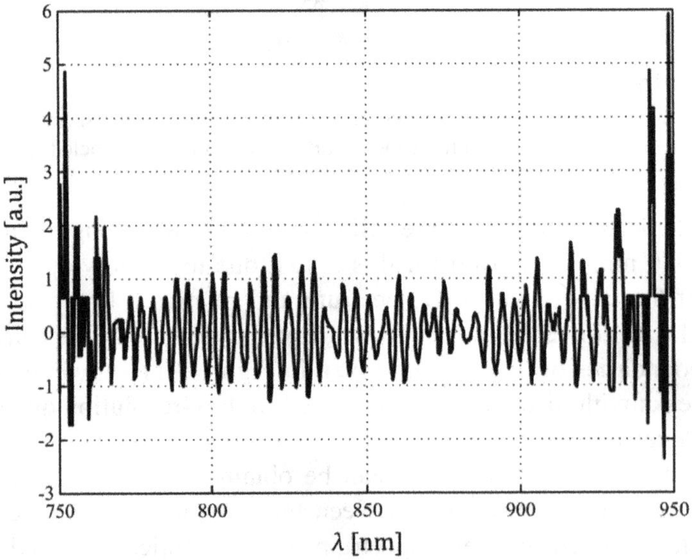

(b)

FIGURE 7.11 (a) Captured optical PSDs of interest. Solid gray line (—) presents the SLD optical PSD and solid black line (—) presents the normalized "high frequency" part of the recorded channeled spectrum. (b) Ratio between the normalized "high frequency" part of the recorded channeled spectrum and SLD optical PSD.

this range. By limiting the frequency range, that is, by filtering this signal, we will obtain the limited resolution in finding the RS heights PDF. For the roughness resolution δz, we have according to the Nyquist–Shannon sampling theorem: $\delta z \approx c/(4B)$, where B is the optical frequency bandwidth of the captured signal and is given with $B = c/(1/\lambda_L - 1/\lambda_H)$ and $\lambda_L = 750$ nm and $\lambda_H = 950$ nm are lower and higher wavelength of the signal, respectively. Finally, for the resolution of the RS heights PDF, we have:

$$\delta z \approx \frac{\lambda_H \lambda_L}{4(\lambda_H - \lambda_L)}. \tag{7.95}$$

By substituting the numerical values into eq 7.95, we obtained for the resolution of the RS heights PDF the following value: $\delta z \approx 0.89$ μm. This is relatively low resolution of the measurement, but still if we want to cover the entire range of the RS heights (range of the heights that will cover 99.9% of all possible heights is equal to approximately 6.6 R_q), we will have approximately 33-independent points, which is sufficiently large to estimate the PDF or to fit with, for example, Gaussian distribution function to obtain all relevant parameters, such as surface roughness. If we made simple inverse Fourier transform of the signal given in Figure 7.11(b), as a solution we will obtain wanted probability PDF convoluted with the impulse response of an ideal band-pass filter (boxcar truncated function). For that reason, we need to apply some of the apodization functions to suppress oscillation of the inverse Fourier transformation function. There are two types of superimposed oscillations. The period of the first one is proportional to the bandwidth of the PSD signal given in Figure 7.11(b) and the period of the second one is proportional to the central wavelength of the OSA captured PSD. To suppress the slow oscillation, we will multiply such obtained shifted PSD with the apodization function. Further, to eliminate the fast oscillation of the inverse Fourier transform function due to the central wavelength of the OSA captured PSD different from zero, we will translate this PSD multiplied by the apodization function in the range between 0 and 200 nm without influencing the final result. For the optimal apodization function $a(\omega)$ we will take Gaussian function given in the following equation:

$$a(\omega) = \begin{cases} \exp\left[-\frac{\pi^2}{\ln 2}\left(\frac{\omega - \omega_C}{2\omega_{EB}}\right)^2\right], & |\omega - \omega_C| \le \omega_{EB} \\ 0, & |\omega - \omega_C| > \omega_{EB} \end{cases}, \tag{7.96}$$

where $\omega_{EB} = 2\pi B$ is the equivalent bandwidth of the PSD signal given in Figure 7.11(b), and ω_C is its central angular frequency. In our case, for the numerical value of the equivalent bandwidth ω_{EB} we have $\omega_{EB} \approx 5.29 \times 10^{14}$ rad/s wand for the central angular frequency: $\omega_C \approx 2.25 \times 10^{15}$ rad/s. Such defined Gaussian apodization function provides us very good side lobes suppression. The ratio between the central and the first side peak is +0.4%. With the gray dashed line in Figure 7.12, it is given the filtered normalized inverse Fourier transform of the apodized and shifted signal given in Figure 7.11(b) that represents the measured RS heights PDF. In the same diagram, it is depicted the fitted Gaussian PDF with the black solid line. It can be noticed very good matching of the measured and fitted PDF. According to the fitted PDF parameters, we obtained the surface roughness to be: $R_q \approx (2.46 \pm 0.01)$ µm with the goodness factor of $R^2 = 0.9646$. The measured distance of the line of average height is estimated to be: d = (80 ± 0.01) µm, which is in very good agreement with the data obtained from the micrometer.

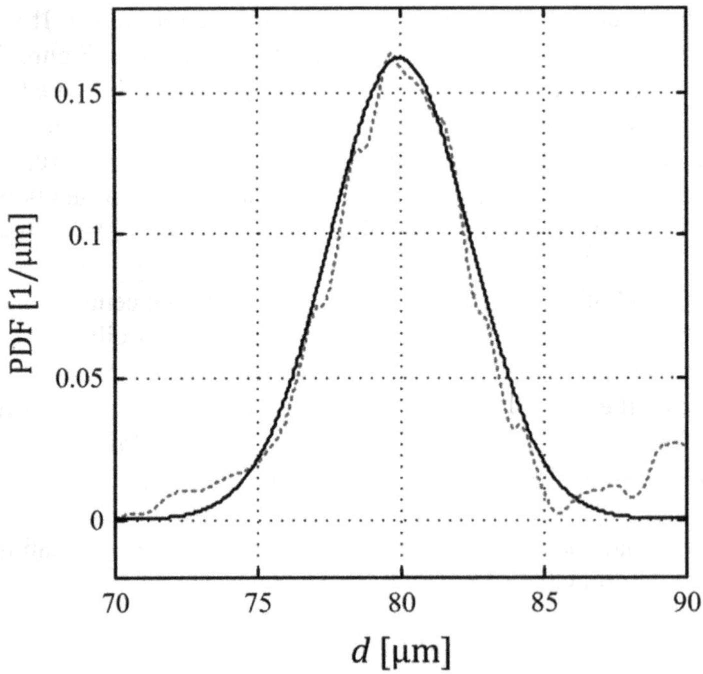

FIGURE 7.12 Measured RS heights PDF is presented with the gray dashed line (- -) and fitted Gaussian PDF is presented with the black solid line (—).

Measured surface roughness of the used lapping tape is slightly smaller than mathematically estimated surface roughness based on the average grain size. The reason for this can lie in the fact that we measured the surface roughness only in a relatively small part of the RS area due to the small dimensions of the light spot onto the RS, thus collecting the data only from the limiting number of the RS points. In general, we need to perform several measurements from different parts of the RS to obtain the better estimate of the RS parameters. However, measurement of the RS parameters, we performed in this section, was just the proof of concept of the mathematical apparatus given in the previous section.

To confirm the result given by eq 7.78, a short analysis will be conducted. Therefore, according to the Cauchy–Schwarz–Bunyakovsky inequality, we have the following relation:

$$\frac{\left|\int_{-\infty}^{+\infty} X(\omega)Y(\omega)\,\mathrm{d}\omega\right|^2}{\int_{-\infty}^{+\infty}\left|X(\omega)\right|^2 \mathrm{d}\omega} \le \int_{-\infty}^{+\infty}\left|Y(\omega)\right|^2 \mathrm{d}\omega, \tag{7.97}$$

where the equality is valid if the following condition is fulfilled: $X(\omega) = \alpha Y(\omega)$, where α is an arbitrary and real constant and $X(\omega)$ and $Y(\omega)$ are real functions of ω. If we take the following relations, according to eq 7.77:

$$X(\omega) = \frac{R_{\mathrm{RS}}(\omega)}{V_{\mathrm{R}}(\omega)} \text{ and} \tag{7.98}$$

$$Y(\omega) = V_{\mathrm{R}}(\omega)\cos(\omega t_0), \tag{7.99}$$

than we have the equality condition fulfilled, that is, we obtained the maximal SNR if we have the following relation fulfilled for the optimal optical PSD shape $V_{\mathrm{R}}^{\mathrm{opt}}(\omega)$:

$$\frac{R_{\mathrm{RS}}(\omega)}{V_{\mathrm{R}}^{\mathrm{opt}}(\omega)} = \alpha V_{\mathrm{R}}^{\mathrm{opt}}(\omega)\cos(\omega t_0). \tag{7.100}$$

or finally:

$$V_{\mathrm{R}}^{\mathrm{opt}}(\omega) = \sqrt{\frac{1}{\alpha}\frac{R_{\mathrm{RS}}(\omega)}{\cos(\omega t_0)}} = K\sqrt{\frac{R_{\mathrm{RS}}(\omega)}{\cos(\omega t_0)}}. \tag{7.101}$$

To obtain the relation between the average grain size and the surface roughness of the lapping tape, that is, to estimate the surface roughness parameters, if we know the distribution of the grain size, we will perform a simplified analysis. In Figure 7.13, it is presented the closer look at the surface of the lapping tape. Each grain is approximated with a small sphere with an average radius of $\langle r \rangle$. We will assume Gaussian distribution of the grain sizes, so for the grain size variance, we can approximately take $\sigma_r \le \langle r \rangle/3$ is valid, because grains have always positive dimensions and in the range $\langle r \rangle \pm 3\sigma_r$ we have approximately 99.9% of all grains.

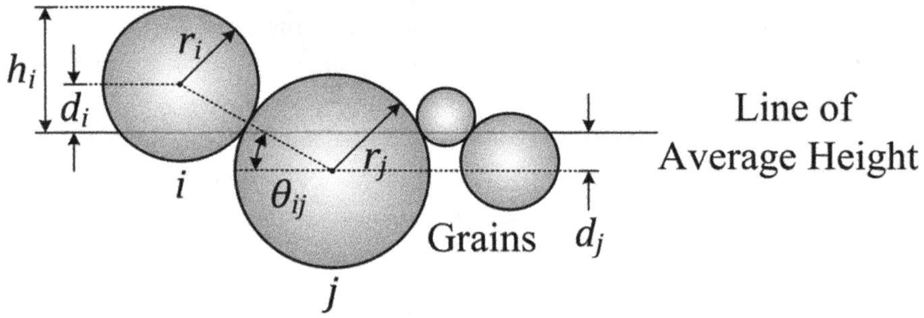

FIGURE 7.13 Closer look at the surface of the lapping tape.

In our analysis, we took the distance d_i between the ith grain center and the line of the RS average height to be positive if the center of the grain is above the line and in the opposite case to be negative. For the height of the ith grain, we have $h_i = r_i + d_i$, where r_i is the ith grain radius. As we are measuring grain heights from the line of the average height, we have $\langle h \rangle = \langle r \rangle + \langle d \rangle$ and $\langle d \rangle = -\langle r \rangle$, respectively. The distribution of the RS heights is a result of the large number of layers of the grains deposited uniformly (lapping tape serves for polishing the fiber end) onto the substrate, so according to the central limit theorem, we can make an assumption that stochastic variable h_i has a Gaussian distribution. Both of the stochastic variables, h_i and r_i have the Gaussian distribution, so the stochastic variable d_i also must have the Gaussian distribution. Further, for the standard deviation $\sigma_h = R_q$ of the RS heights, we have the following: $R_q^2 = \sigma_r^2 + \sigma_d^2$, where we assumed both stochastic processes to be independent. According to the geometry of two neighboring grains given in Figure 7.13, we have the following:

$$d_i - d_j = \left(r_i + r_j\right)\sin\theta_{ij}, \tag{7.102}$$

where d_i and d_j are the distances between the centers of two neighboring grains and the line of average heights, r_i and r_j are their radiuses and θ_{ij}

is the angle between the line of average height and line connecting grains centers. Assuming that all these stochastic processes are independent, we have $\langle\theta_{ij}\rangle = 0$. From eq 7.102, we also have following identity:

$$\langle(d_i - d_j)^2\rangle = \langle(r_i + r_j)^2\rangle\,\langle\sin^2\theta_{ij}\rangle. \tag{7.103}$$

Further from eq 7.103, we have:

$$\langle(d_i - d_j)^2\rangle = \langle d_i^2\rangle - 2\langle d_i\rangle\langle d_j\rangle + \langle d_j^2\rangle = 2\langle d^2\rangle - 2\langle d^2\rangle, \tag{7.104}$$

where it is valid: $\langle d_i^2\rangle = \langle d_j^2\rangle = \langle d^2\rangle = \langle d^2\rangle + \sigma_d^2$. By substituting this last value into eq 7.104, we have: $\langle(d_i - d_j)^2\rangle = 2\sigma_d^2.$.

Similar, we have:

$$\langle(r_i + r_j)^2\rangle = \langle r_i^2\rangle + 2\langle r_i\rangle\langle r_j\rangle + \langle r_j^2\rangle = 2\langle r^2\rangle + 2\langle r\rangle^2, \tag{7.105}$$

where is valid: $\langle r_i^2\rangle = \langle r_j\rangle = \langle r^2\rangle = \langle r\rangle^2 + \sigma_r^2$. By substituting this last value into eq 7.105, we have: $\langle(r_i + r_j)^2\rangle = 4\langle r^2\rangle + 2\sigma_r^2$. Further, for the average value of the sine function of the angle θ_{ij}, we have:

$$\langle\sin^2\theta_{ij}\rangle = \frac{1}{2\pi}\int_{-\pi}^{+\pi}\sin^2\theta\,\mathrm{d}\theta = \frac{1}{2}, \tag{7.106}$$

where we assumed uniform distribution of the stochastic variable θ_{ij} in the range $\theta_{ij} \in [-\pi, +\pi]$. At the end, we have the following:

$$\sigma_d^2 = \langle r\rangle^2 + \frac{\sigma_r^2}{2} \approx \langle r\rangle^2, \tag{7.107}$$

and finally for the standard deviation of the RS heights, we have: $R_q \approx \langle r\rangle$, that is, $R_q \approx 4.5$ μm, where we took $\sigma_r^2 \ll \sigma_d^2$.

7.3 WIDE-DYNAMIC RANGE LOW-COHERENCE INTERFEROMETRY

Since there are no moving mechanical parts present in the sensing systems, the electronically scanned low-coherence interferometry represents a compact, rigid, stable, and fast interrogation interferometric technique. Nevertheless, this technique has a problem with low SNR and low sampling rate. Therefore, to obtain subpixel resolution, a special attention should be paid to the signal processing of the captured interferogram. However, it is still difficult to reach subnanometer resolution with typical SNR values ranging from 20 to 40 dB. To interrogate the low-finesse fiber-optic Fabry–Pérot interferometric sensors, typically the electronically scanned low-coherence interferometric technique has been employed. Such technique is capable of measuring the optical paths difference in the range of 30 μm with the resolution of 0.9 nm thus providing the sensor dynamic range of 90.5 dB. However, to obtain this subnanometer resolution, a birefringent wedge together with two linear polarizers and focusing optics must be employed.

Typical electronically scanned low-coherence interferometry-based sensing system is composed of collimation lens, wedge (Fizeau interferometer), and linear photodetector array. Due to the relatively low number of pixels of the linear array, such sensor suffers from low resolution. To increase the sensor resolution, one of the ways is to decrease the wedge angle. However, in this way, the sensor dynamic range is also decreased. The other way implies the use of the linear arrays with larger number of pixels. This further implies the use of long, custom-made, and expensive linear arrays and wedges that on the other hand, due to the thermal drift and ambient vibration, significantly deteriorate the sensor performances such as stability and accuracy.

A typical camera has approximately three orders of magnitude larger number of pixels when compared with a linear array. Although the camera has a large number of pixels, the main problem of involving cameras in low-coherence interferometric measurement chain is the sensors geometry, that is, their two-dimensionality. To overcome this problem, the wedge and camera axes must be set in the way that their axes are mutually inclined in a horizontal plane for a small angle. Such a wedge and camera configuration gives the following wedge thickness $t(x, y)$ above the camera:

$$t(x,y) = t_0 + x\cos\varphi\tan\theta + y\sin\varphi\tan\theta \approx t_0 + \theta x + \varphi\theta y, \qquad (7.108)$$

where x and y are the coordinates in the camera plain, t_0 is the wedge thickness at the coordinate system origin, φ ($\varphi \ll 1$) is the angle between the wedge and camera axes, and θ ($\theta \ll 1$) is the wedge angle. Based on the camera captured two-dimensional low-coherence interferogram, an algorithm has been developed aimed at finding the central fringe maximum position with high resolution that consists of two steps. The first step is based on the analysis of the camera row signal. Therefore, it is necessary to extract a low-coherence interferogram from a single row, for example, from a row with the maximal fringe visibility. The wedge thickness along this row positioned at $y = y_0$ is given by:

$$t(x, y_0) \approx t_0 + \varphi\theta y_0 + \theta x. \tag{7.109}$$

To obtain an estimation of the central fringe maximum position, one can use one among many algorithms for analyzing low-coherence interferograms. Due to the low SNR and low sampling rate, this estimation isn't sufficient for obtaining high-resolution measurements. Nevertheless, the efficient algorithm will identify at least the position close to the central fringe. When finding the position, that is, the pixel in the analyzed row that is in the close proximity of the central fringe maximum one can switch to the second step of the proposed algorithm. The second algorithm step includes the analysis of the camera column that contains this particular pixel. Along this column, the wedge thickness is given by:

$$t(x_{\mathrm{CF}}, y) \approx t_0 + \theta x_{\mathrm{CF}} + \varphi\theta y, \tag{7.110}$$

where x_{CF} represents the position of the pixel in the close proximity of the central fringe that has been identified in the first step. By comparing eqs 7.109 and 7.110, one can notice that wedge thickness changes very slowly along the column providing us an equivalent wedge with a much smaller angle than in the first step. This allows us to achieve much higher pixel resolution in the central fringe maximum positioning. To reach maximum positioning resolution, angle φ should be chosen as small as possible but still large enough so that each column covers just a couple of fringes. When we find the pixel position y_{CFM} along the observed column that is closest to the central fringe maximum that is further closest to the position $y = y_0$ of the pixel identified in the first algorithm step, the corresponding thickness t_{CFM} above this particular pixel is given by:

$$t_{\mathrm{CFM}} \approx t_0 + \theta x_{CF} + \varphi\theta y_{\mathrm{CFM}}. \tag{7.111}$$

The presented wedge thickness directly corresponds to the optical path difference of the low-coherence sensing interferometer. Based on eq 7.111, the single-pixel resolution in measuring low-coherence sensing interferometer optical paths difference is given by:

$$\delta L \approx 2np\varphi\theta, \tag{7.112}$$

where n is the wedge index of refraction and p is the distance between the centers of the adjacent camera pixels.

To get a better insight in all relevant geometrical parameters of the wedge and camera setup, in Figure 7.14, it is presented the top view of the corresponding setup, together with both side views of the setup cross sections along both relevant axes. Figure 7.14 shows the active zone of the camera that represents the rectangle with maximal dimensions on the camera surfaces that is covered by the wedge. Only the images captured by this active zone will be taken into consideration for further signal processing.

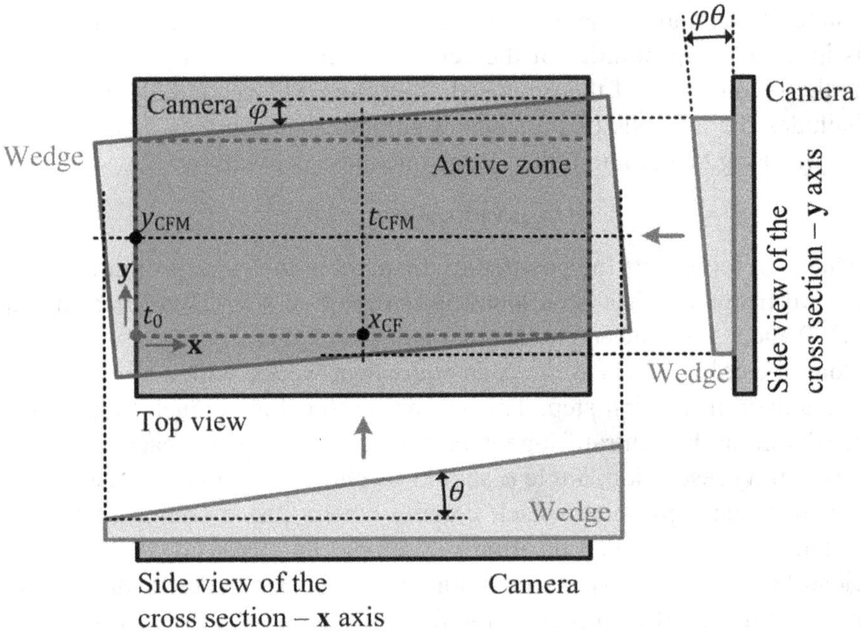

FIGURE 7.14 Top view of the setup, together with both side views of the setup cross sections along both relevant axes.

To test the capabilities of the proposed low-coherence interferometric interrogation technique, a simple low-coherence fiber optic–based absolute position sensor has been built. The layout of the test setup that is based on a single-mode 2 × 2 FOC is given in Figure 7.15. A low-finesse Fabry–Pérot interferometer, which is formed between the tip of the SF thus forming the RM and SM, forms the sensing interferometer. To avoid parasitic interference the distance D between the RM and SM, that is, the cavity length must be greater than one half of the used LCLS coherence length L_C ($D > L_C/2$). A pigtailed SLD is used as the LCLS, which is driven by the commercial LDD that has current and temperature controller. SLD emits infrared light with the central wavelength of 842 nm and with the FWHM spectral bandwidth of 22 nm. The overall optical power launched by the SLD in a single-mode fiber is approximately 1.4 mW. To maximize the interferogram fringe visibility the inactive FOC arm is immersed into the IMG. The back-reflected light is used for further wedge (W) illumination. The low-coherence interferogram is captured by the camera (C) and transferred to the PC for further signal processing.

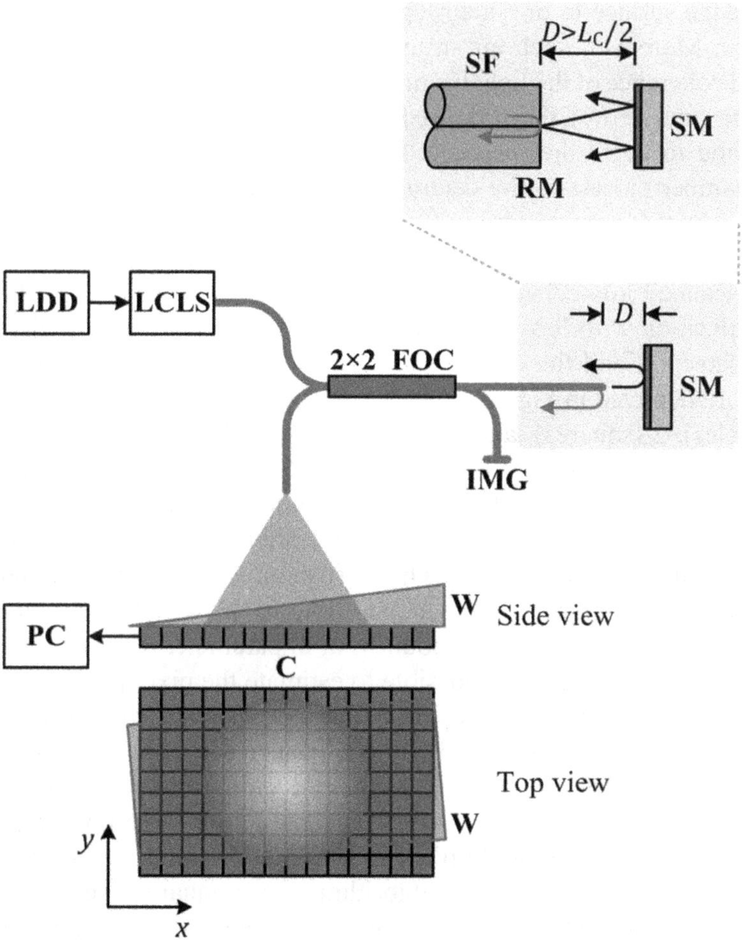

FIGURE 7.15 Layout of the experimental setup.

The images captured by the camera for three distances between RM and SM: (1) $D = 362$ μm, (2) $D = 370$ μm, and (3) $D = 392$ μm, are presented in Figure 7.16. These distances were chosen to obtain the highest fringe visibilities of the sensing interferometer. The presented two-dimensional low-coherence interferograms are obtained for the case: $\varphi \approx 1.84°$ and $\theta = 0.5°$. To provide the highest possible level of the interferometric signal, both sides of the wedge are partially reflective with 30% of reflectivity. One can clearly notice in Figure 7.16 the edge of the wedge in the bottom part of each image. Since there is no any coupling optics the FOC fiber that illuminates the wedge is positioned at a relatively large distance from the wedge surface to provide roughly uniform illumination of the camera surface. Moreover, such an arrangement also provides a relatively high spatial coherence of the light beam illuminating the camera's active surface.

The first step of the proposed algorithm requires a single row to be extracted to be further processed. To be able to extract the pixel position/number closest to the central fringe maximum the raw data from a single row that are depicted in Figure 7.17(a) should be further processed. There are typical steps in the signal processing chain of the electronically scanned low-coherence interferograms such as band-pass filtration, normalization with low-pass filtered signal component, and finally envelope detection and the corresponding fitting. All these signal processing steps are depicted in Figure 7.17(b), (c), and (d), respectively, where lines 1 (circles), 2 (squares), and 3 (triangles) mark the corresponding signals for the following RM-SM distances: $D = 362$, $D = 370$, and $D = 392$ μm, respectively. To suppress the influence of nonuniform irradiance distribution over the camera surface, the normalization step has been performed. Usually, SLD has Gaussian spectral distribution. Therefore, the Gaussian function has been used for envelope fitting in the last step. Based on the fitting parameters, where the goodness of fit parameter R^2 was greater than 0.98 in all three cases, it is possible to estimate the pixel position/number closest to the central fringe position. Based on eq 7.110, the corresponding wedge index of refraction $n = 1.51$ and distance between the centers of the adjacent camera pixels of $p \approx 2.2$ μm, and taking into consideration that sensing interferometer cavity length is equal to the corresponding wedge optical thickness, the single-pixel resolution of central fringe maximum position estimation in the first algorithm step is equal to the $\Delta D \approx np\theta \approx$ 29 nm that represents a relatively low resolution which is on the other side still sufficient for identifying the central fringe.

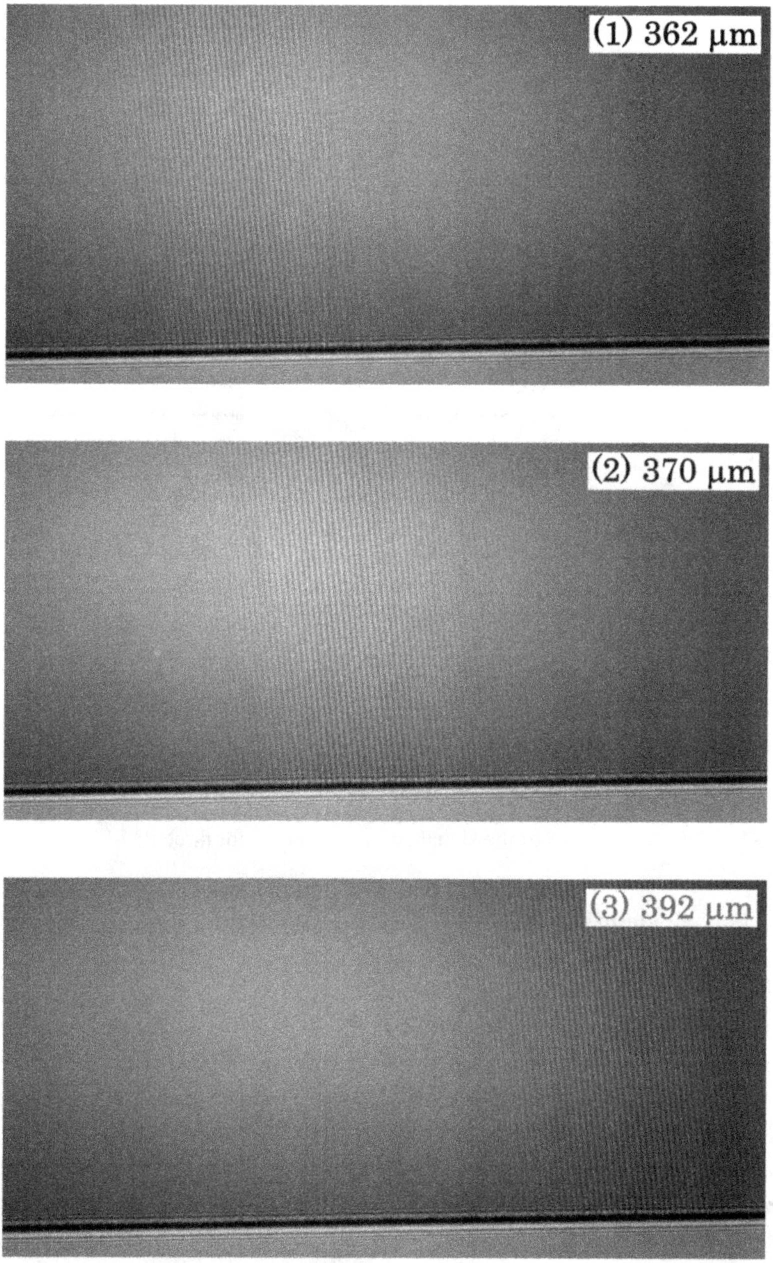

FIGURE 7.16 Two-dimensional low-coherence interferogram captured by the camera.

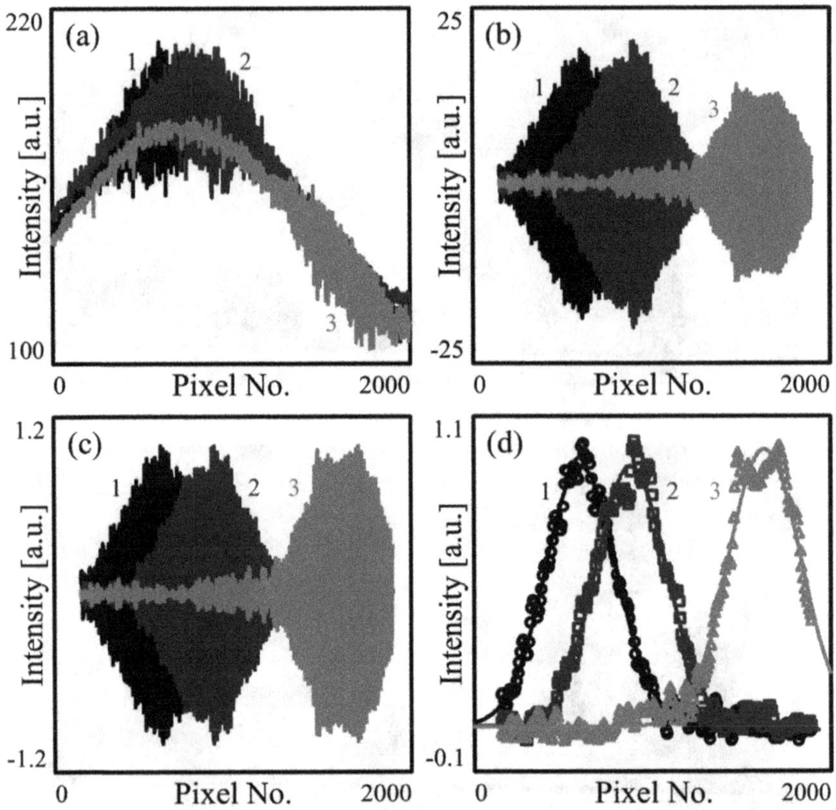

FIGURE 7.17 Three steps of the signal processing chain for three RM-SM distances.

When we identify the corresponding pixel number in the first algorithm step, we switch on the second step. The band-pass filtration and normalization have been performed over each image row to suppress the influence of the nonuniform irradiance distribution. To eliminate the potential errors the rows in the lower part of the images depicted in Figure 7.17, which contain pixels in the close proximity to the wedge edge, are omitted from further processing. After filtration and normalization of the camera row signal, the corresponding column signal that contains the aforementioned pixel should have the cosine shape. The exact position of the central fringe maximum is obtained according to the above-presented algorithm. Based on the values of the experimental setup parameters, the estimated overall single-pixel resolution in central fringe maximum position measurement is $\delta D = \delta L/2 \approx 0.93$ nm.

Based on a simple experiment, we estimate the overall performances of the proposed algorithm. In the presented experimental setup, where SM was positioned at the distance of approximately 370 μm from RM, SM was slightly shifted with the help of calibrated high-precision piezo actuator. The images that are obtained for the initial and shifted SM positions were processed according to the proposed algorithm. In the case when SM shift was 5 nm, we have presented the results in Figure 7.18. As the result of the first step of the algorithm, the Gaussian fitting parameters are obtained, where envelope data and Gaussian fits are presented in Figure 7.18(a) and (b) for 1 (circle) initial and 2 (square) shifted SM position where Figure 7.18(b) represents the enlarged view of Figure 7.18(a) in the close proximity to the maximum value of the fitted function. Based on the fitting parameters, central fringe maximum is positioned with the standard deviation of approximately 1.7 pixels, that is, with the resolution of approximately 48 nm, where signal-to-noise rtaio of the row signal was 24.7 dB. Nevertheless, this is still sufficient for identification of the central fringe, but insufficient to measure the shift of 5 nm. Moreover, the SM shift measured with the first step of the proposed algorithm is in the opposite direction than the real one. However, after rounding both of the corresponding fitting parameters refer to the same pixel.

When we identify the pixel in the corresponding row closest to the central fringe maximum we switch on the second step of the algorithm. The signals of the corresponding column are presented in Figure 7.18(c). It is easy to notice that both signals are overlapping due to the very small SM shift. Therefore, an enlarged view of Figure 7.18(c) central part is presented in Figure 7.18(d) in the close proximity of the central fringe maximum, where lines 1 and 2 represent the initial and shifted SM position, respectively, and lines 3 and 4 represent their corresponding fitting curves with the cosine shape, where the goodness of fit parameter R^2 was greater than 0.987 in both cases. Due to the fact that RM and SM have different phase shifts, 0 and π, respectively, after the light beam reflection instead of central fringe maximum, the central fringe minimum value has been presented in Figure 7.18(d). It is evident from Figure 7.18(d) that the column signal shifts toward the first column pixel that may be confusing if compared to the row shift obtained in the first algorithm step. However, this is expected as the wedge thickness decreases along the inspected column from the top to the bottom and the first pixel is located at the top of the captured image. Finally, based on the fitted parameters, the measured shift is 5.5 nm and its standard deviation is 0.28 nm (0.31 pixels), which, at the same time, represents the overall sensing system subpixel resolution.

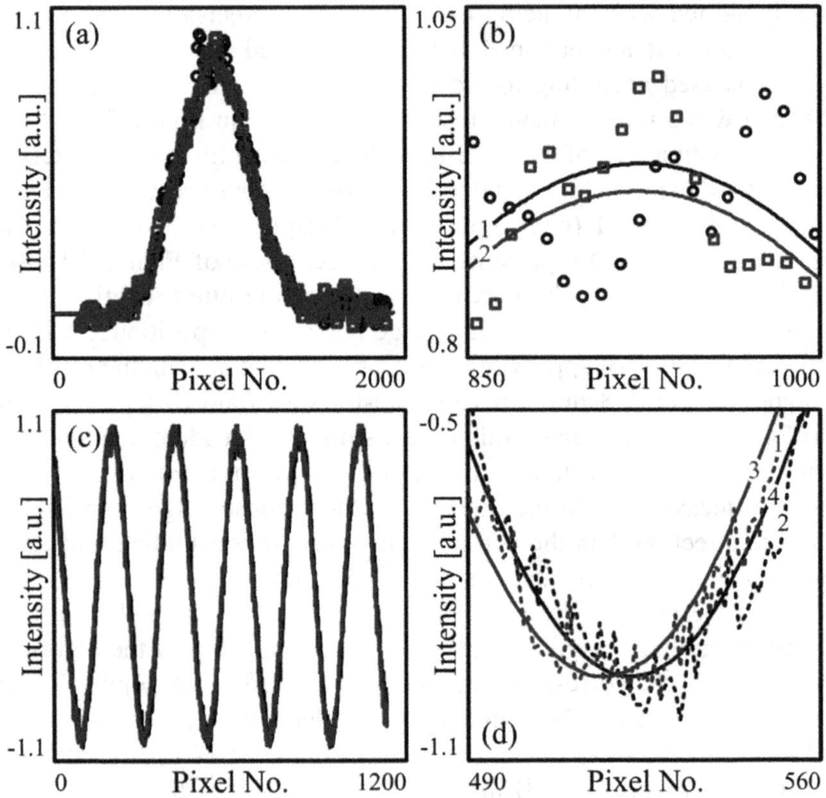

FIGURE 7.18 Estimation process for the SM small shift measurement.

To estimate the dynamic range of such a built sensing system, it will be assumed that the maximally measured cavity length range corresponds to the maximal wedge optical thickness difference along the camera surface. The maximum RM-SM distance that can be measured with the presented sensing system is given by $D_M \approx npN_x\theta \approx 59$ μm, where N_x is the row number of pixels, for which for the used camera we have $N_x = 2048$. The dynamic range of any measurement system is defined as the ratio of the measurement range and resolution. Since SNR changes only for 1.1 dB when the RM-SM distance changes in the range from 362 to 392 μm the resolution of 0.28 nm, which was estimated for the distance of 370 μm, will be taken as the relevant one for the whole measurement range. Taking these values into consideration, we obtain the dynamic range of approximately 106 dB that represents relatively high dynamic range that can be increased if the angle between the wedge and camera decreases, but there is a limit in decreasing of this angle. To cover the entire measurement range with the highest resolution, each column must cover at least the range of single row pixel, that is, $N_y\,\delta D \geq \Delta D$ must be fulfilled, where N_y is the column number of pixels. In our case, we have $N_y = 1536$, where $\varphi \geq 1/N_y \approx 0.037°$ is valid. Consequently, the maximum theoretically achievable dynamic range of such a composed sensing system is given by $20\log_{10}(D_M/\delta D) \approx 20\log_{10}(N_x N_y) \approx 130$ dB, where the single-column-pixel resolution has been considered as the relevant resolution of the complete measurement setup. If compared with the single-pixel dynamic range of the proposed sensing system of 96 dB, where the achieved single-pixel resolution of $\delta D \approx 0.93$ nm has been taken into account, one can notice that there is a space in the sensing system performance improvement. However, there can be difficulties since the angle between the wedge and the camera must be very small thus reaching higher noise levels and consequently lower resolution. One can easily notice this in Figure 7.19(a) where the PSD of the analyzed raw column signal (arrow marks the corresponding carrier signal) is presented. If the angle between the wedge and the camera decreases the carrier signal will move toward the lower frequencies where we have higher noise levels due to the flicker noise.

The PSD of the analyzed raw row signal is also presented in Figure 7.19 (b). In this case, the carrier signal is clearly distinguishable and marked with an arrow. If the angle between the wedge and the camera is kept very small ($\varphi \ll 1$), there will be no significant change in the carrier signal position. One can change the position of the carrier signal only if one chose the wedge with a different angle. However, one can notice that wedge angle has been chosen appropriately as the carrier signal is positioned away from the high level of the flicker noise.

7.4 OPTICAL COHERENCE TOMOGRAPHY TECHNIQUE WITH ENHANCED RESOLUTION

Optical coherence tomography (OCT) is an optical technique that allows to determine the internal structure of weakly absorbing and scattering objects. OCT relies on an inspection of a beam of low-coherent light back-reflected from the internal interfaces of the layered structure of an object that internal structure is to be determined. To find the locations of these interfaces along the path of the penetrating beam, OCT uses interferometry. The back-reflected light from the layered object is combined with the light back-reflected from an RM and then is further analyzed by a detection system. Consecutive measurements with the beam in adjacent transversal positions provide the longitudinal cross-section of the object structure.

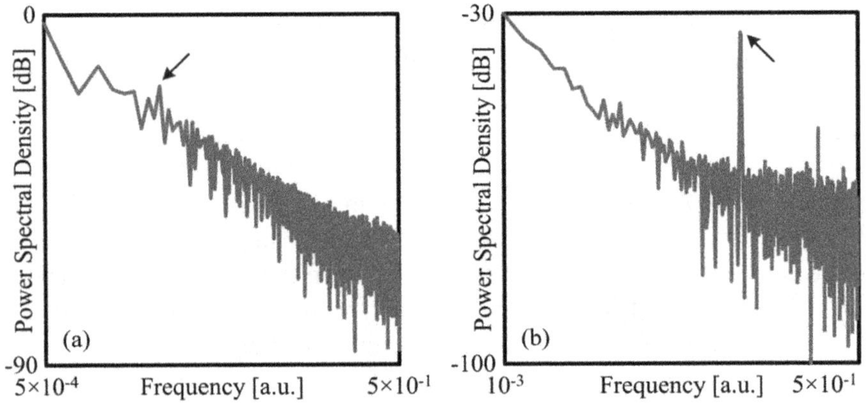

FIGURE 7.19 Power spectral densities of the raw (a) column and (b) row signal.

Based on how the depth information is obtained, the OCT is divided typically into two major classes: the time-domain OCT (TD-OCT) and the Fourier domain OCT (FD-OCT). SLDs are used for the TD-OCT, as an LCLS because the broadband spectra of the SLDs give a short coherence length and hence good depth resolution. In the case of the FD-OCT, the depth information is acquired by taking a Fourier transform of the captured spectral interferogram. FD-OCT can be further divided into two classes based on the type of the used light source and the detection methods. These FD-OCT classes are the spectral interferometric technique OCT and the wavelength tuning FD-OCT. A broadband SLD, a grating, and an optical detectors' array for the detection of spectral interference fringes are used with the spectral interferometric technique. A broadly tunable laser as the light source and a normal detector for detection of the interference fringes are used with the wavelength-tuning technique. The FD-OCT has shown to provide significant advantages in terms of acquisition time and sensitivity when compared to conventional TD-OCT.

Contrary to the conventional microscopy, the mechanisms that govern the axial and transverse image resolutions are independent in OCT. The OCT imaging axial resolution is determined by the coherence length of the used LCLS, and high axial resolution can be achieved independently of the beam-focusing conditions. According to the Wiener–Khinchin theorem, the envelope of the optical signal electrical field autocorrelation function is equivalent to the Fourier transform of the corresponding optical power spectrum. Therefore, the autocorrelation function width, or equivalently the axial resolution, is inversely proportional to the power spectrum width. Since the axial resolution is inversely proportional to the spectral width of the used light source, the broadband optical sources are required to achieve high axial resolution. Typically, SLDs are the most commonly used light sources in the OCT systems since they have high irradiance and are not too expensive. However, the typical coherence lengths of SLDs of approximately 10 – 30 μm are not short enough to achieve the resolution required for many applications. The edge-emitting light-emitting diodes, with the coherence lengths half as long as the SLDs coherence length, are available commercially at low cost, but their emission powers are smaller by an order of magnitude when compared with the SLD's emission power. To overcome the trade-off between the source power and spectral width is to synthesize a broadband source by combining the outputs of several SLDs with different center wavelengths. Multiple quantum-well devices achieve this synthesis by coupling the output of several sources on a single substrate.

To maintain a given SNR, one needs to increase the light source power with the increase of the scanning speed. To meet the demands of the latest generation of OCT systems with scan rates that approach the television video rate, mode-locked $Ti:Al_2O_3$ and Cr^{4+}:forsterite lasers have been employed. These lasers make them attractive sources for fast, high-resolution OCT imaging since they have high power and wide spectral width. However, their lack of portability limits their use of the specialized applications.

To obtain high-resolution OCT system, there are, also, some other techniques such as spectral shaping, iterative deconvolution, dispersion manipulation, destructive interference, and chirped quasiphase matching. All these techniques imply the use of relatively expensive equipment or where an extensive postprocessing of the recorded OCT signal is needed. The technique presented in this chapter is based on the simple algorithm of the deconvolution by division of two spectrums: the channeled spectrum at the interferometer output and the used LCLS spectrum. The sensing system is realized as a two-beam interferometer, where one of the beams illuminates the object that should to be characterized and the other one that illuminates the RM. By capturing both optical spectrums of the used LCLS and of the interferometer channeled spectrum, the high-resolution axial cross section of the object can be obtained as the inverse Fourier transform of the ratio of two optical spectrums.

A simple sensing test setup for the axial sample characterization that we will use in our theoretical analysis is presented in Figure 7.20. A broadband light from the LCLS that is driven by a commercial LDD that keeps constant current and constant temperature of the LCLS is launched into one of the input arms of the 2 × 2 single-mode FOC. One of the output arms of the FOC, which also serves as an SF, is directed toward the object to be characterized. The other output arm is immersed into the IMG to suppress spurious back-reflections. After reflecting from the SF end, which forms the RM and the backscattering from the object, both beams are launched into the OS with the help of the second input arm of the FOC. Having a very small surface area, the cleaved single-mode fiber core of the SF will strictly define the reference phase front of the reference beam. At the same time, the single-mode fiber serves as a modal/spatial filter of the light that is backscattered from the sample and coupled back into the SF. Moreover, to suppress any thermal and/or polarization drifts and/or vibrations that can occur along the SF both beams travel along the same SF. Finally, the OS is connected to the PC for data capturing and further signal processing.

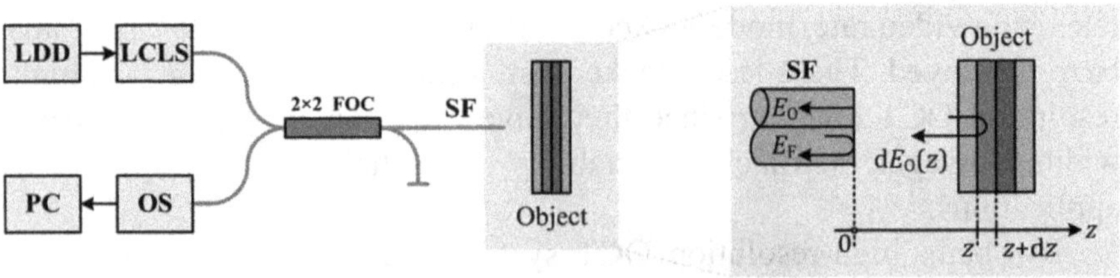

FIGURE 7.20 Block schematic of the experimental setup: LDD—laser diode driver, LCLS—low-coherence light source, PC—personal computer, OS—optical spectrometer, FOC—fiber-optic coupler, and SF—sensing fiber.

For the electrical field $E_{CS}(t)$, which we have back coupled into the SF, that is, at the input of the OS, we can write the following equation:

$$E_{CS}(t) = E_F(t) + E_O(t) = E_F(t) + \int_{L_O} dE_O(z,t), \tag{7.113}$$

where t is time, $E_F(t)$ is the electrical field obtained as a back-reflection from the SF end (reference beam), $E_O(t)$ is the overall backscattered electrical field obtained from the object and coupled back into the SF, L_O is the object length, z is the distance between the SF end and the infinitesimal thin layer in the object, and $dE_O(z, t)$ is the infinitesimal backscattered electrical field obtained from the object layer with the position $(z, z + dz)$ and coupled with the SF. For both electrical fields from eq 7.113, we can write the following:

$$E_F(t) = \sqrt{R_F}\, E_I(t) \text{ and} \tag{7.114}$$

$$dE_O(z,t) \approx E_I\left(t - \frac{2z_O}{c}\right) dR_O(z), \tag{7.115}$$

where $E_I(t)$ is the SF incident electrical field, R_F is the reflection coefficient of the glass–air interface of the SF end, zo is the single pass OPD (one half of the overall OPD) between the SF end and the layer at the position z in the object, c is the speed of light and $dR_O(z)$ is the infinitesimal electrical field coupling coefficient of the layer at the position z in the object. Based on eqs 7.113, 7.114, and 7.115 for the Fourier transform $E_{CS}(j\omega)$ of the captured electrical field $E_{CS}(t)$, we have:

$$E_{CS}(j\omega) = E_I(j\omega)\left[\sqrt{R_F} + \int_{L_O} \exp\left(-j\omega\frac{2z_O}{c}\right) dR_O(z)\right]. \tag{7.116}$$

Further we will take into consideration that our OS consists of a monochromator and a photodetector linear array. The optical detectors of the linear array in the OS are sensitive only on the time average value of the light power impinging onto the detectors, so the OS measures the average PSD of the captured optical signal. Based on eq 7.116 for the OS captured optical PSD, we have:

$$I_{CS}(\omega) = I_I(\omega)\left|\sqrt{R_F} + \int_{L_O} \exp\left(-j\omega\frac{2z_O}{c}\right) dR_O(z)\right|^2, \tag{7.117}$$

where $I_I(\omega) = |E_I(j\omega)|^2$ is the incident optical PSD. By assuming that along the object layers we have approximately a constant axial coupling coefficient of the backscattered light with the SF, we have $dR_O(z) = r(z)$ dz, where $r(z)$ is the axial coupling coefficient. The scattering coefficient of the material in the layer, its absorption, and the distance between the SF end and the layer influence the axial coupling coefficient. Further, assuming that the distance between the SF end and the object is much larger than the object thickness (object length L_O), we can neglect the dependence of the axial coupling coefficient $r(z)$ on the distance due to the geometrical parameters of the complete setup. The absorption of the object will be also neglected, so, finally we can take into consideration that only the layer scattering coefficients of the object material influences axial coupling coefficient.

Because the overall coupling coefficient between the object and the SF is very small, due to the very small aperture, a relatively small acceptance angle of the single-mode SF and low scattering coefficients of the object material, we can make a solid assumption that the optical power of the backscattered light from the object, which is coupled into the SF, is much lower than the optical power back-reflected from the SF end. In this case, the overall spectral fringe visibility much smaller than unity. Having this in mind, we can neglect the squared term of the integral term in eq 7.117. So, the following can be written for the captured optical PSD:

$$I_{CS}(\omega) \approx I_F(\omega)\left[1 + \frac{2}{n_O\sqrt{R_F}}\int_{L_O}\cos\left(\frac{2\omega}{c}z_O\right)r(z_O)\,dz_O\right], \tag{7.118}$$

where $I_F(\omega) = |E_F(j\omega)|^2$ and n_O is the object refraction index, for which it is assumed to be approximately constant along the sample, that is, $n_O(z) \approx n_O$ and where $dz_O = n_O\,dz$ is valid. Finally, from eq 7.118, we have:

$$I_{CS}(\omega) \approx I_F(\omega)\left[1 + K\int_{L_O} r(t)\cos(\omega t)\,dt\right], \tag{7.119}$$

where parameter $K = c/\left(n_O\sqrt{R_F}\right)$ has a constant value and $t = 2z_O/c$ is the time delay between the light beam reflected from the SF end and the light backscattered from the object layer. The integral term in eq 7.119 represents the real part of the Fourier transform $R(j\omega)$ of $r(t)$, so eq 7.119 can be written as:

$$I_{CS}(\omega) \approx I_F(\omega)\left\{1 + \frac{K}{2}\left[R(j\omega) + R(-j\omega)\right]\right\}. \tag{7.120}$$

After rearranging eq 7.120, we have:

$$R(j\omega)+R(-j\omega)\approx\frac{2}{K}\frac{I_{CS}(\omega)-I_F(\omega)}{I_F(\omega)}. \tag{7.121}$$

Finally, after making an inverse Fourier transforms of both sides of eq 7.121, we have the following for the scattering coefficient $r(z_O)$ along the sample:

$$r(z_O)\propto\mathcal{F}^{-1}\left\{\frac{I_{CS}(\omega)-I_F(\omega)}{I_F(\omega)}\right\}\Bigg|_{t=\frac{2z_O}{c}}, \tag{7.122}$$

where $\mathcal{F}^{-1}\{\bullet\}$ denotes the inverse Fourier transform for $t=2z_O/c$ for which we took only those values of the inverse Fourier transform in which $z_O\geq 0$ is satisfied.

As it was mentioned earlier, the OCT resolution is limited by the spectral width of the used LCLS and it is approximately equal to the coherence length of the LCLS. After deconvolution the outgoing electrical field autocorrelation function at the interferometer output by applying the spectrum division, in theory, we can gain unlimited improvement of the axial resolution. In practice, the resolution of the sensing system is limited by the SNR of the used OS, especially at the tails of the LCLS spectrum, where the spectrometer captures very low-optical power levels. Furthermore, after digitalizing the output signals from the OS, the quantization resolution of the used A/D converter will also influence the measurement resolution.

To experimentally test the result obtained by the short mathematical analysis given in the previous section, an experimental characterization of two objects with the known characteristics is performed: the aluminum mirror in the first case and the layered object in the second case. In the presented sensing system, we used a very simple fiber optic–based interferometric sensing unit with the commercially available OS. Such a sensing configuration offers a very small sensing head (bear fiber tip) and high disturbances rejections along the common beam path, thus providing the possibility of placing the signal processing unit at a relatively long distance from the sensing head. One of the advantages of such a built sensing system also is that due to the common paths of the reference and backscattered beams, the wavelength dependence of the FOC splitting ratio is compensated by the spectrum division.

In Figure 7.20, it is presented the block schematic of the experimental setup for the object characterization. As an LCLS, a pigtailed SLD

SLD-381-MP2-DIL-SM-PD from Superlum, Ltd. is used. The SLD is connected to one of the input arms of the 2 × 2 single-mode fused silica FOC and it is driven by the commercial LDD multichannel laser diode control system PRO 800 with integrated Temperature module TED 8xxx and Current module LDC 8xxx from Profile Optische Systeme GmbH at constant diode current of 20 mA and constant diode case temperature of +25°C. Under these conditions, the SLD emits infrared light with the central wavelength of λ_0 = 854.4 nm and spectral bandwidth of $\Delta\lambda$ = 24.6 nm (full width at half-maximum, FWHM) thus providing SLD coherence length of about $L_C \approx 9.5$ μm ($L_C = k\lambda_0^2/\Delta\lambda$, where $k = 0.32$ is valid for the Lorentzian spectrum). In the first case, the SF is directed toward the object that represents the aluminum mirror and in the second case to the layered object. For the layered sample, a silicon carbide (SiC) fiber lapping tape (Part. No. M2124-X, 4" No PSA) from Ultra Tec Manufacturing, Inc. is used. The lapping tape has several layers of SiC grains deposited uniformly onto the substrate. The randomly sized SiC grains have an average size of approximately 9 μm, thus providing the surface with the roughness approximately equal to one half of the average grain size. The layered object to be investigated is firmly fixed to the MMPS Z625B 25 mm Motorized Actuator from Thorlabs, Inc., for fine object positioning with respect to the SF fiber tip. To be able to independently estimate the object position with respect to the SF fiber tip, the micrometer Digimatic Indicator ID-C125B from Mitutoyo Corporation with the accuracy of ±500 nm has been mounted. The light beams back-reflected from the SF fiber end and the layered object illuminate the OS S2000 Miniature Fiber Optic Spectrometer from Ocean Optics, Inc. The used OS is sensitive to the light with the wavelengths ranging from 333.35 to 1012.19 nm with in total 2048 resolvable points and with the mean wavelength sampling period of about 0.33 nm. The second input arm of the FOC is connected to the OS. The optical spectrum is captured via the microcontroller-based 12 – bit A/D card ADC1000-USB and OOIBase32™ spectrometer operating software from the same company and transferred to the PC via universal serial bus interface. The corresponding captured channeled spectrum represents an average of 20 successive optical spectrums with the integration time of 6 ms.

To record only the SLD spectrum obtained as the Fresnel reflection from the fiber end at the very beginning of the measurement process, we left SF far away from any object. For proofing experimentally the resolution enhancement of the proposed sensing system, an aluminum mirror at the distance of approximately 120 μm from the SF end has been placed. In Figure 7.21, both of the OS captured optical spectrums are presented. To suppress any changes in the emitted optical power of the SLD during the measurement process, both spectrums were normalized to the same mean value. The recorded SLD spectrum $I_F(\lambda)$ is presented with a light gray line and the recorded channeled spectrum $I_{CS}(\lambda)$ is presented with a dark gray line. In the same diagram, the "high frequency" part of the captured channeled spectrum, that is, $I_{CS}(\lambda) - I_F(\lambda)$ is presented with a black line.

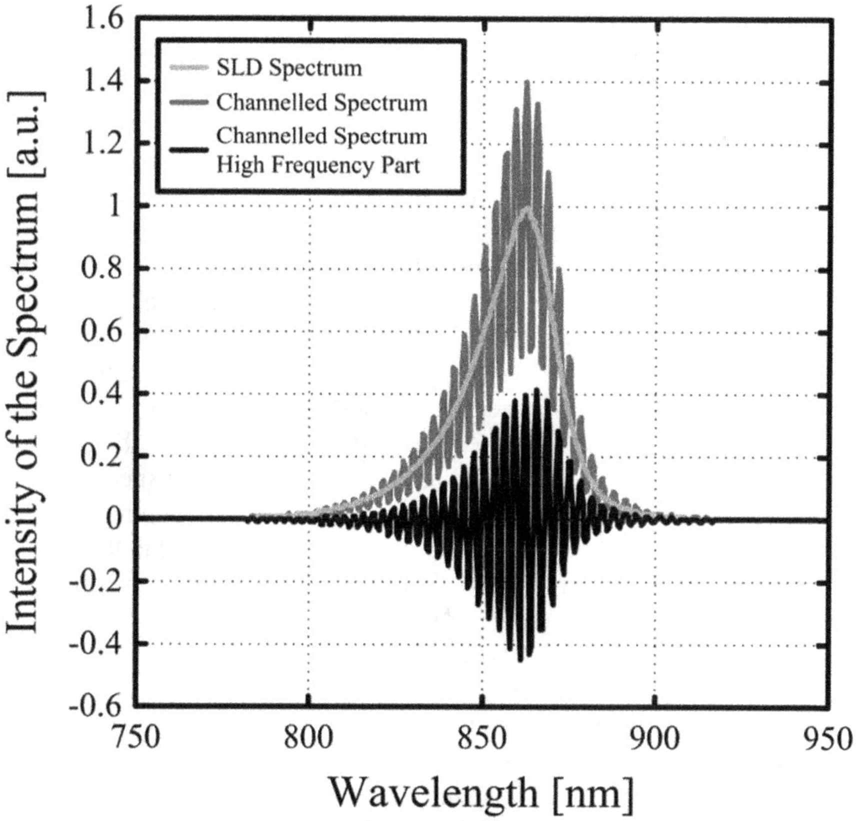

FIGURE 7.21 OS captured optical spectrums, whereas an object the aluminum mirror is used. The recorded SLD spectrum is presented with a bright gray line and the recorded channeled spectrum is presented with a dark gray line. At the same diagram, the "high frequency" part of the recorded channeled spectrum is presented with a black line.

To obtain the scattering coefficient of the aluminum mirror, which in the ideal case should be the Dirac delta function the ratio between the "high frequency" part of the recorded channeled spectrum and the SLD optical spectrum needs to be found. In Figure 7.22, the ratio between the normalized "high frequency" part of the recorded channeled spectrum and the SLD optical spectrum is presented. This ratio of the spectrums is depicted only in the wavelength range from 750 to 950 nm, for example, the rectangular apodization function is applied in this range on the calculated ratio of the spectrums. The rectangular apodization has been applied to suppress the large errors, due to the low-optical power levels at the tails of the SLD spectrum and due to the relatively low resolution and noise of the used A/D converter of the used OS. These errors appear as the result of the division process as it is inherently very sensitive to the input noise especially at the low signal levels. Diagram, given in Figure 7.21, shows this at its boundaries where one has sharp peaks with discrete values obtained by the low-resolution quantization of the low-level signals. However, the use of the rectangular apodization function will only partially suppress the errors caused by the OS noise and the better and narrower apodization function must be used such as narrower Gaussian function to significantly suppress the errors. Moreover, the use of such an apodization function will, also, influence the resolution by reducing it. By using the broadest possible apodization function, one obtains the best possible resolution, but the price that is paid is the increased noise of the ratio of the spectrums and consequently increased noise in the obtained scattering coefficients. One takes that the ratio of these two spectrums has zero value for the wavelength out of this range, that is, we applied a boxcar truncation as an apodization function in the range from 750 to 950 nm. In the case of the aluminum mirror, an overall spectral fringe visibility of about 0.4 is obtained, which is not sufficiently small to simply apply the approximation we made in making the step between eqs 7.117 and 7.118. However, this will not influence the width and the shape of the electrical field autocorrelation function but only its level and thus it will not influence the measured resolution of such a built sensing system.

In Figure 7.23, the normalized inverse Fourier transform of the channeled spectrum $I_{CS}(\omega)$ (marked as IFT of the CS in Fig. 7.23) is presented with a gray solid line and the normalized inverse Fourier transform of the deconvoluted channeled spectrum $[I_{CS}(\omega) - I_F(\omega)] / I_F(\omega)$ (marked as IFT of the DCS in Fig. 7.23) with a black solid line. Based on

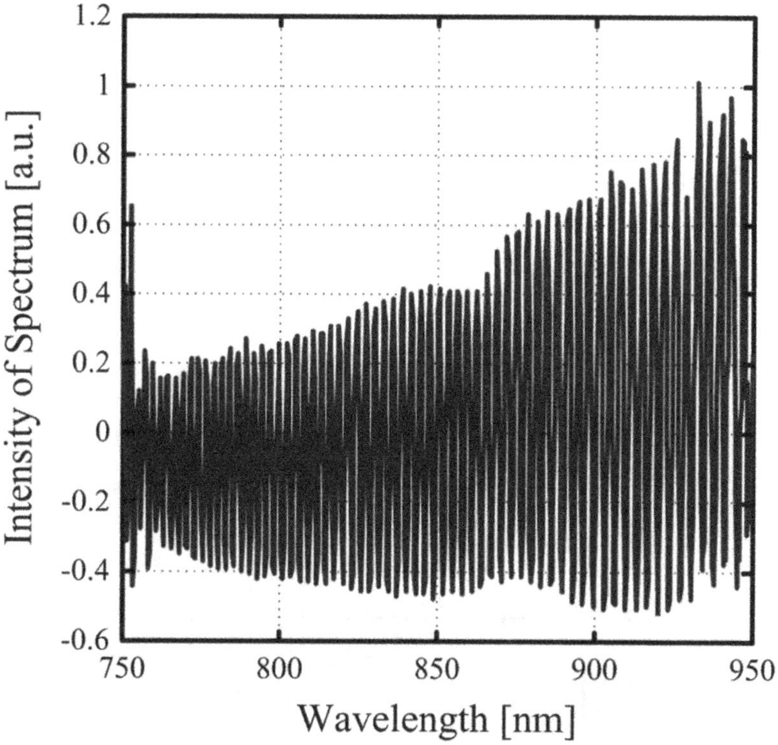

FIGURE 7.22 The ratio between the normalized "high frequency" part of the recorded channeled spectrum and the SLD optical spectrum in the case of the aluminum mirror.

the Wiener–Khinchin theorem, the inverse Fourier transform of the optical spectrum represents the autocorrelation function of the signal of the optical electrical field. The resolution, that is, the minimum distance between two resolvable points in the axial characterization of the inspected object, can be defined as a FWHM width of the optical electrical field autocorrelation function envelope. In Figure 7.23(b), the closer view of the measurement resolutions is depicted, where we approximately obtained 9 μm in the case of standard FD-OCT and approximately 1.3 μm in the case of deconvoluted channeled spectrum. One can obviously notice a sevenfold resolution enhancement. As can be seen, by applying the simple deconvolution by spectrum division, an order of magnitude better resolution than in the case of standard FD-OCT has been obtained. To obtain the best possible resolution, we did not apply any special apodization function. This is the

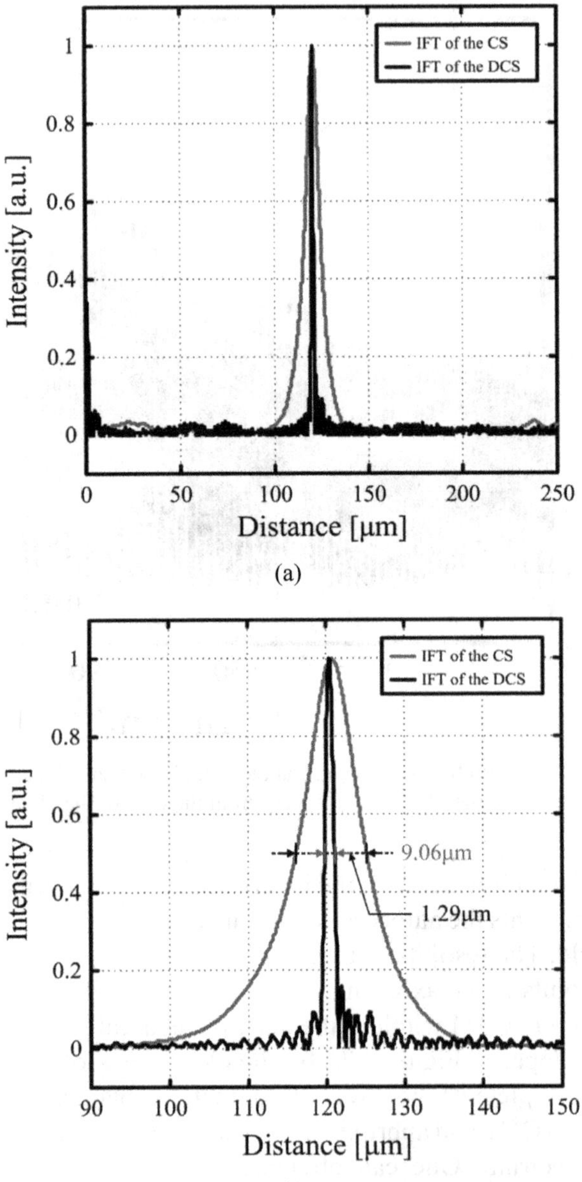

(b)

FIGURE 7.23 (a) Inverse Fourier transform of the channeled spectrum (marked as IFT of the CS) is presented with a gray solid line and the inverse Fourier transform of the deconvoluted channeled spectrum (marked as IFT of the DCS) is presented with a black solid line, (b) the same diagram in closer view in the case of the aluminum mirror.

reason for having relatively small side lobes in the close proximity to the main peak. If, for example, a Gaussian apodization function on the calculated ratio of two spectrums has been applied, the resolution will be slightly decreased and equal to approximately 2.2 μm but a significant suppression of the side lobes will be achieved.

To significantly improve the resolution of standard FD-OCT, deconvolution by spectrum division offers very simple algorithm. One can easily achieve the resolution of today's state-of-the-art OCT systems (approximately 1 μm) where very expensive broadband LCLSs are employed, or where an extensive postprocessing of the recorded OCT signal is performed by using relatively cheap and commercially available LCLSs. For further experimental testing and confirming the capability of resolution enhancement, we performed the characterization of the layered object with the known characteristics. As mentioned earlier, we used fiber lapping tape as the layered object that consists of several layers of the SiC grains with the average size of 9 μm deposited onto the substrate. To provide a sufficiently high power level of the backscattered light and sufficient overall spectral fringe visibility which still satisfies the condition that the overall spectral fringe visibility is significantly smaller than the unity, the object was positioned at a distance of approximately 80 μm from the SF end where in Figure 7.24 both of the OS captured optical spectrums are presented. To suppress any changes in the emitted optical power of the SLD during the measurement process, both spectrums were normalized to the same average value. The captured SLD spectrum $I_F(\lambda)$ is presented with a light gray line and the recorded channeled spectrum $I_{CS}(\lambda)$ is presented with a dark gray line. In the same diagram, the "high frequency" part of the recorded channeled spectrum, that is, $I_{CS}(\lambda) - I_F(\lambda)$ is presented with a black line. One can notice from Figure 7.24 that the overall spectral fringe visibility takes a rather small value of approximately 0.1. The reason for this is in the fact that the optical power of the object backscattered light coupled into the SF is much lower than the optical power obtained as a fiber end back-reflection, which is further in a very good agreement with our assumption given at the beginning of the analysis, where we assumed low values for the sample scattering coefficient of the object.

The scattering coefficient of the layered object, that is, the axial structure of the object can be obtained by making the inverse Fourier transform of the ratio between the "high frequency" part of the recorded channeled spectrum and SLD optical spectrum. We take that the ratio of these two spectrums has zero value for the wavelength out of this range, that is, we also applied a boxcar truncation as an apodization function in the range from 750 to 950 nm, as in the previous case. In this way, we obtained the best resolution but the price to pay is the side lobes that could overlap the real signal.

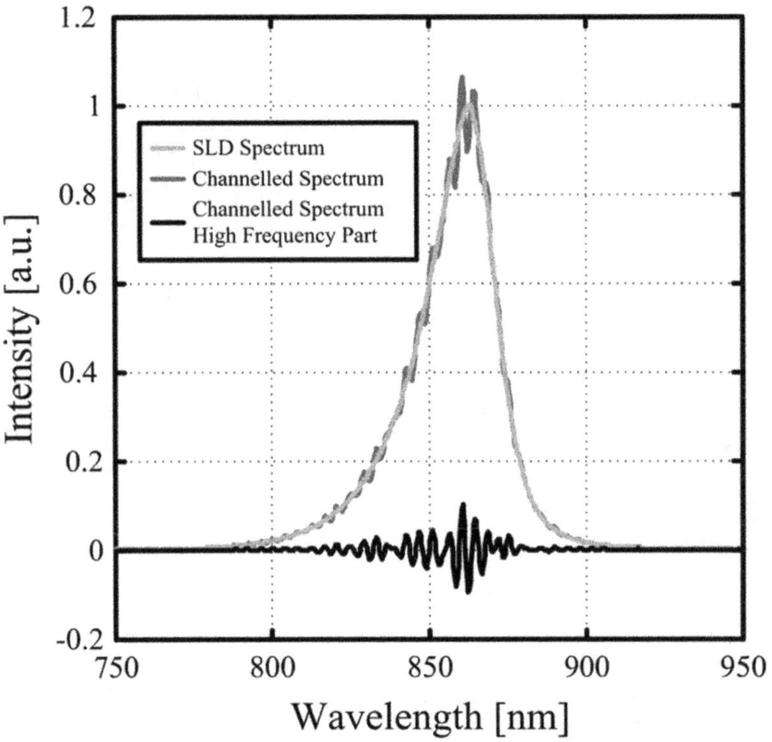

FIGURE 7.24 OS captured optical spectrums in the case of the layered sample (fiber lapping tape). The recorded SLD spectrum is presented with a bright gray line and the recorded channeled spectrum is presented with a dark gray line. On the same diagram, the "high frequency" part of the recorded channeled spectrum is presented with a black line.

In Figure 7.25, the normalized inverse Fourier transform of the channeled spectrum (marked as IFT of the CS) is given with a gray solid line and the normalized inverse Fourier transform of the deconvoluted channeled spectrum (marked as IFT of the DCS) is given with a black solid line. By inspecting the diagrams depicted in Figure 7.25, one can see some hidden details that cannot be seen if we make the simple Fourier transform of the received channeled spectrum without deconvolution it, that is, dividing it with the LCLS spectrum. Therefore, instead of the first single peak, obtained at the position of about 80 μm we have three adjacent peaks. The reason for this can be the lapping tape surface morphology, as the light spot, falling onto the sample, has a diameter of approximately 33 μm, thus illuminating several surface grains. Taking into account that the average grain size is approximately 9 μm, the light spot illuminates on average approximately 13 grains. So, the first three peaks and the next several peaks may originate from the lapping tape surface morphology. The peaks, located at the distance of about 110 μm from the SF end, may originate from the next layer of grains below the tape surface. The index of refraction of SiC is 2.59 for the wavelength range of the SLD. According to Figure 7.25, the optical thickness of the grains is approximately equal to 30 μm (distance between two groups of peaks located at 80 and 110 μm) and taking into consideration the grains index of refraction, we obtained the grain size in the axial direction to be approximately 11 μm which is very close to the mean value of the grains lateral size onto the lapping tape surface. Since grains have similar axial and lateral dimensions, the grains have an approximately spherical shape that is in good agreement with the results presented in literature.

Further, the inspection of Figure 7.25 shows that in the case of the standard FD-OCT, we can only see several peaks due to the convolution process of the lapping tape structure and the autocorrelation function (coherence function) of the SLD electrical field. However, by applying the simple deconvolution by spectrum division, we will suppress the influence of the wide autocorrelation function of the SLD electrical field and many fine details in the axial direction of the object will appear. An almost tenfold resolution improvement is obtained with this simple setup thus confirming the capabilities of this method. The sensing system consists of only relatively cheap and commercially available equipment. In spite of this, we obtained the resolution in the range of 1 μm that can be compared with today's state-of-the-art OCT systems where very expensive broadband light sources are employed. It is worth mentioning that the maximal optical depth, that is, the range that can be achieved with this sensing system is limited by the OS wavelength sampling period and in the case of the used OS, it is equal to 550 μm. If we define the dynamic range of the sensing system as the range-to-resolution ratio, in the case of such a built sensing system, the dynamic range is approximately 53 dB.

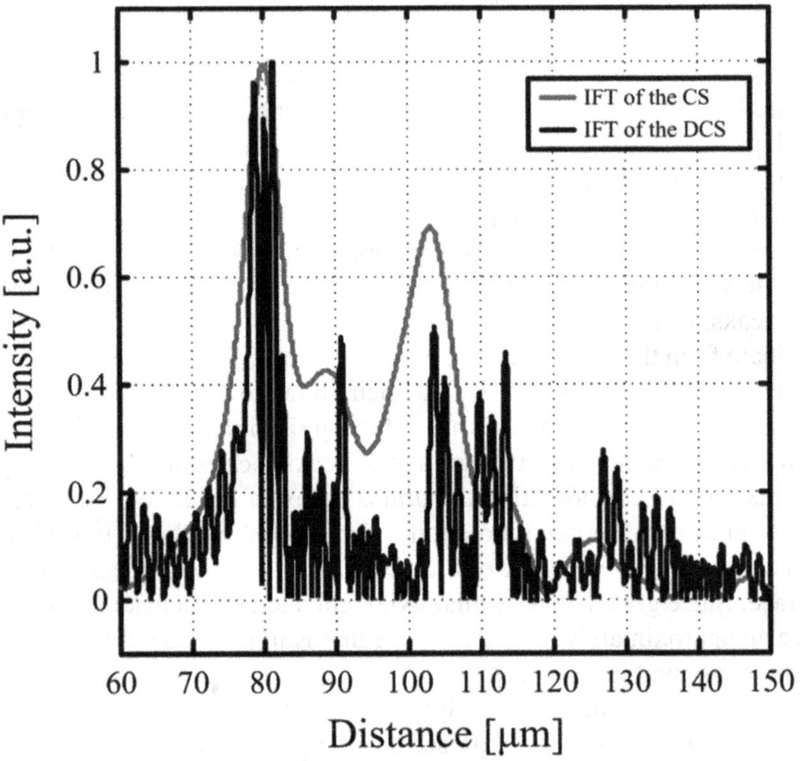

FIGURE 7.25 Inverse Fourier transform of the channeled spectrum (marked as IFT of the CS) is depicted with a gray solid line and the inverse Fourier transform of the deconvoluted channeled spectrum (marked as IFT of the DCS) is depicted with a black solid line, in the second case of the layered sample (fiber lapping tape).

KEYWORDS

- **absolute position measurement**
- **fiber-optic sensor**
- **optical coherence tomography**
- **rough surface height distribution**
- **wide-dynamic range low-coherence interferometry**

REFERENCES

Abdulkadir, A.; Birkebak, R. C. Optical Surface Roughness and Slopes Measurements with a Double Beam Spectrophotometer. *Rev. Sci. Instrum.* **1974,** *45*, 1356–1360.

Arfken, G. *Mathematical Methods for Physicists*, 3rd ed.; Academic Press: Orlando, 1985.

Beckmann, P.; Spizzichino, A. The Scattering of Electromagnetic Waves from Rough Surfaces. Artech House, 1987.

Boulet, C.; Hathaway, M.; Jackson, D. A. Fiber-Optic-Based Absolute Displacement Sensors at 1500 nm by Means of a Variant of Channelled Spectrum Signal Recovery. *Opt. Lett.* **2004,** *29* (14), 1602–1604.

Bouma, B. E.; Tearney, G. J.; Boppart, S. A.; Hee, M. R.; Brezinski, M. E.; Fujimoto, J. G. High-Resolution Optical Coherence Tomographic Imaging Using a Mode-Locked Ti:Al_2O_3 Laser Source. *Opt. Lett.* **1995,** *20*, 1486–1488.

Brundavanam, M. M.; Viswanathan, N. K.; Rao, D. N. Nanodisplacement Measurement Using Spectral Shifts in a White-Light Interferometer. *Appl. Opt.* **2008,** *47* (34), 6334–6339.

Cahill, B.; El Baradie, M. A. LED-Based Fibre-Optic Sensor for Measurement of Surface Roughness. *J. Mater. Process. Technol.* **2001,** *119*, 299–306.

Carrasco, S.; Torres, J. P.; Torner, L.; Sergienko, A.; Saleh, B. E. A.; Teich, M. C. Enhancing the Axial Resolution of Quantum Optical Coherence Tomography By Chirped Quasi-Phase Matching. *Opt. Lett.* **2004,** *29*, 2429–2431.

Chen, S.; Palmer, A. W.; Grattan, K. T. V.; Meggitt, B. T. Digital Signal–Processing Techniques for Electronically Scanned Optical–Fiber White–Light Interferometry. *Appl. Opt.* **1992,** *31*, 6003–6010.

Chen, S.; Palmer, A. W.; Grattan, K. T. V.; Meggitt, B. T.; Martin, S. Study of Electronically-Scanned Optical-Fibre White-Light Fizeau Interferometer. *Electron. Lett.* **1991,** *27*, 1032–1034.

Choma, M. A.; Sarunic, M. V.; Yang, C.; Izatt, J. A. Sensitivity Advantage of Swept Source and Fourier Domain Optical Coherence Tomography. *Opt. Express* **2003,** *11*, 2183–2189.

Dändliker, R.; Zimmermann, E.; Frosio, G. Electronically Scanned White–Light Interferometry: A Novel Noise–Resistant Signal Processing. *Opt. Lett.* **1992,** *17*, 679–681.

Domanski, A. W.; Woliriski, T. R. Surface Roughness Measurement with Optical Fibers. *IEEE Trans. Instrument. Measure.* **1992,** *41*, 1057–1061.

Duplain, G. Low–Coherence Interferometry Optical Sensor Using a Single Wedge Polarization Readout Interferometer. US Patent No. 7259862 B2 (2007).

Farahi, F.; Jackson, D. A. A Fibre Optic Interferometric System for Surface Profiling. *Rev. Sci. Instrum.* **1990,** *61*, 753–755.

Fercher, A. F.; Hitzenberger, C. K.; Kamp, G.; EL-Ziat, S. Y. Measurement of Intraocular Distances by Backscattering Spectral Interferometry. *Opt. Commun.* **1995,** *117*, 43–48.

Gadelmawala, E. S.; Koura, M. M.; Maksoud, T. M. A.; Elewa, I. M.; Soliman, H. H. Roughness Parameters. *J. Mater. Process. Technol.* **2002,** *123*, 133–145.

Gao, F.; Leach, R. K.; Petzing, J.; Coupland, J. M. Surface Measurement Errors Using Commercial Scanning White Light Interferometers. *Meas. Sci. Technol.* **2008,** *19*, 1–13.

Gorecki, C. Interferogram Analysis Using a Fourier Transform Method for Automatic 3D Surface Measurement. *Pure Appl. Opt.* **1992,** *1*, 103–110.

Gradshteyn, I. S.; Ryzhik, I. M. *Tables of Integrals, Series, and Products*, 6th ed.; Academic Press: San Diego, 2000.
Grajciar, B.; Pircher, M.; Fercher, A. F.; Leitgeb, R. Parallel Fourier Domain Optical Coherence Tomography for In Vivo Measurement of the Human Eye. *Opt. Express* **2005,** *13*, 1131–1137.
Griffiths, P. R.; de Haseth, J. A. *Fourier Transform Infrared Spectrometry*, 2nd ed.; John Wiley & Sons, Inc.: Hoboken, 2007.
Hamed, A. M.; El-Ghandoor, H.; El-Diasty, F.; Saudy, M. Analysis of Speckle Images to Assess Surface Roughness. *Opt. Laser Technol.* **2004,** *36*, 249–253.
Hlubina, P. Dispersive Spectral-Domain Two-Beam Interference Analysed by a Fibre-Optic Spectrometer. *J. Mod. Opt.* **2004,** *51* (4), 537–547.
Hlubina, P. Dispersive White-Light Spectral Interferometry to Measure Distances and Displacements, *Opt. Commun.* **2002,** *212*, 65–70.
Hlubina, P.; Gurov, I.; Chugunov, V. White-Light Spectral Interferometric Technique to Measure the Wavelength Dependence of the Spectral Bandpass of a Fibre-Optic Spectrometer. *J. Mod. Opt.* **2003,** *50* (13), 2067–2074.
Horlick, G. Resolution Enhancement of Line Emission Spectra by Deconvolution. *Appl. Spectr.*, *26*, 395–399.
Hsu, I.-J.; Sun, C.-W.; Lu, C.-W.; Yang, C. C.; Chiang, C.-P.; Lin, C.-W. Resolution Improvement with Dispersion Manipulation and a Retrieval Algorithm in Optical Coherence Tomography. *Appl. Opt.* **2003,** *42*, 227–234.
Huang, D.; Swanson, E.; Lin, C. P.; Schuman, J. S.; Stinson, W. G.; Chang, G.; Hec, M. R.; Hotte, T.; Gregory, K.; Puliafito, C. A.; Fujimoto, J. G. Optical Coherence Tomography. *Science* **1991,** *254*, 1178–1181.
Huynht, V. M.; Kurada, S.; North, W. Texture Analysis of Rough Surfaces Using Optical Fourier Transform. *Meas. Sci. Technol.* **1991,** *2*, 831–837.
Jiang, J.; Wang, S.; Liu, T.; Liu, K.; Yin, J.; Meng, X.; Zhang, Y.; Wang, S.; Qin, Z.; Wu, F.; Li, D. A Polarized Low–Coherence Interferometry Demodulation Algorithm By Recovering the Absolute Phase of a Selected Monochromatic Frequency. *Opt. Express* **2012,** *20*, 18117–18126.
Jolic, K. I.; Nagarajah, C. R.; Thompson, W. Non-Contact, Optically Based Measurement of Surface Roughness of Ceramics. *Meas. Sci, Technol.* **1994,** *5*, 671–684.
Kandpal, H. C.; Mehta, D. S.; Vaishya, J. S. Simple Method for Measurement of Surface Roughness Using Spectral Interferometry. *Opt. Lasers Eng.*, *34*, 139–148.
Kingston, R. H. *Optical Sources, Detectors, and Systems*; Academic Press: San Diego, 1995.
Kulkarni, M. D.; Thomas, C. W.; Izatt, J. A. Image Enhancement in Optical Coherence Tomography Using Deconvolution. *Elec. Lett.* **1997,** *33*, 1365–1367.
Kumagai, N.; Yamasaki, S.; Okushi, H. Optical Characterization of Surface Roughness of Diamond by Spectroscopic Ellipsometry. *Diamond Relat. Mater.* **2004,** *13*, 2092–2095.
Lampard, D. Definitions of "Bandwidth" and "Time Duration" of Signals which are Connected by an Identity. *IRE Trans. Circ. Theor.* **1956,** *3*, 286–288.
Lavin, E. P. *Specular Reflection*; Adam Hilger: London, 1971.
Lehmann, P. Surface-Roughness Measurement Based on the Intensity Correlation Function of Scattered Light Under Speckle-Pattern Illumination. *Appl. Opt.* **1999,** *38*, 1144–1152.
Leitgeb, R. A.; Hitzenberger, C. K.; Fercher, A. F. Performance of Fourier Domain Versus Time Domain Optical Coherence Tomography. *Opt. Express* **2003,** *11*, 889–894.

Li, C.; Kattawar, G. W.; Yang, P. Effects of Surface Roughness on Light Scattering by Small Particles. *J. Quant. Spectr. Radiat. Transf.* **2004,** *89*, 123–131.

Lin, C. F.; Lee, B. L. Extremely Broadband AlGaAs/GaAs Superluminescent Diodes, *Appl. Phys. Lett.* **1997,** *71*, 1598–1600.

Lorincz, E.; Richter, P.; Engard, F. Interferometric Statistical Measurement of Surface Roughness. *Appl. Opt.* **1986,** *25*, 2778–2784.

Lu, R.-S.; Tian, G Y. On-line Measurement of Surface Roughness by Laser Light Scattering. *Meas. Sci. Technol.* **2006,** *17*, 1496–1502.

Luk, F.; Huynh, V.; North, W. Measurement of Surface Roughness by a Machine Vision System. *J. Phys. E: Sci. Instrum.* **1989,** *22*, 977–980.

Ma, C.; Wang, A. Signal Processing of White–Light Interferometric Low–Finesse Fiber–Optic Fabry–Perot Sensors. *Appl. Optic.* **2013,** *52*, 127–138.

Manojlović, L. M. A Simple White-Light Fiber-Optic Interferometric Sensing System for Absolute Position Measurement. *Opt. Lasers Eng.* **2010,** *48* (4), 486–490.

Manojlović, L. M. High-resolution Wide-Dynamic Range Electronically Scanned White-Light Interferometry. *Appl. Opt.* **2014,** *53*, 3341–3346.

Manojlović, L. M. Novel Method for Optical Coherence Tomography Resolution Enhancement. *IEEE J. Quantum Electron* **2011,** *47*, 340–347.

Manojlović, L. M.; Živanov, M. B.; Marinčić, A. S. White-Light Interferometric Sensor for Rough Surface Height Distribution Measurement. *IEEE Sens. J.* **2010,** *10* (6), 1125–1132.

Marinescu, I. D.; Uhlmann, E.; Doi, T. K. *Handbook of Lapping and Polishing*; CRC Press, Taylor & Francis Group: Boca Raton, 2007.

Marinescu, I. D.; Uhlmann, E.; Doi, T. K. *Handbook of Lapping and Polishing*; CRC Press, Taylor & Francis Group: Boca Raton, 2007.

Marshall, R. H.; Ning, Y. N.; Jiang, X.; Palmer, A. W.; Meggitt, B. T.; Grattan, K. T. V. A Novel Electronically Scanned White–Light Interferometer Using a Mach-Zehnder Approach. *J. Lightw. Technol.* **1996,** *14*, 397–402.

Myshkin, N. K.; Grigoriev, A. Ya.; Chizhik, S. A.; Choi, K. Y.; Petrokovets, M. I. Surface Roughness and Texture Analysis in Microscale. *Wear* **2003,** *254*, 1001–1009.

Nassif, N.; Cense, B.; Hyle Park, B.; Yun, S. H. In Vivo Human Retinal Imaging By Ultrahigh-Speed Spectral Domain Optical Coherence Tomography. *Opt. Lett.* **2004,** *29*, 480–482.

Neste, R. V.; Belleville, C.; Pronovost, D.; Proulx, A. System and Method for Measuring an Optical Path Difference in a Sensing Interferometer. US Patent No. 6,842,254 B2 (2005).

Ogilvy, J. A. *Theory of Wave Scattering from Random Rough Surfaces*; Adam Hilger: Bristol, 1991.

Park, J.; Li, X. Theoretical and Numerical Analysis of Superluminescent Diodes. *J. Lightwave Technol.* **2006,** *24* (6), 2473–2480.

Park, J.; Li, X. Theoretical and Numerical Analysis of Superluminescent Diodes. *J. Lightw. Technol.* **2006,** *24*, 2473–2480.

Patrikar, R. M. Modeling and Simulation of Surface Roughness. *Appl. Surf. Sci.* **2004,** *228*, 213–220.

Pavliček, P.; Hýbl, O. White-Light Interferometry on Rough Surfaces-Measurement Uncertainty Caused by Surface Roughness. *Appl. Opt.* **2008,** *47*, 2941–2949.

Pavliček, P.; Soubusta, J. Theoretical Measurement Uncertainty of White-Light Interferometry on Rough Surfaces. *Appl. Opt.* **2003,** *42*, 1809–1813.

Pernick, B. J. Surface Roughness Measurements with an Optical Fourier Spectrum Analyzer. *Appl. Opt.* **1979,** *18*, 796–801.

Persson, U. A Fibre-Optic Surface-Roughness Sensor. *J. Mater. Process. Technol.* **1999,** *95*, 107–111.

Persson, U. In-Process Measurement of Surface Roughness Using Light Scattering. *Wear* **1998,** *215*, 54–58.

Pinet, É., Cibula, E.; Đonlagić, D. Ultra-Miniature All–Glass Fabry–Pérot Pressure Sensor Manufactured at the Tip of a Multimode Optical Fiber. *Proc. of SPIE* **2007,** *6770,* 67700U-1-8.

Poole, P. J.; Davies, M.; Dion, M.; Feng, Y.; Charbonneau, S.; Goldberg, R. D.; Mitchell, I. V. The Fabrication of a Broad-Spectrum Light-Emitting Diode Using High-Energy Ion Implantation. *IEEE Photon. Technol. Lett.* **1996,** *8*, 1145–1147.

Poon, C. Y.; Bhushan, B. Comparison of Surface Roughness Measurements by Stylus Profiler, AFM and Non-Contact Optical Profiler. *Wear* **1995,** *190*, 76–88.

Rao, Y. J.; Jackson, D. A. Universal Fiber-Optic Point Sensor System for Quasi-Static Absolute Measurements of Multiparameters Exploiting Low Coherence Interrogation. *J. Lightwave Technol.* **1996,** *14* (4), 592–600.

Recknagel, R.-J.; Notni, G. Analysis of White Light Interferograms Using Wavelet Methods. *Opt. Commun.* **1998,** *148*, 122–128.

Ribbens, W. B. Interferometric Surface Roughness Measurement. *Appl. Opt.* **1969,** *8*, 2173–2176.

Rice, S. O. Mathematical Analysis of Random Noise. *Bell Syst. Tech. J.* **1952,** *23*, 282–332.

S. H. Wang, C. J. Tay, C. Quan, H. M. Shang, and Z. F. Zhou, Laser integrated measurement of surface roughness and micro-displacement, Meas. Sci. Technol., *11*, 454–458.

Sandeman, R. J. Use of Channeled Spectra to Measure Absolute Phase Shift and Dispersion in Two Beam Interferometry. *Appl. Opt.* **1971,** *10* (5), 1087-1091.

Santos, G.; Cywiak, M.; Barrientos, B. Interferometry with Coherent Gaussian Illumination for Roughness and Shape Measurement. *Opt. Commun.* **2004,** *239*, 265–273.

Schmitt, J. M.; Lee, S. L.; Yung, K. M. An Optical Coherence Microscope with Enhanced Resolving Power. *Opt. Commun.* **1997,** *142*, 203–207.

Schnell, U.; Zimmermann, E.; Dändliker, R. Absolute Distance Measurement with Synchronously Sampled White-Light Channelled Spectrum Interferometer. *Pure Appl. Opt.* **1995,** *4*, 643–651.

Shan, S.; Ji-Hua, G.; Jian-Song, G.; Ping, X. Enhancement of Optical Coherence Tomography Axial Resolution by Spectral Shaping. *Chin. Phys. Lett.* **2002,** *19*, 1456–1458.

Silvennoinen, R.; Peiponen, K.-E.; Myller, K. *Specular Gloss*; Elsevier: Amsterdam, 2008.

Smith, L. M.; Dobson, C. C. Absolute Displacement Measurements Using Modulation of the Spectrum of White Light in a Michelson Interferometer. *Appl. Opt.* **1989,** *28* (15), 3339–3342.

Smith, L. M.; Dobson, C. C. Absolute Displacement Measurements Using Modulation of the Spectrum of White Light in a Michelson Interferometer. *Appl. Opt.* **1989,** *28*, 3339–3342.

Spagnolo, G. S.; Ambrosini, D. Diffractive Optical Element Based Sensor for Roughness Measurement. *Sens. Actuat. A* **2002,** *100*, 180–186.

Sprague, R. A. Surface Roughness Measurement Using White Light Speckle. *Appl. Opt.* **1972,** *11*, 2811–2816.

Stover, J. C. *Optical Scattering—Measurement and Analysis*; McGraw-Hill: New York, 1990.

Taplin, S.; Podoleanu, A. Gh.; Webb, D. J.; Jackson, D. A. Displacement Sensor Using Channelled Spectrum Dispersed on a Linear CCD Array. *Electron. Lett.* **1993,** *29* (10), 896–897.

Tay, C. J.; Wang, S. H.; Quan, C.; Shang, H. M. In Situ Surface Roughness Measurement Using a Laser Scattering Method. *Opt. Commun.* **2003,** *218*, 1–10.

Tearney, G. J.; Brezinski, M. E.; Bouma, B. E.; Boppart, S. A.; Pitris, C.; Southern, J. F.; Fujimoto, J. G. In Vivo Endoscopic Optical Biopsy with Optical Coherence Tomography. *Science* **1997,** *276*, 2037–2039.

Thomas, T. R. *Rough Surfaces*, 2nd ed.; Imperial College Press: Singapore, 1999.

Toh, S. L.; Shang, H. M.; Tay, C. J. Surface-roughness Study Using Laser Speckle Method. *Opt. Lasers Eng.* **1998,** *29*, 217–225.

Tomassini, P.; Rovati, L.; Sansoni, G.; Docchio, F. Novel Optical Sensor for the Measurement of Surface Texture. *Rev. Sci. Instrum.* **2001,** *72*, 2207–2213.

Udupa, G.; Singaperumal, M.; Sirohi, Kothiyal, M. P. Characterization of Surface Topography By Confocal Microscopy: I. Principles and the Measurement System. *Meas. Sci. Technol.* **2000,** *11*, 305–314.

Udupa, G.; Singaperumal, M.; Sirohi, R. S.; Kothiyal, M. P. Characterization of Surface Topography by Confocal Microscopy: II. The Micro and Macro Surface Irregularities. Meas. Sci. Technol., *11*, 315–329.

van Ginneken, B.; Stavridi, M.; Koenderink, J. J. Diffuse and Specular Reflectance from Rough Surfaces. *Appl. Opt.* **1998,** *37*, 130–139.

Vorburger, T. V.; Ludema, K. C. Ellipsometry of Rough Surfaces. *Appl. Opt.* **1980,** *19*, 561–573.

Wakaki, M.; Kudo, K.; Shibuya, T. *Physical Properties and Data of Optical Materials*; Taylor and Francis Group: Boca Raton, 2007.

Wang, S. H.; Jin, C. J.; Tay, C. J.; Quan, C.; Shang, H. M. Design of an Optical Probe for Testing Surface Roughness and Micro-Displacement. *Precision Eng.* **2001,** *25*, 258–265.

Washington, C. *Particle Size Analysis in Pharmaceutics and Other Industries*; Ellis Horwood: New York, 2005.

Widrow, B.; Kollár, I. *Quantization Noise*; Cambridge University Press: Cambridge, 2008.

Wojtkowski, M.; Leitgeb, R.; Kowalczyk, A.; Bajraszewski, T.; Fercher, A. F. In Vivo Human Retinal Imaging by Fourier Domain Optical Coherence Tomography. *J. Biomed. Opt.* **2002,** *7*, 457–463.

Yilbas, Z.; Hasmi, M. S. J. Surface Roughness Measurement Using an Optical System. *J. Mater. Process. Technol.* **1999,** *88*, 10–22.

Yun, S. H.; Tearney, G. J.; deBoer, J. F.; Iftimia, N.; Bouma, B. E. High Speed Optical Frequency-Domain Imaging. *Opt. Express* **2003,** *11*, 2953–2963.

Zhang, K.; Butler, C.; Yang, Q.; Lu, Y. A Fiber Optic Sensor for the Measurement of Surface Roughness and Displacement Using Artificial Neural Networks. *IEEE Trans. Instrument. Measure.* **1997,** *46*, 899–902.

Zhang, Y.; Sato, M.; Tanno, N. Resolution Improvement in Optical Coherence Tomography Based on Destructive Interference. *Opt. Commun.* **2001,** *187*, 65–70.

Author Biography

Lazo M. Manojlović, PhD, is currently with the Ministry of Telecommunications and Information Society of the Republic of Serbia, Belgrade, as a Special Adviser to the Minister. He is also with the Zrenjanin Technical College, Serbia, as a Professor. His experience and research interests include interferometry, optical sensing systems, and laser rangefinders. His extensive experience includes roles with Integrated Microsystems Austria GmbH, Wiener Neustadt, Austria, where he was engaged in research on the development of different types of fiber-optic sensors; the Vienna University of Technology, Austria, where he was engaged in research on the development of high-precision piezoactuator-based positioning systems; Pupin Telecom DATACOM, Belgrade, Serbia, where he was involved in designing, installing, and commissioning of the first hybrid fiber coaxial network for the cable distribution systems in Belgrade; the Military Institute of Technology, Belgrade, Serbia, where he was engaged in research on designing and testing several types of ultra-low noise wideband photodiode preamplifiers for quadrant photodetector and for tracking electronics circuits for analog signal processing in pulsed-laser tracking systems for laser guidance missiles; and the Institute Mihailo Pupin, Belgrade, Serbia, as a part of his extended studies, where he was engaged in research on the development of applicative software for text-to-speech synthesis of Serbian spoken language in the C programming language based on diphone concatenation.

Dr. Manojlović received his BSc and MSc degrees in electrical engineering from the Faculty of Electrical Engineering, University of Belgrade, Serbia, and his PhD degree in electrical engineering from the Faculty of Technical Sciences, University of Novi Sad, Serbia.

Author Biography

Index

A

Absolute position measurement, 246–258
Absolute radiometer, 38–40
Algorithms for signal processing, 209
  centroid algorithm, 213–214
  envelope coherence function method, 212–213
  minimum signal-to-noise ratio, 212
  modified centroid algorithm, 217–218
    linear scanning, sensitivity of, 220–228
    optical path difference measurement error, 228–242
  phase-shifted interferograms, 214–215
  threshold comparison method, 210–211
  wavelet transform algorithm, 215–217
Amplitude division method, 111–113
Analog-to-digital (A/D) converter, 281
Antireflective coating, 144
Arbitrary functions, 22
Avalanche photodiodes (APD), 71–72

B

Back-reflected light, 293
Band-pass filtration and normalization, 296
Block schematic, 251
Bose–Einstein distribution, 59

C

Case studies
  absolute position measurement, 246–258
  analog-to-digital (A/D) converter, 281
  C-843 DC-Motor Controller Card, 250
  Cauchy–Schwarz–Bunyakovsky, 274, 287
  CCD and PSD, 262–263
  curve fitting tool, 254
  Fabry–Perot interferometer, 246
  frequency range, 285
  full width at half maximum (FWHM), 248
  Gaussian apodization function, 286
  Gaussian distribution, 270
  Gaussian spectrum shape, 250
  index matching gel (IMG), 250
  laser diode driver (LDD), 250
  LCLS, 276–277
  low-coherence light source (LCLS), 246–247
  Motorized Actuator, 279
  motorized micrometer positioning stage (MMPS), 250
  normalized autocorrelation function, 249
  OSA, 261
  OSA detectors, 277
  PSD signal, 285
  RM back-reflected light, 273
  RS heights distribution, 265
  sensing fiber (SF), 246
  sensing mirror (SM), 246
  signal-to-noise ratio (SNR), 256–257
  Smart Move Operating Software, 250–251
  spectral fringe visibility, 252–253
  superluminescent diode (SLD), 247
  white-light Michelson interferometer, 261
  Wiener–Khintchine theorem, 266–267
Communication system
  (CATV—**C**ommon-**A**ntenna **T**ele**V**ision), 185
  coherent fiber-optic communication systems, 191–194
  distribution networks, 185–186
  fiber distributed data interface (FDDI), 186
  fiber-optic communication system, 188–191
  local area network (LAN), 186–187
  metropolitan-area networks (MANs), 185
  point-to-point links, 183–184
Corpuscular nature of light
  arbitrary functions, 22
  classical electromagnetic theory, 14

continuity equation, 16
cross-section area, 16
Doppler effect, 18
electromagnetic radiation, 14, 22
hypothesis rejection, 14
mathematical relation, 17
moving emitter, 18
photon energy, 15, 17
photon energy–frequency relationship, 22
photon momentum, 19, 21
Planck's constant, 22

D

Dirac delta function, 308
Distribution networks, 185–186
Doppler effect, 13, 18
DWDM (***D****ense* ***W****avelength* ***D****ivision* ***M****ultiplexing*), 154

F

Fabry-Pérot interferometer, 139–140, 246, 292
Fiber-optic sensors
block schematic of, 251
case studies
absolute position measurement, 246–258
analog-to-digital (A/D) converter, 281
C-843 DC-Motor Controller Card, 250
Cauchy–Schwarz–Bunyakovsky, 274, 287
CCD and PSD, 262–263
curve fitting tool, 254
Fabry–Perot interferometer, 246
frequency range, 285
full width at half maximum (FWHM), 248
Gaussian apodization function, 286
Gaussian distribution, 270
Gaussian spectrum shape, 250
index matching gel (IMG), 250
laser diode driver (LDD), 250
low-coherence light source (LCLS), 246–247, 276–277
Motorized Actuator, 279
motorized micrometer positioning stage (MMPS), 250
normalized autocorrelation function, 249
OSA detectors, 261, 277
PSD signal, 285
RM back-reflected light, 273
RS heights distribution, 265
sensing fiber (SF), 246
sensing mirror (SM), 246
signal-to-noise ratio (SNR), 256–257
Smart Move Operating Software, 250–251
spectral fringe visibility, 252–253
superluminescent diode (SLD), 247
white-light Michelson interferometer, 261
Wiener–Khintchine theorem, 266–267
optical coherence tomography (OCT), 299
consecutive measurements, 300
Dirac delta function, 308
FD-OCT, 309
fiber end back-reflection, 311
Fourier domain OCT (FD-OCT), 300
Fourier transform, 304, 308
full width at half-maximum (FWHM), 306
Gaussian apodization function, 311
LCLS, 305
OS captured optical, 303
SLD electrical field, 313
SLD spectrum, 308
SLD, measurement process, 307
Wiener–Khinchin theorem, 301, 308
wide-dynamic range low-coherence interferometry
back-reflected light, 293
band-pass filtration and normalization, 296
Fabry–Pérot interferometer, 292
Fizeau interferometer, 290
FOC arm, 293
low-coherence interferometric measurement, 290
PSD, 299
RM-SM distance, 298
sensing system, 298
SNR, 289
wedge index, 291

Fiber optics
communication system
(CATV—**C**ommon-**A**ntenna **T**ele**V**ision), 185
coherent fiber-optic communication systems, 191–194
distribution networks, 185–186
fiber distributed data interface (FDDI), 186
fiber-optic communication system, 188–191
local area network (LAN), 186–187
metropolitan-area networks (MANs), 185
point-to-point links, 183–184
DWDM (***D**ense **W**avelength **D**ivision **M**ultiplexing*), 154
optical fibers
chromatic dispersion, 174–179
fiber loses, 181–182
geometrical optics and, 157–163
multimode and single-mode fibers, 155–156
polarization mode dispersion, 180–181
wave optics, 163–174
sensing systems
interferometer topologies, 197–202
sensor basic topologies, 195–197
Fizeau interferometer, 290
Full width at half maximum (FWHM), 128, 248

G

Gaussian apodization function, 286

I

Index matching gel (IMG), 250

L

Laser diode driver (LDD), 250
Light
celestial bodies, 127
coherence theory, 118
mutual coherence function, 119–121
quasimonochromatic light, 121
spatial coherence, 122–124
time and length, 123–127
interferometric sensors
laser-Doppler interferometry, 147–148
Michelson's stellar interferometer, 148–150
Rayleigh refractometer, 146–147
vibration amplitudes, 148
multilayer thin films, 140
antireflective coating, 144
electric fields amplitudes, 141
eliminating amplitudes, 142
heat-reflecting and heat-transmitting mirrors, 141
magnesium fluoride, 144
magnetic fields amplitudes, 141
maximum reflectance, 146
reflectance and transmittance, 144
reflection and transmission coefficients, 143
stack for, 145
transfer matrix, 143
wave vectors, 142
multiple-beam interference, 135
Fabry–Pérot fringes, 139–140
incident wave irradiance, 138
plane-parallel plate, 136, 137
reflected rays, 137
sharp fringes of, 139
slab perpendicular, 138
transmitted beam amplitude, 138
transmitted irradiance, 138
two-beam interference, 105–107
amplitude division method, 111–113
Mach–Zehnder interferometer, 115–117
Michelson interferometer, 114–115
Sagnac interferometer, 117–118
wavefront division method, 108–111
white-light interferometry, 127
CCD linear array, 130
channeled spectrum, 131–132
full width at half maximum (FWHM), 128
low-coherence interferometric pattern, 135
measured channeled spectrum, 133
optical coherence-domain reflectometry, 134

optical wedge for, 131
reflectometer/tomography measurement system, 133
spectral interferometry, 132
Twyman–Green interferometer, 128
with electronic scanning, 130
Light, wave nature
Cartesian coordinate system, 4, 8
curl equations, 4
Doppler effect, 13
electric field, 5
electromagnetic field, 3
electromagnetic radiation, 13
electromagnetic spectrum, 8, 9
electromagnetic wave, 10
electromagnetic wave amplitude, 5
electromagnetic waves, 7
Euler representation, 8
force and density, 3
harmonic solution, 5
Lorentz transformation, 14
magnetic induction, 4
Maxwell's equations, 3
momentum, 11
monochromatic wave, 7
one dimensional wave equation, 5
phase velocity, 10
physical quantity, 11
plane wave, equiphase surfaces of, 7
Poynting vector, 13
propagation vector, 6
vacuum permittivity, 3
wavelength λ, 5
Local area network (LAN), 186–187
Low-coherence fiber-optic sensor
algorithms for signal processing, 209
centroid algorithm, 213–214
envelope coherence function method, 212–213
minimum signal-to-noise ratio, 212
modified centroid algorithm, 217–218, 220–228, 228–242
phase-shifted interferograms, 214–215
threshold comparison method, 210–211
wavelet transform algorithm, 215–217
LCLS, 208–209
Michelson interferometers, 206
photodetector (PD), 207
Low-coherence light source (LCLS), 276–277

M

Mach–Zehnder interferometer, 115–117
Metropolitan-area networks (MANs), 185
Michelson interferometer, 114–115
Motorized micrometer positioning stage (MMPS), 250
Multiple-beam interference, 135
Fabry–Pérot fringes, 139–140
incident wave irradiance, 138
plane-parallel plate, 136, 137
reflected rays, 137
sharp fringes of, 139
slab perpendicular, 138
transmitted beam amplitude, 138
transmitted irradiance, 138

O

Optical coherence tomography (OCT), 299
consecutive measurements, 300
Dirac delta function, 308
FD-OCT, 309
fiber end back-reflection, 311
Fourier domain OCT (FD-OCT), 300
Fourier transform, 304, 308
full width at half-maximum (FWHM), 306
LCLS, 305
OS captured optical, 303
SLD electrical field, 313
SLD spectrum, 308
SLD, measurement process, 307
Wiener–Khinchin theorem, 301, 308
Optical detector
Bose–Einstein distribution, 59
fixed observation time, 57
incoherent light, 59–60
photodetection modeling
low-pass filter, 60
optical power (flux), 62
photodetector impulse responses, 61
photodetection techniques, 56
photodetectors (PD)
avalanche photodiodes (APD), 71–72

photoconductors, 65–67
photodiode amplifier circuit, 95–102
photodiodes, 67–70
photomultiplier (PMT), 64–65
position sensing photodiodes, 72–75
quadrant photodiode (QPD), 75–92
typical photodiode, equivalent circuit model of, 92–95
photon counting, 56, 58
photon generation, 57
Poisson distribution, 58–59
Optical power (flux), 62
Optical radiometry, 28
absorption and transmission, 34–36
Kirchhoff's law, 36–38
Lambertian radiator, 30–31
radiant flux, 29–30
radiometric measurements, 31–34
reflection, 34–36

P

Photodetection modeling
low-pass filter, 60
optical power (flux), 62
photodetector impulse responses, 61
Photodetectors (PD), 207
avalanche photodiodes (APD), 71–72
photoconductors, 65–67
photodiode amplifier circuit, 95–102
photodiodes, 67–70
photomultiplier (PMT), 64–65
position sensing photodiodes, 72–75
quadrant photodiode (QPD), 75–92
typical photodiode, equivalent circuit model of, 92–95
Photometry
and realization of candela, 51–52
human eye, spectral response, 47–48
Photomultiplier (PMT), 64–65
Photon energy, 15, 17
Photon energy–frequency relationship, 22
Planck's constant, 22

Q

Quadrant photodiode (QPD), 75–92

R

Radiometric and photometric measurements
detector-based candela realization, 52
elementary area, 28
geometric representation, 28
isothermal cavity, 37
maximal value, 26
measurement techniques in
absolute radiometer, 38–40
integrating sphere, 45–46
radiant flux measurement, 40–44
optical radiometry, 28
absorption and transmission, 34–36
Kirchhoff's law, 36–38
Lambertian radiator, 30–31
radiant flux, 29–30
radiometric measurements, 31–34
reflection, 34–36
photometry
and realization of candela, 51–52
human eye, spectral response, 47–48
radiant flux, 27
solid angle, 26
spectral luminous efficiency function
photopic vision $V(\lambda)$, 49–50
scotopic vision $V'(\lambda)$, 49–50
transfer processes, 27

S

Sagnac interferometer, 117–118
Sensing fiber (SF), 246
Sensing mirror (SM), 246
Signal-to-noise ratio (SNR), 256–257
Smart Move Operating Software, 250–251
Superluminescent diode (SLD), 247

T

Two-beam interference, 105–107
amplitude division method, 111–113
Mach–Zehnder interferometer, 115–117
Michelson interferometer, 114–115
wavefront division method, 108–111
Twyman–Green interferometer, 128

W

White-light interferometry, 127
CCD linear array, 130
channeled spectrum, 131–132
full width at half maximum (FWHM), 128
low-coherence interferometric pattern, 135
measured channeled spectrum, 133
optical coherence-domain reflectometry, 134
optical wedge for, 131
reflectometer/tomography measurement system, 133
spectral interferometry, 132
Twyman–Green interferometer, 128
with electronic scanning, 130
Wide-dynamic range low-coherence interferometry
back-reflected light, 293
band-pass filtration and normalization, 296
Fabry–Pérot interferometer, 292
Fizeau interferometer, 290
FOC arm, 293
low-coherence interferometric measurement, 290
PSD, 299
RM-SM distance, 298
sensing system, 298
SNR, 289
wedge index, 291
Wiener–Khintchine theorem, 266–267

For Product Safety Concerns and Information please contact our EU representative GPSR@taylorandfrancis.com
Taylor & Francis Verlag GmbH, Kaufingerstraße 24, 80331 München, Germany

www.ingramcontent.com/pod-product-compliance
Lightning Source LLC
Chambersburg PA
CBHW060809220326
41598CB00022B/2573

9781774637364